NEUTRON BEAM DESIGN, DEVELOPMENT, AND PERFORMANCE FOR NEUTRON CAPTURE THERAPY

BASIC LIFE SCIENCES

Ernest H. Y. Chu, Series Editor

The University of Michigan Medical School
Ann Arbor, Michigan

Alexander Hollaender, Founding Editor

Recent volumes in the series:

NEUTRON BEAM DESIGN, DEVELOPMENT, AND PERFORMANCE FOR NEUTRON CAPTURE THERAPY

Edited by

Otto K. Harling
Massachusetts Institute of Technology
Cambridge, Massachusetts

John A. Bernard
Massachusetts Institute of Technology
Cambridge, Massachusetts

and

Robert G. Zamenhof
Tufts–New England Medical Center
Boston, Massachussetts

PLENUM PRESS • NEW YORK AND LONDON

Library of Congress Cataloging-in-Publication Data

International Workshop on Neutron Beam Design, Development, and
    Performance for Neutron Capture Therapy (1989 : Massachusetts
    Institute of Technology)
      Neutron beam design, development, and performance for neutron
    capture therapy / edited by Otto K. Harling and John A. Bernard and
    Robert G. Zamenhof.
           p.   cm. -- (Basic life sciences ; v. 54)
       "Proceedings of an International Workshop on Neutron Beam Design,
    Development, and Performance for Neutron Capture Therapy, held March
    29-31, 1989, at the Massachusetts Institute of Technology, in
    Cambridge, Massachusetts"--T.p. verso.
      Includes bibliographical references and index.
      ISBN 0-306-43610-8
      1. Boron-neutron capture therapy--Congresses. 2. Neutron beams-
    -Congresses.   I. Harling, Otto K. II. Bernard, John A., 1949-
    III. Zamenhof, Robert G. IV. Title. V. Series.
    RC271.N48I58 1989
    616.99'40642--dc20   90-42651
                                                              CIP

Proceedings of an International Workshop on Neutron Beam Design,
Development, and Performance for Neutron Capture Therapy,
held March 29–31, 1989, at the Massachusetts Institute of Technology,
in Cambridge, Massachussetts

ISBN 0-306-43610-8

© 1990 Plenum Press, New York
A Division of Plenum Publishing Corporation
233 Spring Street, New York, N.Y. 10013

Printed in the United States of America

WORKSHOP ORGANIZATION AND SPONSORS

Sponsors

US Department of Energy, Human Health and Assessments Division
Callery Chemical Company
Theragenics Corporation
Nuclear Reactor Laboratory, Massachusetts Institute of Technology
Department of Radiation Oncology, Tufts – New England Medical Center

Organizing Committee

J. A. Bernard, Massachusetts Institute of Technology
R. M. Brugger, University of Missouri, Columbia
R. G. Fairchild, Brookhaven National Laboratory
O. K. Harling, *Co-chairman*, Massachusetts Institute of Technology
J. S. Robertson, US Department of Energy
R. G. Zamenhof, *Co-chairman*, Tufts – New England Medical Center

Rapporteurs

R. M. Brugger, University of Missouri
G. Constantine, AERE, Harwell, England, United Kingdom
O. K. Harling, *Chairman*, Massachusetts Institute of Technology
F. J. Wheeler, Idaho National Engineering Laboratory

PREFACE

For this Workshop, the organizers have attempted to invite experts from all known centers which are engaged in neutron beam development for neutron capture therapy. The Workshop was designed around a series of nineteen invited papers which dealt with neutron source design and development and beam characterization and performance. Emphasis was placed on epithermal beams because they offer clinical advantages and are more challenging to implement than thermal beams. Fission reactor sources were the basis for the majority of the papers; however three papers dealt with accelerator neutron sources. An additional three invited papers provided a summary of clinical results of NCT therapy in Japan between 1968 and 1989 and overviews of clinical considerations for neutron capture therapy and of the status of tumor targeting chemical agents for NCT.

Five contributed poster papers dealing with NCT beam design and performance were also presented. A rapporteurs' paper was prepared after the Workshop to attempt to summarize the major aspects, issues, and conclusions which resulted from this Workshop.

Many people contributed to both the smooth functioning of the Workshop and to the preparation of these proceedings. Special thanks are reserved for Ms. Dorothy K. Eichel for her attention to many of the administrative details, to Mrs. Carolyn R. Hinds and Ms. Joice Himawan who typed the edited papers, and to Mr. Ara B. Sanentz who meticulously edited each paper. Others who warrant mention include Mr. Leonard Andexler, Mr. James Bernard, Mr. Kwan S. Kwok, Mrs. Rebecca S. Lau, Mr. Shing Hei Lau, Ms. Susan Tucker, and Ms. Georgia Woodsworth.

OTTO K. HARLING
JOHN A. BERNARD
ROBERT G. ZAMENHOF

REACTOR-BASED NEUTRON BEAMS

ACCELERATOR-BASED NEUTRON BEAMS

DOSIMETRY AND TREATMENT PLANNING

RAPPORTEURS' REPORT

RAPPORTEURS' REPORT

Rapporteurs

    R. M. Brugger
    G. Constantine
    O. K. Harling, Chairman and
    F. J. Wheeler

INTRODUCTION

The four rapporteurs listed above provided oral summaries and comments on what they believed were some of the important aspects of the Workshop. Those comments are provided here in written form. Also included are some issues which were raised during the Workshop and at the rapporteurs session which are felt to be worthy of further attention by the NCT community. Each rapporteur contributed written material in specific areas and these contributions were edited and organized into this report by the Chairman.

DESIGN PHILOSOPHIES TO PRODUCE EPITHERMAL-NEUTRON BEAMS

The goal for epithermal-neutron beam design is to generate a neutron beam with enough intensity to provide therapy while minimizing patient risk and discomfort. The beam should have minimal contamination from fast-neutron, gamma and thermal-neutron components.

The energy range of the neutrons that are desired for a beam to treat deep-seated cancers is roughly greater than 1 eV and less than 20 keV. These limits are not precise – 0.5 eV neutrons are still somewhat penetrating, while neutrons with energies considerably higher than 20 keV may still provide a therapeutic advantage. However, thermal neutrons are not penetrating and are only appropriate for treating shallow cancers. The nonselective dose from the capture of the thermalized neutrons in the phantom or patient, i.e. $H(n,\gamma)$ and $N(n,p)$, sets a lower limit on the purity or selectivity which is achievable in NCT beams.

The general types of neutron sources that are available are reactors, accelerators, and radioisotope sources. Reactors produce neutrons covering the range from thermal to above 10 MeV. The larger reactors produce enough neutrons, but the challenge is to select from all the neutrons only those of the desired energies. Also the core gamma rays and the induced gamma rays must be suppressed. High energy accelerators present most of the same problems as do reactors but with lower intensity. Lower energy accelerators, that employ a specific reaction, i.e. $Li(p,n)$, are now being seriously considered. For these sources, the maximum energy of the neutrons can be about 800 keV with very few gamma rays produced at the target. However, the intensity of neutrons from these sources is currently too low. At the present time radioactive sources appear to be too low in neutron yield to be seriously considered, although very large Cf-252 sources could conceivably be produced with useful intensities. The practicality of such Cf-252 based neutron sources is questionable.

Both the reactors and the accelerators produce neutrons of too high an energy and these must be removed from the beam or shifted to lower energies. There are two ways to

Neutron Beam Design, Development, and Performance for Neutron Capture Therapy
Edited by O. K. Harling *et al.*
Plenum Press, New York, 1990

approach this problem. In the filter concept, a material is placed between the source of neutrons, which is partially moderated, and the patient position. The cross section of the material is such that the filter scatters the high energy neutrons out of the beam while passing the intermediate energy neutrons. The filter should be placed midway between the source and the patient with enough collimation so that with one interaction of a neutron with the filter, the neutron is redirected and it does not reach the patient position.

Some success with the design of filtered beams has been reported. The filtering materials that are now being tried include Al, S, Ti, V, and Ar, which transmit useful energy bands, while B-10, Cd, and Li-6 are used to remove thermal neutrons and to reduce the induced dose at the surface of the phantom or patient.

In the moderator concept, the fast neutrons from the source are reduced in energy by moderation to arrive at the desired energy. The moderator material should remove neutrons from the fast-energy groups and shift these to the intermediate energy groups more effectively than it removes neutrons from these intermediate groups into the thermal groups. Thus there is an enhancing of the number of intermediate energy neutrons compared to fast neutrons. The moderating material should be placed close to the source of neutrons. Less desirable, but acceptable, is to place the material close to the patient position.

Several beams have been designed and several built based on the moderated source concept. The moderator materials now being tried are Al_2O_3, $Al + D_2O$, S, AlF_3, Si, SiO_2, Be, and BeO. Moderated sources are being designed for both reactors and accelerators.

A few comments about geometry might be made. First, a large source area is needed to produce intense beams from the reactors that are available. Unfortunately, most beam tubes at reactors are small in diameter and will not be practical for intense beams. Second, the flight path from the source to the patient position should be short to enhance intensity. Third, from the comments at this meeting, it appears that the beam at the patient position should be large enough to irradiate the full head. Fourth, a tangential beam may not be a net improvement compared to a radial beam tube. Fifth, a directed beam striking the surface of the patient is an improvement over a non-directed flux of neutrons at the surface of the patient. However the use of the later is possible. If the source size and the source-to-patient distance is optimized, even reactors of moderate power, <1 MW, may be useful for NCT with epithermal beams.

Gamma rays from the source as well as gamma rays induced in the structural material of the beams must be considered and reduced. The low energy accelerator produces minimal gamma rays at the source but still produces neutron induced gamma rays. Design of reactor sources must address both. In most designs, some bismuth or lead must be placed in the beam to reduce the source and gamma rays induced in structures. To reduce the number of induced gamma rays, small amounts of Li-6 or possibly B-10 or Cd can be added to the structural material. It is not yet clear whether Bi is an improvement over Pb as a gamma shield when placed in the flux of intermediate neutrons. In either case one must consider impurities in the Pb or Bi that could increase the flux of induced gamma rays.

DETERMINISTIC TECHNIQUES OF BEAM DESIGN

The final result of the design calculations should be a full characterization of the beam through knowledge of the detailed angular fluxes as a function of energy and position at some surface at the exit port. This characterization should be known for the unperturbed beam (no phantom or tissue present) and for the perturbed beam (phantom or tissue present). The beam characterization should be made with the best available information from calculation and/or measurement.

The neutron and gamma transport from source to tissue may be calculated by methods employing Monte Carlo or by deterministic methods including discrete-ordinates, integral-transport, high-order diffusion theory, or other techniques. This calculation may also be accomplished by combinations of properly-coupled techniques or techniques properly normalized to well-measured fluxes. The limits of each method should be carefully evaluated to determine the effect on accuracy. For example, even high-order diffusion theory may be inadequate for deep transport or for properly describing leakage currents from one spatial compartment to another. Independent calculations using different solution methods and different data sets are valuable for final configurations.

The flux may be determined in one step for the entire system or piecewise by compartments from the source to the target where each compartment is coupled by the detailed angular fluxes from the previous compartment. These angular fluxes should be determined with the effects of neighboring compartments accounted for.

The majority of the calculational methods reported in this Workshop used Monte Carlo methods (e.g., the MCNP code, developed at Los Alamos National Laboratory) or discrete-ordinate methods (e.g., the DOT code, developed at Oak Ridge National Laboratory). The Monte Carlo method and the discrete ordinates method can in theory solve the Maxwell-Boltzmann transport equation with numerical error approaching zero. In practice, however, approximations are always required. Monte Carlo has an advantage for precise cross section and geometric representation, but extreme biasing (for variance reduction) may be required since the natural efficiency for transporting source neutrons to target tissue is often 10^{-6} or less. This extreme biasing can result in undersampling of the source (where the source may be the origin of the particle or the net angular flux from one compartment to another). On the other hand discrete ordinates may require approximations in geometry and fineness of spatial, angular, and/or energy mesh.

The final goal of the calculation is to characterize the spatial and energy distribution of flux within tissue. The phantom or tissue volume is the final compartment in the solution process. It is necessary to characterize the beam into this final compartment, both in the presence and absence of target media. Discrete ordinates or perhaps another appropriate deterministic method could offer advantages in this characterization because the detailed angular, spatial, and energy dependence of the flux is naturally carried through the solution. This is also feasible with the Monte Carlo method. However, extensive CPU time is required because this characterization requires the computation of thousands (or tens of thousands) of pieces of data to prevent excessive statistical variance in individual values of the detailed angular flux.

MONTE CARLO TECHNIQUES OF BEAM DESIGN

Following are some observations of beam performance using Monte Carlo techniques. Monte Carlo has advantages and disadvantages compared with other possibilities such as discrete ordinates, collision probability codes, and other deterministic methods. The overwhelming advantage of Monte Carlo for most of the applications we are discussing lies in the ability to model complex geometry. In general almost any geometry can be defined by combinations of planes, cones, ellipsoids, spherical and cylindrical surfaces, etc. Using combinatorial geometry the required regions in space can be unambiguously specified. A very useful facility in the powerful and versatile MCNP code is the angular transformation option, in which a part of the model can be defined fairly simply, e.g. as a cylinder parallel to the x-axis. Then to represent, for example, a slanting beam tube, whose equation would be much more complicated, the cylinder is rotated by specifying a simple angular transformation matrix. It is important to carry out careful checking to assure that the geometry is what the user intended. Many codes have facilities for plotting out the intersection of the model with user-specified planes as a series of maps which can then be scrutinized for consistency.

Monte Carlo's obvious disadvantage lies in the fact that individual neutron or photon histories are calculated sequentially and the accuracy of the results depends critically on the number of particles scored in the regions investigated. To reduce the standard deviation by a factor of 2 requires a 4-fold increase in the number of neutrons tracked and the computing costs involved. Methods for accumulating the wanted results more quickly abound in Monte Carlo codes, but considerable care must be exercised in applying them. The best-known are the complementary splitting and Russian roulette in which the problem geometry is divided into regions of different importance weighting by splitting surfaces. As neutrons being tracked approach closer to the part of the model we are interested in (a spectrum shifter just outside the core, for example), they cross importance boundaries and are split into fragments and tracked separately. Conversely, a neutron fragment crossing a boundary into a region of lower importance is increased in weight but subjected to Russian roulette, i.e. killed or allowed to continue, the choice being made with appropriate probabilities. By tracking a larger number of subparticles, improvements in the variance are gained, at the expense of course of more distant parts of the model, where sampling is sparse. Various other techniques can be applied – in particular, in beam geometry, angular biasing can be an extremely powerful tool in improving statistical accuracy.

Many of the earlier Monte Carlo codes followed the practice of the deterministic methods in dividing the neutron population into energy groups and ascribing a single value for a given cross section to all neutrons within any one group. A suitable average value had to be calculated for each group. While satisfactory for narrow groups and constant or slowly varying cross sections, this is less adequate in the NCT field, where we are concerned with attenuation through materials characterized by resonances and windows in their total cross sections. To consider a whole series of resonances and anti-resonance windows would be beyond the scope of such a code. More recently written codes such as MCNP employ point energy data so that each neutron is tracked using interaction cross sections appropriate to its actual energy. Thus the physics is properly represented. Cross sections are obtained by interpolation from extensive point energy tabulations, typically several thousand (E,σ) pairs for absorption, total, fission, etc. cross sections, as well as differential data. Subdivision into energy groups is only carried out at the scoring stage.

Scoring in Monte Carlo is the means by which we obtain our results. In MCNP a very flexible set of scoring or tally options is available. The number of neutrons crossing boundaries, number of collisions and pathlengths within given volumes can all be used to obtain neutron fluxes and spectra together with their variances. Interaction probabilities, i.e. specified cross section times the flux integrated over the whole spectrum can be accumulated during the tracking.

An extremely valuable capability common to many codes is the point detector. Here we specify a dose point (such as the patient location, for example) where we want to know the spectrum. In the beam tube configuration this is the most useful tally type. During the tracking procedure, each time a neutron makes a collision, MCNP computes its probability of contributing to the flux at the dose point. This is the product of three factors – first, the probability of scattering in the right direction; secondly, an inverse square law effect for the distance involved; and, thirdly, the attenuation factor due to intervening materials.

Contributions accumulated at the dose point can be modified in different ways and stored in separate tallies. Thus they can be multiplied by an energy dependent function, for instance an activation cross section or a dose KERMA to obtain a reaction rate or neutron dose rate, respectively. Alternately, a groupwise multiplier can be applied such as inverse lethargy width to give a spectrum in terms of flux per unit lethargy rather than flux per energy group.

MCNP has another option which conserves CPU time that is invaluable for filter design work, namely the ability to further modify the point detector contributions by specifying extra intervening materials and their thicknesses (i.e. a filter). Account is taken of the attenuation of the individual neutron contribution by total scattering while passing through such a filter. Neutrons are not being actively tracked through this filter so there is no allowance for multiple scattering. Thus this should be used with caution when the actual filter terminates close to the patient.

The MCNP code, as emphasized above, provides neutron doses in typical NCT geometries and in a similar manner provides information about photon transport and dose. Included in this code's capabilities is the generation of neutron-induced doses in complicated geometries.

FIGURES OF MERIT FOR BEAM PERFORMANCE AND QUALITY

Figures of merit for beam performance and quality can be formulated in a number of ways. One such approach is discussed by Clement et al. and Zamenhof et al. in these proceedings. In those papers, beam performance and quality are based upon measured and/or calculated depth-dose distributions, in phantoms, of all radiation components. Three criteria or figures of merit are used to describe NCT beams. These criteria are summarized below:

Advantage Depth – The advantage depth (AD) provides a measure of the maximum depth for a therapeutic benefit. It is defined as the depth for which the total dose or total therapeutic dose to tumor, from all components of the radiation field, would equal the maximum dose to healthy tissue. Beyond this depth, tumor would be expected to receive less dose than the maximum delivered to healthy tissue.

Advantage Ratio – The advantage ratio (AR) provides a measure of the ratio of total therapeutic dose to the total background dose integrated over a given depth. The depth is taken from the surface to the AD.

The AD and AR can be considered primarily as figures of merit for beam quality. It is essential that the NCT beam deliver a larger dose to tumor than healthy tissue (related to AR) and have adequate penetrability (related to AD). However, quality alone is not sufficient for useful NCT therapy. Therefore a measure of intensity is also used to characterize beam performance.

Advantage Depth Dose Rate – The advantage depth dose rate (ADDR) is the total therapeutic dose rate at the AD. This is the therapeutic dose rate available to treat tumor at the maximum useful depth for the NCT beam. What would be considered a useful ADDR is dependent upon many factors, including the use of fractionation, the patient load, cost, etc. For initial clinical trials of NCT, beams of relatively low intensity, requiring a total of several hours of irradiation in a fractionated treatment plan may be very useful. If NCT therapy becomes an established treatment modality, it is clear that higher intensity beams, which can treat patients in minutes, could be very desirable.

The concepts of AD, AR, and ADDR are based on measured and calculated detailed dose-depth distributions which are obtained with the best available KERMA factors. These KERMAs may be modified in the future, e.g. as we obtain a better understanding of the microdosimetry in NCT. Such improvements could easily be incorporated into the above approaches to figures of merit for NCT beams.

EXPERIMENTAL VERIFICATION OF BEAMS / NEUTRON SPECTRAL MEASUREMENTS

Confirmation of calculations is vital to ensure that we can build an epithermal NCT facility that can do what is required of it, both in minimizing the unwanted components of proton recoil damage and contamination by gammas, and in achieving adequate intensity at the patient position.

Dealing with the neutron spectrum determination first, we have a variety of methods available. In general, candidate beams for NCT have very broad spectra and several methods must be employed to give good coverage of the whole range.

The ^{3}He detector relies on the ^{3}He(n,p)T reaction, which releases 764 keV plus the energy of the incident neutron. Limits on energy resolution restrict its use to the upper part of our range of interest, i.e. from several keV upwards, since 764 keV must be subtracted from the measured overall reaction energy.

A rather more tractable system has been the proton recoil spectrometer. Here the lower energy limit is in the region of 10 keV. The principle lies in measuring the energy distribution of the protons recoiling from the hydrogen filling gas of a small spherical chamber. For a particular neutron energy, this will be a continuous distribution from zero to the full neutron energy. The spectrum of proton recoil energies must therefore be unfolded to obtain the original neutron spectrum. Corrections must be made for the wall effect to account for proton recoils that deposit part of their energy in the chamber wall. To obtain an understanding of this and to increase the range of measurement, several chambers filled to different gas pressures are used, and sophisticated computing techniques are employed to unfold their responses simultaneously. High gamma fields cause problems in the acquisition of the proton spectra, especially at the lower energies, though this can be ameliorated by pulse rise time discrimination. A limit on the data acquisition/processing of ~1000 pulses per second makes spectrum measurements very time consuming.

Bonner spheres, polyethylene spheres of various diameters containing a ^{3}He detector at the center, or a thermal-neutron sensitive foil if the beam is very intense, are other variants. Their energy responses tend to be rather flat so that unfolding a spectrum from a range of different size spheres is a bit difficult. They cannot resolve a sharply varying neutron spectrum such as that emerging through a cross-section window. Since their spectral response is based on uniform illumination by the neutrons, the range of useable sphere sizes is determined by the beam diameter.

Activation detectors have many adherents and, over the energy region where they have a good response, adjusting a calculated spectrum by multi-foil analysis techniques, well tried in reactor dosimetry circles, can be used as confirmatory evidence.

For the energy region between thermal and 10 keV, a subsidiary filter technique can be used, comprising a range of different thickness ^{10}B disks which can be interposed in the beam to modify the spectrum, cutting out lower energy neutrons, with a progressively increasing "threshold" as the disks get thicker. Measuring the resulting spectra with BF_3 chambers for a 1/v weighting, and a detector sensitive to the total neutron current, such as a vanadyl sulphate bath, gives a series of responses that can be unfolded in the same way as the activation detectors. The energy dependence of the attenuation of the ^{10}B disks can best be measured by pulsed neutron time-of-flight techniques.

EXPERIMENTAL VERIFICATION OF BEAM PERFORMANCE IN PHANTOMS

The final proof of NCT beam performance, short of actual patient treatment, is the measurement of beam performance in phantoms which are similar to human heads/brains. Such measurements are also very useful in evaluating NCT performance during beam development. Measurements and calculations of beam performance in air only are not adequate to determine NCT beam performance and may mislead in that inadequate attention may be paid to the major dose components such as capture gamma rays from light hydrogen. Trade-offs in beam design need to be made with an emphasis on all major dose components and this can best be done by calculating and measuring performance in representative head phantoms.

Phantoms made from polyethylene and water have been used extensively for beam studies. Tissue equivalent plastic materials are now available and should be considered for this use. The shapes and sizes of phantoms which have been used by the various NCT research groups vary significantly. Right circular cylinders, elliptical shapes, and human skull shapes have been used. The differences in material, shape, and sizes of phantoms which are being utilized make it difficult to easily compare the performance of neutron beams among different NCT groups. It would be useful therefore to try to agree upon phantom size, shape, and material and to move toward the use of a "standard phantom" for NCT work. A committee under the aegis of the NCT Society could be appointed to develop recommended standards for head phantoms.

Evaluating beam performance requires careful in-phantom measurements of all radiation field components. Thermal flux profiles can easily be measured, using an accepted ASTM procedure, with small non-perturbing detectors such as gold foils or wires, and a measure of the epithermal flux can be obtained by making these measurements with and without cadmium covers. The measured thermal flux can then be used to calculate doses for the B-10(n,α) and N-14(n,p) reactions. Gamma dose and fast-neutron dose measurements are more difficult to accomplish. Arguably, the best approach for this mixed field dosimetry is to use the paired ion chamber method. A neutron sensitive chamber, such as a tissue equivalent ion chamber, is paired with a neutron insensitive version with a magnesium or graphite wall. From these two measurements of the mixed radiation field, the fast-neutron and gamma-ray components can be separated and usually determined to acceptable levels of accuracy.

Measured detailed depth-dose distributions in phantoms are essential for benchmarking computational codes needed to design and optimize NCT treatment beams. Computations, e.g. with Monte Carlo codes such as MCNP, can provide separately all the components of the radiation field in the head phantom. Measurements generally cannot separately determine such components as core gammas, gammas induced in structure and gammas induced in phantom. The combination of good computational tools and detailed measurements in realistic head phantoms is recommended for efficient NCT beam design and development.

CURRENT STATUS OF NCT BEAMS

There are a considerable number of neutron sources currently available or under development worldwide which have potential applicability for NCT. To provide a rough summary of the current status of these sources for NCT, the following figure has been devised. This figure shows the known status of the NCT beams and gives a measure of intensity based on the estimated time to deliver 2000 RBE (cGy) at advantage depth of 6-7 cm in brain, with 30 μg/g of B-10 in tumor and 3 μg/g B-10 in blood. It should be understood that this simplified presentation does not provide enough information to compare the details of beam quality for the various facilities. Currently there are two epithermal beams actually available, one at Brookhaven and one at MIT. These two beams are undergoing testing and further optimization. Several other beams shown in the figure may be available in a year or so. For any unintended errors in the figure the rapporteurs apologize in advance.

ISSUES RAISED IN THE RAPPORTEURS' SESSION

<u>Neutron Energies Suitable for NCT</u>

There was considerable and lively discussion concerning the maximum neutron energies which would be suitable for epithermal NCT beams. From Monte Carlo, MCNP based, ideal beam studies (clean neutron beams at specific energies without any gamma ray contamination), it appears (Clement et al., these proceedings) that there is a wide range of neutron energies from about 1 eV to over 20 keV which can be utilized to produce high quality epithermal NCT beams. The figures of merit used in these ideal beam studies are the advantage depth and advantage ratio. It should also be pointed out that neutron energies above 20 keV need not be avoided quantitatively, e.g. for an ideal neutron beam at 70 keV, MCNP calculates a centerline in-phantom advantage depth and advantage ratio of 5.8 cm (in polyethylene) and 3.5 respectively for a 17.8 cm diameter beam incident upon a 17.8 cm diameter cylindrical phantom (assuming 30 ppm B-10 in tumor and 0 ppm in blood). Such ideal beam studies also show that as the energy is increased further the figures of merit for the beam rapidly deteriorate so that at energies much higher than 70 keV there is zero advantage depth and zero advantage ratio. The current ideal beam calculations, reported in these proceedings, should be augmented by further studies of the biological effect, on cell cultures, of keV energy neutrons under conditions more relevant to clinical radiotherapy conditions.

In addition, it should be pointed out, there is no single ideal beam for NCT, since the criteria (AD and AR) project different ideal neutron beam energies depending upon the boron-10 distributions and concentrations assumed. For example, the "standard" 30/3 (T/B) distribution, recommended for this Workshop for the purposes of intercomparison uniformity, is approximately representative of the "second generation" of boron

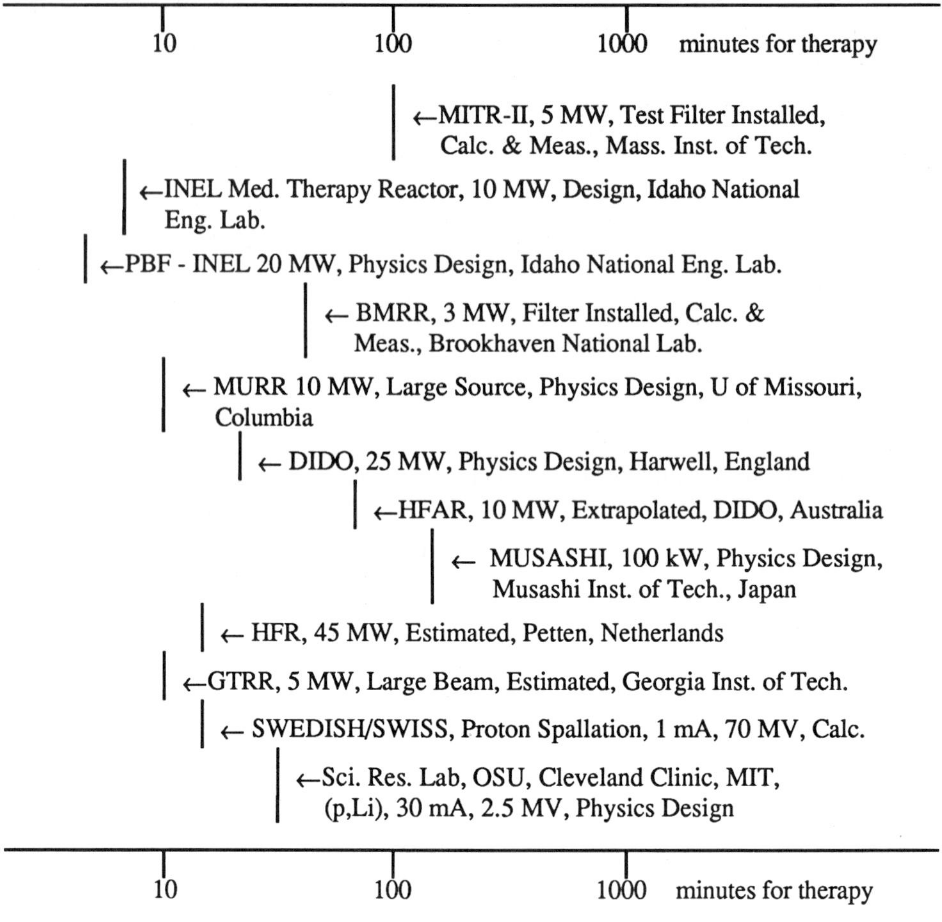

compounds, such as BSH monomer and dimer. However, it is possible that if antibody mediated boron delivery becomes a reality, very different boron distributions, both macroscopic and microscopic, may be observed. If boron concentrations in tumor were in the 5-10 ppm range with (T/B) ratios of 20-50, for example, neutron beams lower in energy than those mentioned above might be optimal.

<u>Dose Delivered to Sensitive Organs During NCT</u>

Several participants pointed to a need for the NCT community to examine carefully such issues as irradiation effects in sensitive organs such as the optic nerve, pituitary, and the brain stem. Such issues will have to be dealt with as new clinical trials are planned. Consequently an essential feature of any clinical treatment planning procedure should be the capability of prospectively and retrospectively evaluating radiation doses to such tissues. With the ability to correlate NCT dose levels with known dose tolerances of cranial tissues, or with observed radiation sequelae, a clearer understanding of these issues will be gained. We hope that these comments will encourage some of the NCT community members to investigate these concerns more thoroughly.

<u>Scalp Dose</u>

One potential problem that must be evaluated very carefully with neutron irradiation of patients through the intact scalp and skull is the possibility of serious delayed radiation injury to the scalp and its underlying tissues. Although moderate acute skin reactions, such as dry or moist desquamation, do not usually pose serious hazards for the patient, delayed skin reactions, such as radiation necrosis, potentially might, and may indeed be life-threatening. Delayed skin reactions typically occur from four months to a year following irradiation, although the latency is a function of both the dose level and of the area of skin irradiated. Hatanaka has reported, at this Workshop, the occurrence of serious acute effects arising from neutron interactions within scalp tissue during BNCT. The occurrence of late effects (e.g., vasculitis) should also be considered since one anticipates some long-term survivors among NCT patients. Rubin (ONCOVAIL '89) has reported $TD_{5/5} = 15$ Gy and $TD_{50/5} = 20$ Gy (single dose, low LET) for scalp vasculitis, which can be a serious, and even fatal, late complication of radiation therapy. Von Essen (cited in W.T. Moss, W.N. Brand, and H. Battifora, "<u>Radiation Oncology: Rationale, Techniques, Results</u>," 5th ed., chapter 3, St. Louis, C.V. Mosby Company, 1979) shows that using superficial X-ray irradiation of a 30 cm^2 area of skin in one fraction, a 3 percent incidence of delayed skin necrosis is observed at an exposure level of 1,900 R. Since the RBE of superficial X-ray irradiation of skin (for late reactions) is believed to be associated with an RBE significantly greater than 1, the cobalt-60 equivalent dose to this value would most likely be in excess of 2,000 cGy.

Making the conservative assumption that the entire neutron capture therapy scalp dose would be delivered in a single fraction (i.e., assuming that radiation damage repair is non-existent), a total scalp RBE-dose of up to 2,000 cJ/kg would most likely be well tolerated by the scalp with little likelihood of serious delayed radiation injury. The component of boron-10 dose to the scalp would depend on the distribution of the boron compound employed, but if the boron-10 were confined intraluminally, its contribution should be equivalent to what it would be in normal brain. With the current sulfhydryl boron compounds, evidence shows that after administration 95 percent or more of these compounds become rapidly bound to plasma globulins, and hence would not be expected to diffuse out of the blood. Therefore, any protective factors based on geometric absorbed fractions that have been proposed for evaluating boron-10 dose to cerebral vasculature would also be expected to apply to scalp.

Surface-sparing characteristics of epithermal-neutron beams are, therefore, desirable, although limiting total macroscopic RBE-dose to scalp to approximately 2,000 cJ/kg would alone most likely provide adequate protection against delayed radiation necrosis of the scalp. If boron compounds were to be employed in which extraluminal boron-10 concentrations would be expected to be high (such as, possibly, some of the melanoma seeking compounds), a careful reanalysis of these issues would need to be carried out.

Relevant information pertaining to scalp and brain tolerance to neutron capture therapy irradiations has been provided by R. Zamenhof in a private communication. The study was performed to investigate the tolerance dose of normal canine brain to neutron capture therapy performed through intact scalp and skull. A dose escalation protocol was established by maintaining the thermal-neutron fluence, from the MIT reactor, at a depth of 1 cm from the scalp surface at a constant value (1.0×10^{13} cm^{-2}) and manipulating the pharmacological dose of boron-10 compound (95% B-10 enriched sodium-mercaptoundecahydrododecaborate) administered and the time interval between end of compound administration and initiation of thermal-neutron irradiation. Irradiation was carried out through a 2.5 cm diameter collimator aperture to the left cerebral hemispheres of six colony bred beagle dogs of age 6 months.

At 8.5 months post-irradiation all dogs were sacrificed and gross and microscopic pathologic examination was carried out on the brains. The macroscopic (from blood-borne B-10) doses ranged from 1248 cGy to 3977 cGy. The corresponding macroscopic RBE-weighted doses ranged from 2245 to 8522 cJ/kg. None of the neutron irradiated dogs exhibited any abnormal clinical signs. Pathology indicated no radiation injury to either the scalp or brain tissue of any of the neutron irradiated dogs. Four additional beagle dogs were irradiated with 120 kV (peak) X-rays as controls, with doses ranging from 2000 to 4000 cGy. One of the X-ray irradiated dogs developed seizures and ataxia in the ninth month. On pathology, two of the X-ray irradiated dogs manifested classical signs of delayed radiation injury. It is concluded that a superficial cortical thermal-neutron fluence of 1.0×10^{13} cm^{-2} together with a blood boron-10 concentration of 39 ppm (boron-10 by weight) during 50-minute thermal-neutron irradiation of the canine brain through intact scalp and skull is easily tolerated when the background radiation components are adequately controlled.

CONCLUSION OF THE RAPPORTEURS

It is the consensus of this Workshop's rapporteurs that the technology currently available for epithermal beam design, construction, and dosimetric characterization is sufficient to make the neutron irradiation component of NCT potentially safe and effective for clinical therapy. A few facilities already possess existing epithermal beams suitable for clinical applications, while other facilities are in the course of designing and/or constructing such beams.

Overviews of Clinical Experience, Clinical Considerations, and Tumor Targeting

CLINICAL RESULTS OF BORON NEUTRON CAPTURE THERAPY*

H. Hatanaka

Professor of Neurosurgery
Teikyo University
Tokyo

The early American clinical trials of Boron Neutron Capture Therapy (hereinafter BNCT) for malignant brain tumors at Brookhaven National Laboratory (BNL) and at the Massachusetts Institute of Technology (MIT) ended in 1961 without having achieved the objectives of the therapy. The aim had been a highly selective destruction of tumor cells while preserving normal organ which harbors the tumor. The last series, 17 cases treated at the MIT Research Reactor, yielded an average survival of 5.7 months. Contrary to expectation, the survival was short and damage to the normal brain (including vascular damage and coagulation necrosis of the brain in the worst case) was found upon autopsy of these patients. BNCT was totally discontinued after 1961. However, it is important to note that the early clinical experience obtained between 1951 and 1961 and described in the literature of that period contributed to my revised therapy conducted after 1968 in Japan, although I had never taken part in the earlier American effort.

When I arrived in the United States for the first time in 1964, I was very much inspired by Dr. William H. Sweet, then the Chief of the Neurosurgical Service at Massachusetts General Hospital and concurrently Professor of Surgery at Harvard Medical School. Although Dr. Sweet was no longer active with BNCT, Dr. Gordon L. Brownell, the Chief of the Physics Research Laboratory at Massachusetts General Hospital and later Professor of Nuclear Engineering at MIT, and Dr. Albert H. Soloway, Associate Chemist in the Neurosurgical Service of Massachusetts General Hospital, were still studying the basic problems of BNCT. Besides these three people who had been engaged in the actual treatment of patients until 1961, a young physicist, Dr. Ralph G. Fairchild at Brookhaven, proposed the use of epithermal neutrons for BNCT so that deep-seated tumors could be easily reached by thermal neutrons. I studied boronated uracil to observe its selective incorporation into tumor nucleic acids in order to see if it would be a feasible idea to use such nucleic acid pseudo-precursors. I reproduced a vascular injury by irradiating cat's brain at the MIT Reactor after injecting one of the boron compounds originally used in the 1960-61 MIT series. Then, by electron microscopy, I demonstrated the remarkable protective effect of adrenocorticosteroid upon the injured cerebral vascular wall even though such a primitive boron compound had been used. Next, I studied the feasibility of using boron sulfhydrides. I continued the basic studies of such boron sulfhydrides back in Japan in 1967 and 1968. The representative form of such boron sulfhydrides was the well known $Na_2B_{12}H_{11}SH$ which was originally made by DuPont, and later supplied to the author by Shionogi Pharmaceutical Company in Osaka. As a neurosurgeon who has to face every day the sorrowful fate of patients with such malignant brain tumors, my conscience did not allow me to wait too long. On August 20, 1968, I treated the first patient in Tokyo.

---

* This paper was presented as an after-dinner talk at the reception for the Workshop on the evening of March 29, 1989 at MIT.

Before clinical trials could be started, there was a lot of red tape involved in my obtaining permission from the Japanese government. I had to persuade a committee composed of several renowned Japanese radiologists and physicians from other specialties. They gave me permission mainly for two reasons: 1) There was a patient who was destined to die and who was waiting helplessly for some new treatment for a tumor that had never responded to repeated surgery, chemotherapy, or conventional radiotherapy and 2) I had strong supporting evidence for the potential of the treatment including good tumor uptake of the boron compound then available to me, autoradiographic findings of selective distribution of the compound in the tumor cells, minimum toxicity, and the curative and selective effect of BNCT on animals carrying tumors [1]. Initially each proposal to treat a patient had to be reviewed by the Atomic Energy Commission and its subcommittee. Also, each treatment result had to be reported and again discussed. After I had treated 10 patients, the Japanese government relaxed the regulations and ceased requiring that I submit a pre-operative application for each treatment. This requirement was replaced by pre-operative and post-operative reports. Even now, after 20 years, each patient's name, age, sex, nationality, tumor and its location must be reported to the Science and Technology Agency under the Prime Minister.

In fact, the patients treated in the early days were mostly those with recurrent glioblastomas for whom all other conventional therapies had failed. Patients with such a history are usually excluded from any American or European cooperative studies for comparison of therapeutic results. In cooperative studies conducted in Europe or the United States, the criteria for the selection of patients are: 1) histological diagnosis established by surgical biopsy or removal, 2) full alertness of the patient before a concerned treatment, 3) no other chemotherapy or radiotherapy before such clinical studies, and 4) no hematological or hepatic function disorders. The purpose of these cooperative clinical studies is solely to assess the clinical effect of the treatment and not always to give actual aid to those patients in despair and distress. As a consequence, proponents of these studies advocate the so-called prospective randomized studies in which half of the enrolled patients may undergo an ineffective placebo treatment.

In my series before 1986, any patient who arrived asking for the treatment was accepted without regard to the severity of the patient's condition. Even patients who had had repeated surgical or radiological treatments were not excluded. For example, a 60-year-old man without spontaneous respiration was accepted and treated for his grade IV glioblastoma involving the brain stem (medulla) which contains the respiration center. Amazingly, he recovered his breathing 3 days following BNCT. A 3-year-old child with a recurrent astrocytoma in the pons recovered from a 3-month coma and developed sufficiently, physically and mentally, to attend elementary school. Critics of my BNCT program unanimously quote the fact that I do not conduct a randomized study and say that I may be treating only selected, favorable cases. But the truth is that most of my patients are in the terminal stage. Physicians who send their patients to me have a tendency to send only those patients with tumors growing in the technically untouchable areas of the brain. In other words, these patients' tumors are mostly in much deeper locations than where thermal neutrons can produce the best result. In addition, half of the patients in my series have been previously treated by radiotherapy and chemotherapy, thus lowering their tolerance to further radiation therapies. Furthermore, previous skin incisions and craniotomies by other surgeons are frequently not the best ones for BNCT. I prefer very large openings in the skin and skull so that thermal neutrons can be more evenly distributed in a deep tumorous area. It is true that the study I am conducting is certainly not randomized, but I am treating patients who are too gravely ill to be included in popular randomized cooperative studies.

I have treated 98 patients as of the end of March, 1989. These 98 present quite a variety of tumors with various histories. In this paper, I will only discuss grade III-IV gliomas which had never received previous chemotherapy or radiotherapy. Grade III-IV gliomas have traditionally been called "glioblastomas". But now there is a new tendency to call only grade IV a glioblastoma. Grade III gliomas are sometimes referred to as "anaplastic gliomas" and grade IV gliomas are termed "highly malignant tumors". To avoid confusion of nomenclature, I have made it a rule to use only the term "grade III-IV gliomas" in accordance with the World Health Organization (WHO) standard. Between 1968 and 1989 there were 49 such patients out of the total of 98 patients. These 49 patients can be easily compared with other people's statistics because grade III and IV gliomas are the most widely used as material for such European or American cooperative studies. (Note: Two patients who died within a month after surgery were excluded, thus making the total 49.)

There is another reason for choosing grade III-IV gliomas. These are extremely malignant tumors that kill people within a year or two, and thus can be used for a rapid assessment of the effect of a new therapy.

In the same period, I also treated 46 patients with grade III-IV gliomas by conventional therapies, including a combination of surgery, conventional radiotherapy, and chemotherapy.

Figure 1 shows data from four different groups of patients. Group 1 consists of patients treated by a combination of conventional treatments. It is marked as PHOTON because the most important part of the treatment was conventional radiotherapy with photons. Group 2 consists of all 38 patients with grade III-IV tumors who were treated by BNCT between 1968 and 1985, excluding preirradiated cases. These patients were treated without regard to location or depth of the tumor. Group 3 comprised the 12 patients sorted out from Group 2 because of the superficial location of their tumors. These tumors involved the superficial layer of the brain; in other words, they were within 6 cm of the cortical surface. As any physicist will note, these tumors are within the limit of the maximum therapeutic depth [2] of the current regimen of BNCT. Group 4 consisted of the 11 patients treated after 1985, without any exclusions.

All of the 'photon' patients of Group 1 died within 7 years of their treatment. Their 5-year survival rate was less than 3%. Group 2 patients, that is all of the BNCT patients between 1968 and 1985, had a 5-year survival rate of 19%. This rate will never satisfy me. However, the Group 3 patients, whose tumors were within the limits of maximum therapeutic depth, had a very satisfying rate of 58%. Some people have criticized me for selecting these "good" cases out of all my patients. However, Group 4 patients who constitute all of the patients treated for grade III-IV gliomas after 1985 have shown a result very comparable to Group 3. The reason is that after 1985 I adopted a new strategy for treating patients with deep-seated tumors. It entails using deuterium water to facilitate deeper penetration of thermal neutrons [3]. Patients with superficially located tumors are, of course, still welcomed. It can be safely said that the results of this new strategy are reinforcing the conclusion drawn from the Group 3 results.

By looking at these results, one must not be dismayed at the gloomy destiny of patients whose tumors are not within 6 cm of the surface because, by now, we know that the replacement of water in the brain with deuterium water and/or the use of epithermal neutrons will definitely increase the rate of survival of these patients.

I agree that a randomized study is important and useful for testing drugs. At the base of this philosophy there is the concept that the means of the therapy can be constant or, in other words, "standardized" without regard to differences among practicing physicians. However, it is nonsense to apply randomization to evaluate surgery. Surgical operations depend heavily not only on tools but also on individual surgeons.

Boron-neutron capture therapy for brain tumors is essentially radiosurgery, because BNCT is a type of surgery using radiological means as a tool. But how to use it depends on surgeons and the result depends on how it is used. You cannot expect the same result from different physicians using BNCT because they use the same means differently.

You can standardize BNCT as to method, but the neutron dose and where and how to deliver the neutrons are entirely variable, depending on each tumor, as in surgery. The decision about how to deliver the neutrons depends heavily on each surgeon's skill, experience, and knowledge of the disease. So there cannot be a randomized study. Post-

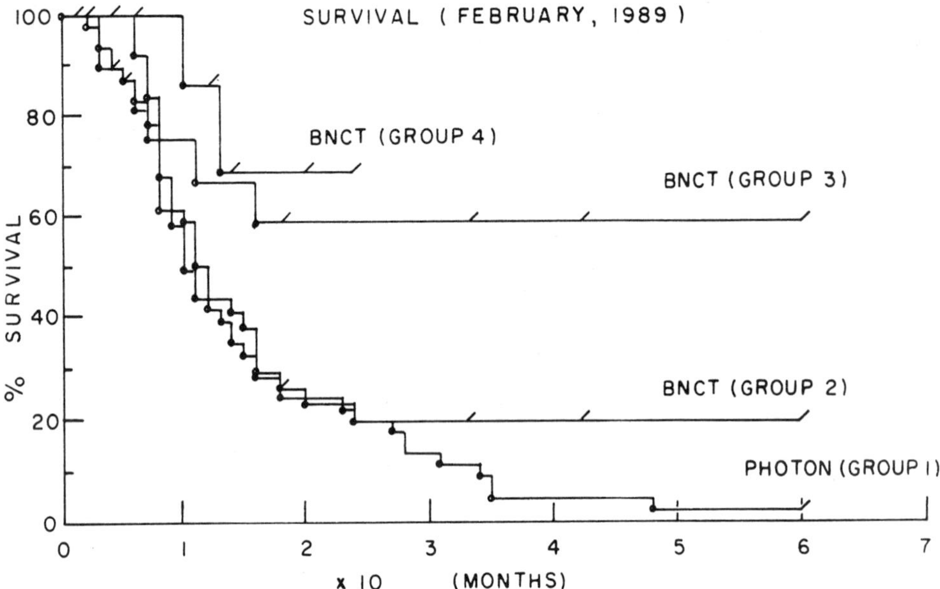

Figure 1. Abscissa is the number of months after the operation. For example, 6 (x10) is 60 months. The ordinate is % survival. Tumors are grade III-IV gliomas, excluding pre-irradiated cases.

Photon (Group 1):   All of the photon-plus-chemotherapy patients (linear accelerator or cobalt). (46 cases)

BNCT (Group 2):   All of the BNCT patients without regard to depth of the tumor, treated between 1968 and 1985. (38 cases)

BNCT (Group 3):   All of the BNCT patients with tumors found within 6 cm from the brain surface, treated between 1968 and 1985. (12 cases sorted out from 38 cases)

BNCT (Group 4):   All of the BNCT patients treated in the past 3 years. (11 cases)

It should be emphasized that the future of deep-tumor treatment is quite promising because we will be able to use deuterium water to facilitate the penetration of thermal neutrons into tissue and because epithermal facilities will become available.

operative care and treatment also require utmost foresight and experience. This means that physicians have to be trained in BNCT.

In fact, if you want to test my words, the easiest thing to do is to send me several patients rather than, as is now often the case, only one patient. Send me patients who fall within the criteria of patient selection that are accepted by many cooperative studies. I mean you should not send me only those patients who are far more terminally ill than those in cooperative studies. Also, you should not forget that I have only a thermal neutron beam facility with which I can satisfactorily treat tumors within only 6 cm of the surface of the brain except with the use of deuterium water.

After that, you should carefully follow the patient exactly as I do by discarding the misbelief that a glioblastoma is not curable. This adamant misbelief among physicians providing follow-up care has caused the loss of a number of patients who suffer from degenerating tumor tissue which gives some toxic effect to the surrounding brain matter. At this stage, dying tumors are often mistaken for recurrent tumors [4]. There was even a patient who committed suicide when he was told by a local radiologist that he had a recurrence.

I will now show you a case where a post-BNCT tumor eventually caused second-look surgery. Fortunately in this case, after the second-look surgery, although the patient lost some normal motor cortex tissue and became hemiparetic, she remains well otherwise. The so-called "recurrent" tumor tissue turned out to be highly necrotic tumor tissue. (Figures 2 and 3)

DISCUSSION AND CONCLUSION

Clinical results as a whole clearly indicate the need to improve delivery of neutrons to the tumors [5]. From my experience and by a simple calculation, it is concluded that if some 17 boron-neutron capture nuclear reactions occur within a single tumor cell, the cell will be destroyed. To attain this goal, one has to deliver a minimum of 2.5×10^{12} neutrons per gram of tumor when the average concentration of boron-10 is 25 μg ^{10}B/g. Factors that may assist in achieving this objective are:

1) Epithermal neutrons are an already well-known means to improve penetration of neutrons in the tissue.

2) The aperture of the neutron beam is important. If the aperture of the craniotomy bone window is large, the neutrons will reach deeper. A surgeon should bear this in mind before fashioning a skull opening.

3) Deuterium water, when it is used to replace tissue water or the cerebrospinal fluid in the spaces of the brain, will reduce attenuation of thermal neutrons and thus will facilitate a satisfactory penetration of neutrons. The price of deuterium water is by now reasonably inexpensive as compared to anti-cancer drugs. Acute toxicity in terms of LD_{50} measured by my group was exactly the same as water.

In conclusion I would like to say that physicians involved in malignant brain tumor treatment should start treating patients without waiting for the construction of an ideal epithermal beam because 500,000 people in the world are dying each year of malignant brain tumors and at least one-third of them could be saved without waiting for an epithermal beam.

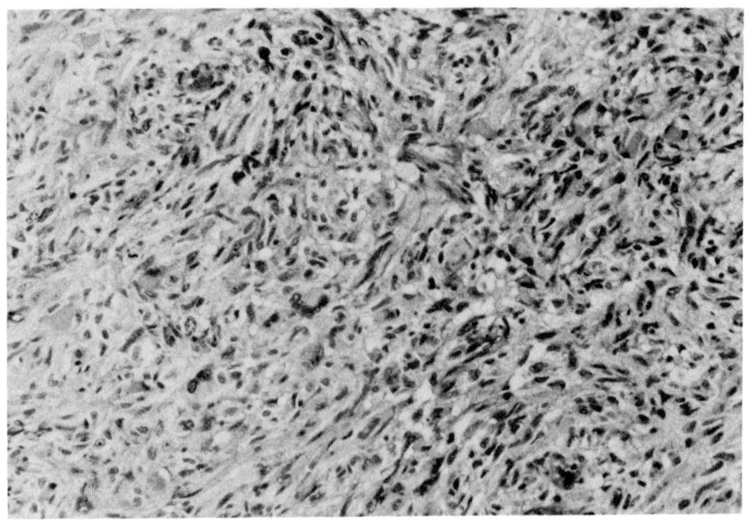

Figure 2. A Caucasian woman's glioblastoma microscopic specimen (200X magnification).

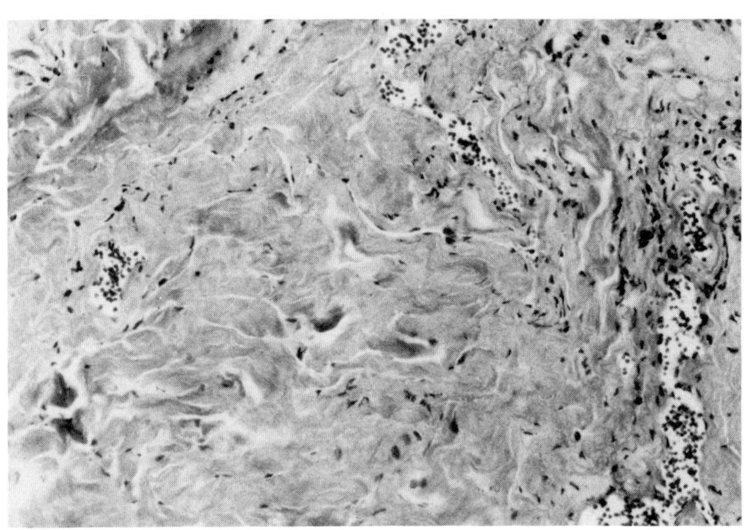

Figure 3. A "tumor" specimen obtained during second-look surgery 14 months after BNCT when clinically the patient suffered from symptoms resembling "recurrence." No viable tumor cells are recognized. Fibers indicating a degenerating process are seen together with small dark-stained (pyknotic) cells. Some of these cells are lymphocytes.

REFERENCES

1.   "Boron-Neutron Capture Therapy for Tumors," ISBN-4-89013-052-7, H. Hatanaka, ed., Nishimura & Co., 1-754-39 Asahimachi-dori, Niigata, Japan (1986).

2.   H. Hatanaka et al., "Pathological Studies of Maximum Therapeutic Depth by the Current Standard Technique of BNCT," <u>Third Int. Symp. on Neutron Capture Therapy</u>, Bremen, FRG, 31 May–June 3, 1988.

3.   T. Kadosawa, A. Takeuchi, T. Nakaichi, T. Matsumoto, T. Nozaki, O. Aizawa, M. Takagaki, M. Moritani, K. Amano, H. Hatanaka, Y. Hayakawa, and T. Inada, "Basic Research in Neutron Capture Therapy (Part I) - Preliminary Study on the Use of D_2O for Boron-Neutron Capture Therapy," <u>Annual Report of Joint Studies at Musashi Institute of Technology Reactor</u>, 12:9 (1987).

4.   H. Hatanaka, "Experience of Boron-Neutron Capture Therapy for Malignant Brain Tumours - with Special Reference to the Problem of Postoperative CT Follow-Ups," <u>Acta Neurochirurgica.</u>, Suppl., 42:187 (1988).

5.   G. L. Brownell, "A Reassessment of Neutron Capture Therapy in the Treatment of Cerebral Gliomas," in <u>Proc. Seventh National Cancer Conf.</u>, p. 827 (1973).

CLINICAL CONSIDERATIONS FOR NEUTRON CAPTURE THERAPY OF BRAIN TUMORS*

H. Madoc-Jones,[1] D. E. Wazer,[1] R. G. Zamenhof,[1] O. K. Harling,[2] and
J. A. Bernard, Jr.[2]

[1] Department of Radiation Oncology
Tufts – New England Medical Center
Boston, MA

[2] Nuclear Reactor Laboratory
Massachusetts Institute of Technology
Cambridge, MA

INTRODUCTION

In welcoming this Workshop on Neutron Capture Therapy to Boston, it would be appropriate to reflect upon the fact that, at the turn of the century, Boston was a major center for the development and use of x-rays to cure cancer. This was summarized by Dr. Francis H. Williams in his book entitled, "The Roentgen Rays in Medicine and Surgery as an Aid in Diagnosis and as a Therapeutic Agent," published in 1901. Dr. Williams published some of the first photographs documenting the curative potential of the then newly discovered x-rays for the treatment of skin cancers. In the ensuing decades, the science and technology of ionizing radiation has been developed and expanded to effectively treat a wide variety of tumors. For many, radiation therapy as a primary or adjuvant treatment either leads to cure of their disease or is a highly effective means of palliation. However, the results of even modern radiotherapy for the management of primary brain tumors and tumor metastases to the brain from primaries arising elsewhere in the body have been extremely disappointing. In spite of significant advances in our understanding of tumor biology and technical advances in radiation delivery, the results of radiotherapy for these tumors have not changed significantly during the past twenty years. We are, therefore, very excited at the potential clinical impact of neutron capture therapy (NCT).

As we begin this Workshop, I would like to acknowledge that today, the primary clinical impetus for NCT is thanks to the early pioneering work of Professor William Sweet, in the United States in the 1950s and 1960s, and to the subsequent work of Professors Hiroshi Hatanaka and Yutaka Mishima in Japan since the late 1960s. In addition, recognition must be afforded to the many other investigators in various fields of science and engineering who collectively built a solid scientific foundation for NCT in the intervening years.

---

* This paper was presented as the opening address for the Workshop.

Neutron Beam Design, Development, and Performance for Neutron Capture Therapy
Edited by O. K. Harling *et al.*
Plenum Press, New York, 1990

PRIMARY BRAIN TUMORS

In the United States, there are approximately ten thousand new primary brain tumor cases annually [1]. Most of these tumors are derived not from neural tissue but from the supporting stroma and are classified as astrocytomas. Therapy for these primary brain tumors depends upon the histologic grade and their location within the brain. The role of surgery is based on maximal cytoreduction and ranges from a simple biopsy to complete resection, although one can never be certain that any such resection is really complete. The role of radiation therapy is dependent upon the extent of the surgical resection and also the histologic grade. Because few tumors, if any, can be completely resected with surety, most patients receive post-operative radiation therapy. The exceptions are patients with very low-grade tumors which have a long natural history and, therefore, may simply be followed carefully after surgical resection.

Standard radiotherapy for the treatment of high-grade brain tumors following a biopsy or subtotal resection is to give external beam radiation with high-energy x-rays (4 or 6 MV) to a dose of approximately 60 Gy in fractions of 1.8 or 2.0-Gy daily, five days a week. Recently, there have been a number of series published in which high dose interstitial radiation has been added to classical external beam radiation to provide localized tumor doses up to 160 Gy [2,3]. Unfortunately, despite these high doses, a clear therapeutic benefit has not yet been proven. Average results of treatment show that for anaplastic astrocytomas, the median survival is approximately twenty-seven months, whereas for glioblastoma multiforme it ranges from eight to fourteen months [1]. Many years ago, Dr. Juan Taveras (Chairman Emeritus of Diagnostic Radiology at the Massachusetts General Hospital) demonstrated that untreated glioblastoma multiforme results in a median survival of approximately three months. Hence, the effect of the external beam radiation delivered at classical doses is to add only approximately six months to the survival of the patient.

One of the main problems in any form of treatment for brain tumors is the difficulty in defining the tumor border because there may be a tendency for disease to infiltrate diffusely into surrounding normal brain tissue. Figure 1 shows a CT scan of a patient with an astrocytoma in the left posterior parietal area. It is difficult to see any clear margins despite the use of contrast material. This illustrates the difficulty that the radiotherapist has in deciding how to draw a field to include the main body of the tumor for a high dose boost. Figure 2 compares another astrocytoma viewed by CT scanning (right) with a magnetic resonance image (left). It can be seen that the magnetic resonance image shows a somewhat different picture than the CT scan. This T2-weighted MR image shows a large area of edema around the main disease. This edema is not clearly seen on the CT scan and it is not obvious whether this area should be encompassed in the high dose boost volume or not. Both CT and magnetic resonance images only show a picture of gross disease without an accurate assessment of the potential microscopic infiltration of tumor cells beyond the main mass. At the Mayo Clinic, Daumas-Duport et al. [4] have proposed a classification of astrocytomas based on the presence of isolated tumor cells around the periphery of the main tumor mass as illustrated in Figure 3. This classification scheme identifies three types of tumor growth: the first has few isolated cells away from the main bulk of disease; the second has a central area of densely located tumor cells with some areas of infiltrating fingers containing isolated tumor cells; and the third is a tumor where there is no real central core but the tumor cells are diffusely distributed. Clearly, unless edema is associated with microscopic distribution of cells, both CT and MRI images may underestimate the extent of tumor in both types 2 and 3.

These data suggest that if high dose irradiation is limited to a small volume encompassing only the contrast-enhanced area noted on the CT scan, then a subset of patients may have tumors that fall beyond this target volume. An example of this is depicted in Figure 4. Shown is a patient who was treated with a combination of external beam and interstitial implant radiation to a dose of 160 cGy for a right parietal tumor which recurred away from the primary in both the right and left frontal lobes.

It has been suggested in one series that serial stereotactic biopsies may be performed [4] to assess the distribution of microscopic disease. However, this could raise

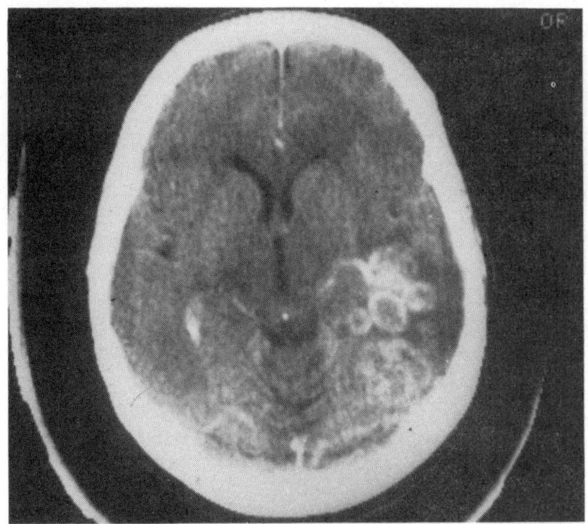

Figure 1. Contrast-enhanced CT scan of a patient with a high-grade glioma of the left parietal/occipital lobes.

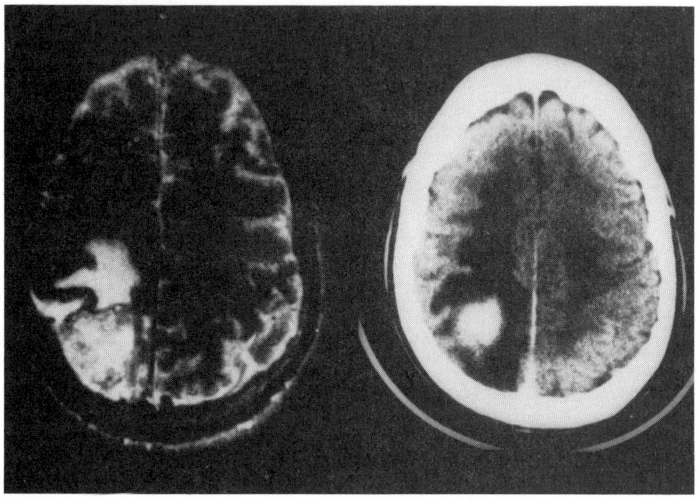

Figure 2. Comparison of MR (left) and CT (right) images in the same patient with a high-grade glioma of the right parietal/occipital lobes.

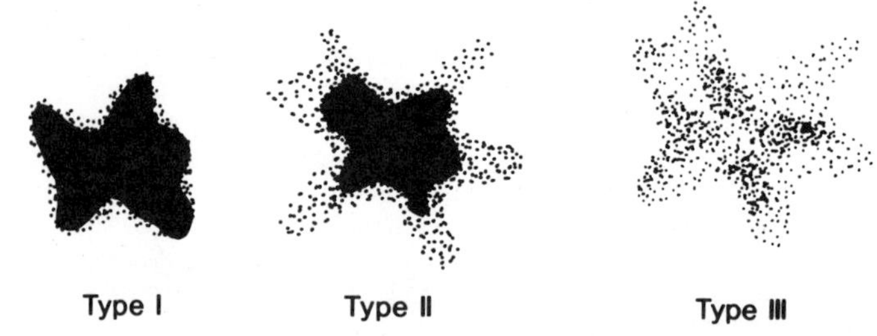

Figure 3. Classification scheme for the pathologic micro-anatomy of astrocytomas. Adapted from Daumas-Duport et al. [4].

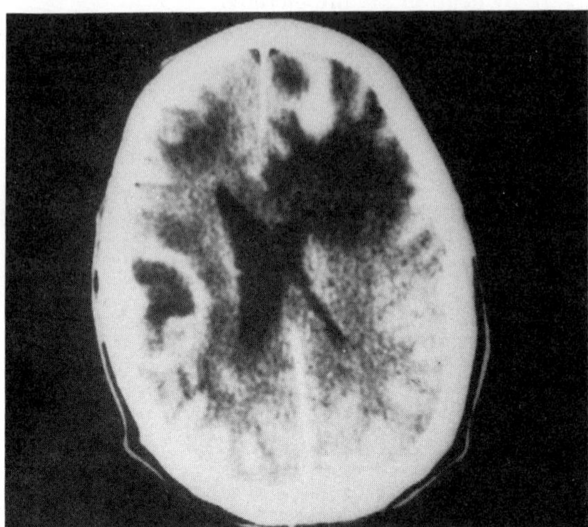

Figure 4. Contrast-enhanced CT scan of a patient with a high-grade tumor presenting in the right parietal lobe who failed in both frontal lobes after high-dose irradiation.

morbidity, would be very expensive, and has not as yet been shown to be of benefit. Therefore, we are left with the necessity of using a so-called best estimate by the clinician based upon CT and MRI images. Despite these limitations, clinical practice forces us to rely upon CT and T2-weighted magnetic resonance images in order to define the target volume for classical radiotherapy and we would need to do the same for neutron capture therapy unless we decide to treat the entire brain volume.

MELANOMA BRAIN METASTASES

In addition to the need to improve radiotherapy for primary brain tumors, there is an equally important need to improve therapy for brain metastases. While metastases from lung or breast tumors are most common, melanoma is the third most common tumor found to metastasize to brain. This is of particular interest to those involved in neutron capture therapy programs because Professor Mishima in Japan has demonstrated the principle of treating melanoma metastatic to other sites (extremities, neck) using boronated phenylalanine as a compound capable of incorporating boron preferentially into melanoma cells.

In Australia, the incidence of melanoma metastatic to the brain is actually greater than the incidence of primary brain tumors. In the United States, there are approximately 4,200 symptomatic presentations of metastatic melanoma in brain annually [5]. Of patients with cutaneous melanoma, six to eleven per cent will eventually develop clinical evidence of CNS metastases. Moreover, autopsy studies suggest that the real incidence is much higher, ranging from thirty-six to ninety per cent. Standard therapy for melanoma brain metastases is usually whole brain irradiation which only provides effective palliation. There may be a limited role for surgical resection of solitary peripheral metastases, followed by whole brain prophylactic irradiation. The median survival of patients treated conventionally for melanoma brain metastases is only four months, generally worse than for high-grade primary brain tumors. More than half of the patients treated for metastatic brain melanoma die as a direct consequence of the progression of the disease in the central nervous system [5]. At this time, many fractionation regimens for classical radiotherapy have been tried but none has demonstrated any clear superiority. As a result, the most commonly used dose is 30 Gy in ten fractions. Because melanoma brain metastases are frequently present at multiple sites (Figure 5), it is necessary to treat the whole brain by conventional radiation or by neutron capture therapy.

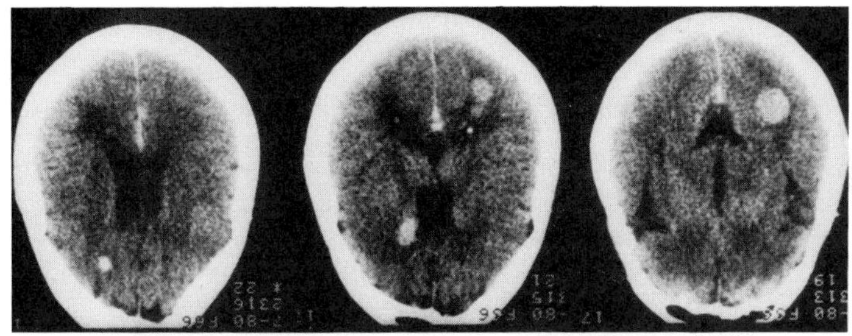

Figure 5. Contrast-enhanced CT scan at multiple levels of a patient with melanoma brain metastases.

NEUTRON CAPTURE THERAPY FOR GLIOBLASTOMA AND CNS MELANOMA

When considering neutron capture therapy for either primary brain tumors or metastatic melanoma in brain, the first question to discuss is how to deliver the neutron beam to the target tissue while sparing the intervening normal tissues. It is necessary to consider the effects of radiation on the intervening skin, bone, dura, and normal brain tissue. The failure of the first boron neutron capture clinical trial conducted by Professor Sweet and others in Boston and Long Island in the 1950s and 1960s was in part the result of radiation injury to the normal tissues of the scalp. One way to limit damage to the scalp, bone, and dura is to surgically reflect a large piece of calvarium and scalp. This was done for most of the patients in the U.S. trials and it was subsequently done by Professor Hatanaka in his personal series in Japan. Another approach to this problem is to use the superior depth-dose characteristics of epithermal neutron beams instead of thermal beams. For obvious reasons, it is this latter approach which would have the most practical impact because it would spare the patient the necessity of the neurosurgical procedure used to reflect scalp and bone. This in turn means avoidance of the attendant risks of infection and hemorrhage. The ability to treat through the scalp also makes available more options for

Table I

Summary of macroscopic (blood) and endothelial doses and RBE-doses (absorbed fraction of 1/3 applied to B-10 dose to give endothelial dose) for neutron irradiated dogs (R1-R6). Doses evaluated at 1 cm from scalp surface. Thermal neutron fluence: $1.0 \cdot 10^{13}$ cm^{-2} (50.5 minute irradiation time). RBE factors are: 2.3 for B-10 reaction, 1.6 for neutrons, and 1.0 for gammas.

Dog #	Macro-Dose (cGy)	Macro-RBE-Dose (cJ/kg)	Endo-Dose (cGy)	Endo-RBE-Dose (cJ/kg)
R1	1248	2245	782.5	1175
R2	1596	3047	899.0	1442
R3	2521	5173	1207	2151
R4	2800	5814	1300	2365
R5	2983	6236	1361	2505
R6	3977	8522	1692	3267

the design of optimal irradiation geometries and provides the capability to fractionate the treatment.

DOSE SCENARIOS FOR NEUTRON CAPTURE THERAPY

One question that needs to be resolved prior to a new U.S. clinical trial of NCT is what total radiation dose should be used. Ideally, one would need to know the maximum dose tolerated by normal brain tissue and also the effective tumoricidal dose. Because neither of these questions can be clearly answered at the present time, the design of a clinical NCT treatment protocol should incorporate a dose escalation schedule. Apart from questions regarding the optimal dose for NCT, should NCT treatment be delivered in a single irradiation, or should fractionation be employed – and if so, how many fractions? As we ponder these questions, we have in mind that classical radiotherapy with low-LET radiation is based upon the principle of fractionation in order to protect normal tissues and possibly, in some circumstances, to maximize the lethal effect of the radiation on tumor tissue.

There are few existing studies which can guide us in determining the tolerance of normal brain to NCT. One such study was performed by Dr. Zamenhof and his colleagues at MIT and the Harvard Medical School [6]. Table One is taken from the report of this study in which six healthy beagle dogs were irradiated to the left hemisphere of the brain with a single fraction of thermal neutrons from the MIT Research Reactor, MITR-II, after being injected into the carotid arteries with BSH boron compound (95% B-10 enriched).

The dogs were euthanized after nine months and detailed histological examinations of their brains were performed. This study demonstrated that, for the beagle brain, a total macroscopic dose by NCT of 3977 cGy delivered in one fraction produces no observable radiation effect at nine months, either clinically or histologically. Employing the RBE factors specified in the caption of Table One, this tolerance dose corresponds to an RBE dose of 8522 cJ/kg. The results of this study provide us with at least a starting point to evaluate what constitutes a "safe" dose in NCT.

Despite recognized similarities in the radiation response of human and canine brain, which would encourage us to use the results of the beagle study described above to determine NCT brain and scalp tolerance doses for humans, there are some important differences between the conditions of the beagle irradiations and those of our proposed protocol for a clinical trial of NCT.

First, the beagle brains were exposed to a thermal neutron beam, while we propose to treat patients with an epithermal neutron beam. For similar brain geometry, the higher integral, or volume-averaged, dose resulting from the epithermal beam irradiation would reduce the brain's tolerance. This would occur because the epithermal beam is more penetrating and hence affects a greater volume of the brain. However, the linear dimensions of a beagle brain are approximately three times smaller than those of a human brain. Consequently, we have estimated that the integral doses in the two cases (normalized to equal maximum dose) are approximately equal, if single field irradiation is employed. Given that for our clinical trial, we propose to treat the whole brain by parallel-opposed fields, the question arises as to whether the integral dose in the latter case would be increased. If we normalize single-field and parallel-opposed irradiations to equal tumor dose at midline, there is no significant difference between the integral dose to normal brain in the two cases.

Second, the beagle brains were exposed to single-fraction irradiation, while we propose to treat patients by fractionation. However, on the assumption that only the low-LET dose components will be sensitive to fractionation, we can incorporate this difference by applying a pseudo-RBE factor of 0.5 (for 4-6 fractions) to the low-LET dose component in developing our treatment plan.

These results in beagles give us some sense of the single fraction tolerance of normal brain tissue exposed to thermal neutrons. However, the cytotoxic components of the beam are not restricted to the densely ionizing particles that are a consequence of the boron-10 reaction. A large portion of the dose (as much as 50%) is due to low-LET contaminating photons. It is the presence of this low-LET dose component and its potential contribution to tumor cell kill as well as to normal tissue injury that highlights the question of fractionation.

Conventional radiotherapy with low-LET photons requires a fractionated course of treatment to establish a "therapeutic ratio" whereby normal tissue is spared in the process of killing tumor cells. The effect of fractionation on normal skin tissue is illustrated in Figure 6 whereby the effect of single large doses is compared to equivalent doses given in ten daily fractions. We can see that the S-shaped dose-response curve is shifted to the right as a consequence of fractionation, meaning that a higher total dose can be given with the same degree of skin reaction.

Clinically, normal tissue reactions are divided into "acute" and "late"-reacting tissues. Acute or early-reacting normal tissues are usually those with a short turnover time that manifest injury during or shortly after a course of treatment. Examples include the skin, mucous membranes, and lining of the gut. Late-reacting normal tissues tend to have a prolonged turnover time whereby radiation injury may not be manifest for months to years after a course of radiotherapy. Examples of this type of tissue include vasculature, fibrous connective tissue, and the supporting cells of the central nervous system (glia).

The shapes of the dose-survival curves for acute and late-responding tissues appear to be different and could account for their different sensitivities to fractionated treatment. As shown in Figure 7, the survival curve for late-responding tissues tends to be "curvier" with a broadened initial shoulder, followed by a steep exponential fall-off. Visual inspection of these curves suggests that late-responding tissues are more severely injured than acute-responding tissues when fraction sizes are very large.

The relative "curviness" of these survival curves can be more precisely described by fitting the data with a linear quadratic equation whereby surviving fraction is given by the expression $\alpha D + \beta D^2$, where D is the magnitude of the dose fraction. Figure 8 illustrates the type of dose curve obtained with this equation for low-LET radiations. It is postulated, in this model, that there are two components of cell killing: one (α) is proportional to dose (D) while the other (β) is proportional to the square of the dose (D^2). The relative influence of the αD and βD^2 terms can be described by the ratio α/β. This ratio tends to be much larger for acute-responding tissues than for late-responding ones.

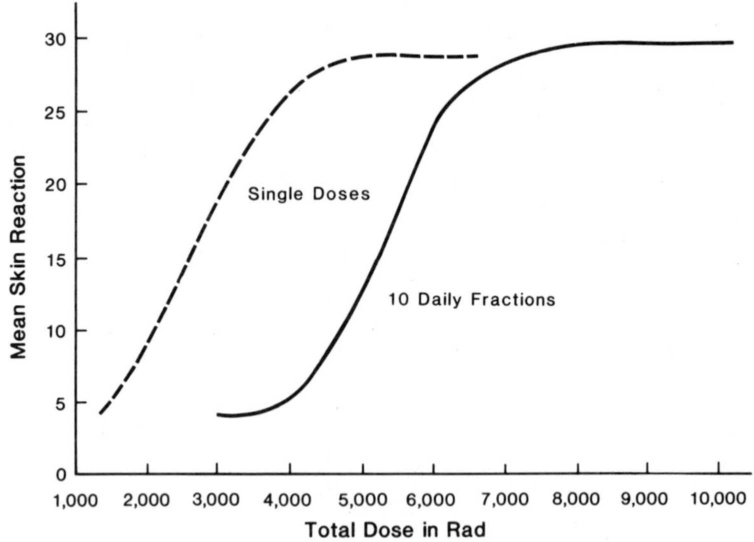

Figure 6. Dose-response curves for mouse skin reaction comparing single dose to ten daily fractions. Adapted from Hall [9].

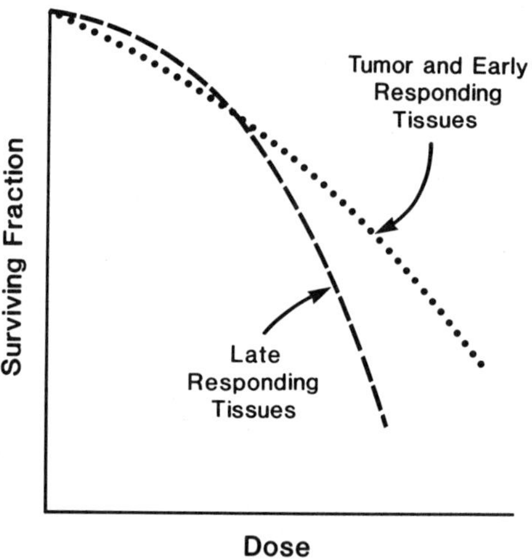

Figure 7. A comparison of the shapes of dose-survival curves for early (acute) and late-responding tissues. Adapted from Hall [9].

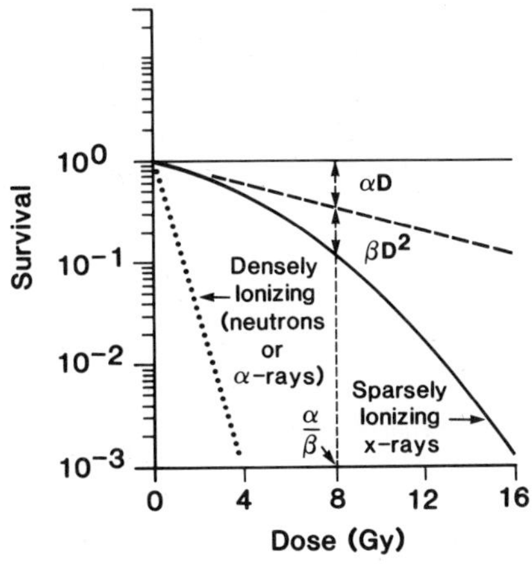

Figure 8. Dose-survival curve shape as described by the linear quadratic model. Adapted from Hall [9].

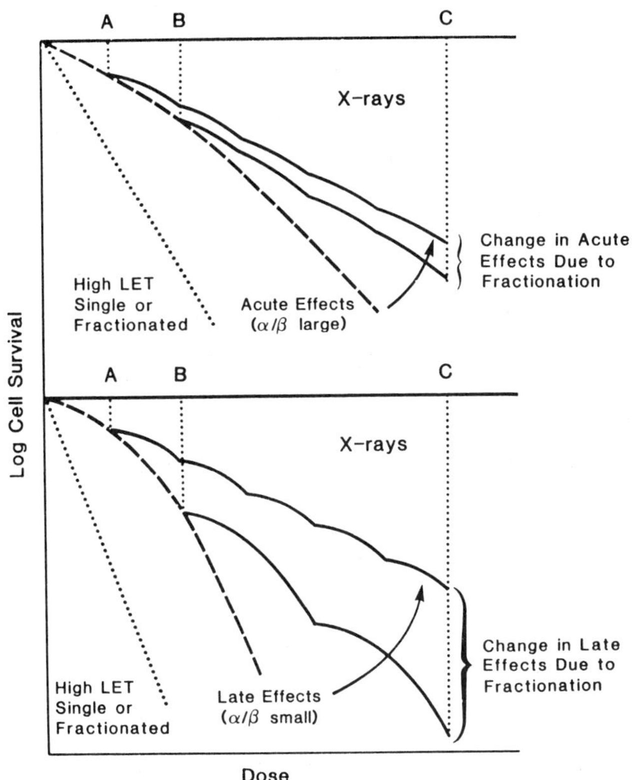

Figure 9. A comparison of survival curves, single dose and fractionated for early (acute) and late-reacting tissues. Adapted from Hall [9].

The relevance of these findings to clinical fractionation is shown in Figure 9. A fractionated course of treatment with fraction size A is compared to a course with a larger fraction size B. In those tissues with a large α/β ratio, the change in log cell survival with fraction size at total dose C is minimal when compared to the difference seen in tissues with a small α/β ratio. This suggests that acutely-responding tissues are much less sensitive to the effects of fractionation than are late-responding tissues. This finding is quite consistent with clinical observations. However, it needs to be emphasized that these relationships are true only for low-LET radiation. As shown in Figures 8 and 9, high-LET radiations, such as fast neutrons, have a purely log-linear dose-survival curve and are minimally influenced by fractionation.

These data indicate that the risk of injury by the low-LET component of a mixed-LET radiation field (as encountered in NCT) to the supporting stroma of the brain and overlying scalp and bone would be minimized by a fractionated course of treatment. However, one must also consider the potential contribution of the low-LET component to tumor kill and whether this would be affected by fractionation. The answer to this is less clear.

Most tumors encountered in clinical practice tend to exhibit radiobiologic parameters that are similar to acute-responding tissues, i.e., with a high α/β ratio. However, this can be variable. In Figure 10, using data from Deacon et al. [7], the surviving fraction at 2 Gy is compared for a variety of tumor types which have been grouped into five categories relative to their clinical radioresponsiveness. We see that there are clear differences in the range of surviving fractions when comparing groups A through E. A separate analysis reveals a striking difference in the α/β ratios of group A versus group E with mean values of 60.4 and 5.77, respectively [8]. This is further illustrated in Figure 11 (left) where the dose-survival curves have been plotted for different tumor types. These curves suggest that glioblastomas and melanomas are less radiosensitive to low-LET radiation than is adenocarcinoma. However, these data must be interpreted with extreme caution because such survival curves represent a mean of multiple determinations. Figure 11 (right) gives a more accurate view of the broad heterogeneity in radiosensitivity that can be encountered within a single histologic category. This makes any sweeping generalizations about the radiosensitivity of specific tumor type relative to normal tissue hazardous.

We conclude that a clinical trial of NCT should be fractionated to minimize the risk of complications from the low-LET component of the beam. Ideally, the fraction size chosen should allow for a low-LET dose component in the range of 2.3 - 3.0 Gy per fraction. This would approximate the dose used in conventional radiotherapy and thereby exploit empirically-derived experience in optimizing low-LET fractionated treatment. It is unclear whether any sparing of tumor would occur with this approach. As demonstrated above, fractionation should have minimal effect on the cytotoxic effect of the high-LET dose component, such as the dose due to boron-10.

In order to develop objective treatment planning criteria for NCT, it is necessary to estimate the different RBE factors for the various radiation components comprising the dose field. In the case of the low-LET gamma component of dose (and the low-LET fractions of other dose components), the RBE factor is generally assumed to be equal to 1. However, with fractionation, the low-LET tolerance of brain increases and the RBE factor (relative to a single-fraction reference) decreases. From experience gained over many years of conventional radiotherapy of brain, the delivery of a given dose in four to six fractions versus a single fraction increases brain tolerance with respect to late effects by approximately a factor of 2. Thus, the _effective_ RBE of the low-LET dose components may be considered to be 0.5. As mentioned above, this makes sense if our reference condition is the delivery of the dose in a single fraction. But why should one reference one's RBE assessments to a single fraction treatment? Because the most important dose component in NCT is the very high-LET B-10 dose, which to all intents and purposes is uninfluenced in its biological effect by fractionation, it is logical to reference any

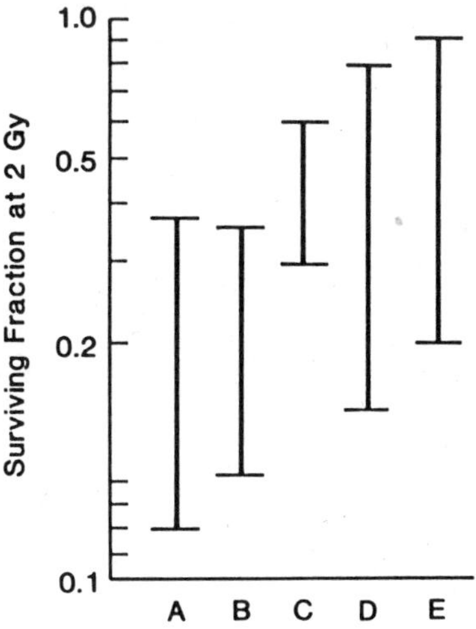

Classification of Human Tumors according to Clinical Radioresponsiveness

A. Neuroblastoma, lymphoma, myeloma

B. Medulloblastoma, small-cell lung carcinoma

C. Breast, bladder, cervix carcinoma

D. Pancreas, colorectal, squamous lung carcinoma

E. Melanoma, osteosarcoma, glioblastoma, renal carcinoma

Figure 10.   A comparison of surviving fractions at a dose of 2 Gy for various tumor types grouped according to radioresponsiveness. Error bars depict range of values observed. Adapted from Deacon et al. [7].

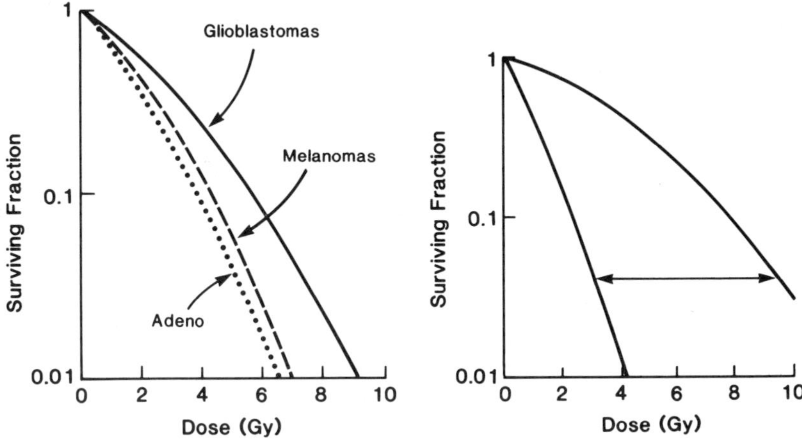

Figure 11.   <u>Left</u> – Average *in vitro* survival curves for glioblastoma, melanoma, and adenocarcinoma cell lines. <u>Right</u> – Range of survival curves obtained for 36 separate melanoma cell lines. Adapted from Fertil and Malaise [10].

fractionation-mediated effects (such as the biological effects of the low-LET dose) to the high-LET dose component which behaves **as though** it were delivered in a single fraction.

For the intermediate-LET dose components, such as protons generated following thermal neutron capture by nitrogen in the brain and recoil protons generated by the collisions of epithermal and fast neutrons with hydrogen nuclei, the RBE factor that is frequently agreed upon is 1.5 to 2. Of course, it is well known that RBE decreases with increase in dose, but at typical brain-tolerance doses, the RBE for intermediate to high-LET radiations has essentially "bottomed out."

For the high-LET dose components, specifically the B-10 dose, the assessment of RBE is much more problematic. Unlike the case of the low and intermediate-LET dose components, which produce a relatively homogeneous ionization distribution at the cellular level, the ionizations from the alpha and Li-7 fragments of the B-10 reaction are generally inhomogeneous, because of the inhomogeneous distribution of boron in tissue at the cellular and sub-cellular levels. For instance, with regard to normal brain, it is believed that the tissue most at risk from B-10 dose is the microvasculature, because with the current boron compounds of interest for NCT, blood concentrations of boron are known to be much higher than parenchymal brain tissue concentrations. For the response of microvasculature to B-10 dose, RBE factors of 2-3 have often been assumed. However, in the case of tumor cells, the intracellular mechanisms for the incorporation of boron strongly influence the biological response of the cell. Boron compounds such as sodium-mercapto-undecahydro-dodecaborate (BSH) or para-borono-phenylalanine (BPA), which do not appear to be selectively incorporated in tumor cell nuclei, would be expected to have a lower RBE for the B-10 dose than compounds such as porphyrins, which are expected to manifest greater degrees of nuclear incorporation, and hence bring the critical DNA ionization targets closer to the locations of the B-10 disintegrations. RBE factors of up to 7 have been measured for melanoma cells exposed to melanin precursors labeled with B-10.

PROPOSED CLINICAL PROTOCOL AT TUFTS – NEW ENGLAND MEDICAL CENTER AND MIT

Based upon the above considerations, we have developed the outline of a clinical protocol using the epithermal beam currently under development at the MIT Research Reactor. First, for primary brain tumors, the tumor would initially be resected as completely as possible as part of a debulking strategy. After a 2–3 week delay to allow for healing, the patient would be infused with a boron compound. After a 12-24 hour delay to allow tumor/blood B-10 ratios to be optimized (to be determined by blood and tissue kinetic measurements of boron concentration), the patient would be irradiated in 4–6 fractions to a total normal brain RBE dose of 1000-2000 RBE-cGy (cJ/kg), using parallel-opposed fields, through the intact scalp and skull. The irradiation time is anticipated to be approximately 15–30 minutes. The main consideration in wishing to pursue a fractionated regimen, albeit a minimal fractionation compared with conventional x-ray therapy, is the need to spare the normal tissues from the late-effects of the low-LET gamma components of the radiation field.

SUMMARY

The radiotherapeutic management of primary brain tumors and metastatic melanoma in brain has had disappointing clinical results for many years. Although neutron capture therapy was tried in the United States in the 1950s and 1960s, the results were not as hoped. However, with the newly developed capability to measure boron concentrations in blood and tissue both quickly and accurately, and with the advent of epithermal neutron beams obviating the need for scalp and skull reflection, it should now be possible to mount such a clinical trial of NCT again and avoid serious complications. As a prerequisite, it will be important to demonstrate the differential uptake of boron compound in brain tumor as compared with normal brain and its blood supply. If this can be done, then a trial of boron neutron capture therapy for brain tumors should be feasible. Because boronated phenylalanine has been demonstrated to be preferentially taken up by melanoma cells through the biosynthetic pathway for melanin, there is special interest in a trial of boron neutron capture therapy for metastatic melanoma in brain. Again, the use of an epithermal beam would make this a practical possibility. However, because any epithermal (or thermal) beam must contain a certain contaminating level of gamma rays, and because even a pure neutron beam causes gamma rays to be generated when it interacts with tissue, we think that it is essential to deliver treatments with an epithermal beam for boron neutron capture therapy in fractions in order to minimize the late-effects of low-LET gamma rays in the normal tissue.

I look forward to the remainder of this Workshop, which will detail recent progress in the development of epithermal, as well as thermal, beams and new methods for tracking and measuring the uptake of boron in normal and tumor tissues.

ACKNOWLEDGMENTS

This research was supported by Grant No. DE-FG02-87ER-6060 from the U.S. Department of Energy, Office of Health and Human Assessments.

REFERENCES

1. V. Levin, G. Sheline, and P. Gutin, "Neoplasms of the Central Nervous System," in Cancer: Principles and Practice of Oncology, 3rd Ed., V. DeVita, S. Hellman, and S. Rosenberg, eds., Lippincott, Philadelphia, p. 1557 (1989).

2. M. Chun, P. McKeough, A. Wu, D. Kasdon, D. Heros, and M. Change, "Interstitial Iridium-192 Implantation for Malignant Brain Tumors," Brit. J. Radiol., 62:158 (1989).

3. S. Liebel, P. Gutin, W. Wara et al., "Survival and Quality of Life after Interstitial Implantation of Removable High-Activity Iodine-125 Sources for the Treatment of Patients with Recurrent Malignant Gliomas," Int. J. Radiat. Oncol. Biol. Phys., 17:1129 (1989).

4. C. Daumas-Duport, B. W. Scheithauer, and P. J. Kelly, "A Histologic and Cytologic Method for the Spatial Definition of Gliomas," Mayo Clinic Proc., 62:435 (1989).

5. M. Amer, M. Al-Sarraf, L. Baker, and V. Vaitkevicius, "Malignant Melanoma and Central Nervous System Metastases," Cancer, 42:660 (1978).

6. R. G. Zamenhof, W. S. Schoene, G. L. Brownell, G. R. Wellum, H. Hatanaka, A. Takeuchi, and M. Shalev, "An Investigation of the Tolerance of Canine Brain to Thermal Neutron Capture Therapy." (In Preparation.)

7. J. Deacon, M. J. Peckham, and G. G. Steel, "The Radioresponse of Human Tumors on the Initial Slope of the Cell Survival Curve," Radiother. Oncol., 2:317 (1984).

8. M. M. Elkind, "The Initial Part of the Survival Curve: Does It Predict the Outcome of Fractionated Radiotherapy?," Radiat. Res., 114:425 (1988).

9. E. J. Hall, Radiobiology for the Radiologist, 3rd Ed., Lippincott, Philadelphia, PA (1988).

10. B. Fertil and E. P. Malaise, "The Mean Inactivation Dose: Experimental Versus Theoretical," Radiat. Res., 108:222 (1986).

TUMOR TARGETING AGENTS FOR NEUTRON CAPTURE THERAPY

A. H. Soloway, F. Alam, R. F. Barth, A. K. M. Anisuzzaman
and B. V. Bapat

The Ohio State University
Columbus, OH

The development of compounds for use as delivery agents in the treatment of cancer by neutron capture therapy has largely focused on the use of boron-10 even though there are other nuclides having higher capture cross section values for thermal neutrons. The major reasons are: (1) it is nonradioactive and comprises approximately 20% of naturally occurring boron; (2) boron-10 enriched materials are readily available commercially; (3) the products of the fission reaction are largely high LET recoiling ^{7}Li and ^{4}He (alpha) nuclei; (4) the pathlength of these particles is such that they are confined to a radius approximating the diameter of a single cell; and (5) the chemistry of boron is such that it may be incorporated into a multitude of different stable chemical structures.

Current boron chemistry concepts and target cell affinity as applied to neutron capture therapy are now beginning to become more clearly delineated. Compounds are now being synthesized which have a biological basis for tumor accretion. Even so, our knowledge as to the basis by which particular compounds may be sequestered by tumor and yet at the same time either be excluded from normal structure or else be present with markedly reduced concentration in these structures remains rudimentary and fragmentary at best.

In order to simplify this presentation, we shall divide the substances into two broad categories. Those which are administrated as low molecular weight substances and those substances which may be classed as biopolymers or higher molecular weight materials. Secondly, the types of malignancies where these compounds are to be used will be indicated.

The first group of small molecules are the sulfhydryl-containing polyhedral boranes and their oxidation products. Nearly twenty-five years ago, two of these compounds, synthesized by chemists at E. I. Dupont, Inc. were shown to have a capability to achieve a significant concentration differential between gliomas and normal brain/blood in rodents [1]. The compounds shown in Figure One were the ones used. In all likelihood, the preparations based upon the methods of synthesis and handling contained disulfide analogues as well as

SULFHYDRYL-CONTAINING POLYHEDRAL BORANES

$$Na_2B_{12}H_{11}SH$$

$$Na_2B_{10}Cl_8(SH)_2$$

Figure 1.

Neutron Beam Design, Development, and Performance for Neutron Capture Therapy
Edited by O. K. Harling *et al.*
Plenum Press, New York, 1990

Table One

Tumor and Blood Distribution of $B_{12}H_{11}SH^{2-}$

| Micrograms of Boron / g | | Tumor : Blood Ratio |
Tumor	Blood	
17.2	4.4	3.9
21.8	5.0	4.4
6.5	1.7	3.8
4.4	2.1	2.1
33.0	6.2	5.3
6.0	2.2	2.7
37.6	2.1	17.9
19.9	3.9	5.1
4.2	1.1	3.8
5.0	1.8	2.8
13.3	2.0	6.6
7.5	0.9	8.3
23.4	3.4	6.9
18.7	3.5	7.7
		Average: 5.7

Dose: 35 µg Boron/g mouse/day

Total Dose: 175 µg Boron/g

the mercapto entity. As a result of these initial promising animal studies [2], shown in Table One, Dr. Hiroshi Hatanaka with the important involvement of chemical researchers at Shionogi Research Laboratories began synthesizing and evaluating the chemically pure sodium mercaptoundecahydro-<u>closo</u>-dodecaborate in animals [3] and in man [4]. A recent study has confirmed its high incorporation in a human malignant astrocytoma [5]. Using this sodium salt of $B_{12}H_{11}SH^{2-}$, Hatanaka has been treating patients with malignant brain tumors by boron neutron capture therapy (BNCT) over the past 20 years [6]. His clinical results have indicated that this binary system may have great potential not only for brain tumors but for the treatment of a variety of different malignancies. However, the biochemical and physiological bases by which these compounds localize and achieve significantly elevated concentrations in neoplasms vis-à-vis normal structures still remain unclear. Certainly there is a major difference physiologically between the $B_{12}H_{12}^{2-}$ anion and its mercapto counterpart. Various researchers have shown that the mercapto compounds do have the potential for forming mixed disulfides with disulfide moieties on various plasma proteins [7]. Such a covalent linking of boron groups to proteins may be cleaved by mercapto compounds like dithiothreitol. This indicates that attachment is probably via the formation of a new disulfide group. However, it is uncertain whether the capability to become incorporated in plasma proteins is the basis for accretion of malignant cells. Researchers have speculated that such boron-containing proteins are endocytosed possibly in a manner analogous to the internalization of antibodies into malignant cells. Those studies have now demonstrated a significantly greater accretion and diminished degradation of these proteins by neoplastic cells in comparison with their corresponding normal cells [8]. Thus, the attachment to plasma proteins, such as albumin, has been viewed as a possible basis for the more selective concentration of these compounds in tumor. However, such studies remain to be carried out in order to determine the validity of this hypothesis. Recently, there has been increasing interest in the disulfide or dimer and its oxidation product, $B_{12}H_{11}S(O)SB_{12}H_{11}^{4-}$ (Figure Two). Studies by Joel, Slatkin, and Fairchild at Brookhaven National Laboratory have indicated that the disulfide appears to achieve higher concentration in tumor than does the monomer [9]. At the same time, there appears to be increased toxicity as demonstrated by

1. $B_{12}H_{11}SH^{2-}$; $(B_{12}H_{11}S-)_2^{4-}$; $B_{12}H_{11}S(O)SB_{12}H_{11}^{4-}$

2. $B_{10}Cl_8(SH)_2^{2-}$; $B_{10}H_9SH^{2-}$

3. Potential New Sulfhydryl Compounds:

$B_{20}H_{17}SH^{4-}$; $B_{20}H_{17}SH^{2-}$

Figure 2.

elevated liver enzyme levels. The mechanisms by which the dimer achieves higher concentrations in tumor remains obscure. One possibility could be related to the facile generation of a stable free radical by homolytic cleavage of the disulfide [10] and its incorporation into proteins; an alternative is the disulfide's direct reaction with sulfhydryl-containing biopolymers. Certainly the mechanism by which these sulfur-containing polyhedral boranes achieve differential tumor localization is important, not only per se, but as the basis for the design of other boron structures which may achieve great selectivity for brain tumors and other malignancies. The lability of these sulfur-containing boranes has posed problems in developing analytical procedures designed to measure the various components in solution and in biological materials.

Another use of low molecular weight compounds in BNCT relates to the treatment of malignant melanoma. Mishima first proposed the incorporation of boron into melanin or its precursors as the basis for treating melanomas [11]. The joint Japan-Australia Workshop, spearheaded by Mishima and Allen, has given impetus to the development of boron compounds for this tumor. At Brookhaven National Laboratory, Fairchild, Glass, Coderre, and Micca have also contributed [12]. One of the compounds with which these three groups have been actively involved is p-boronophenylalanine. This structure was first synthesized by H. R. Snyder et al. [13]. The basis for its use is the high avidity which melanomas possess for aromatic amino acids and their subsequent conversion to melanin. The synthesis of other boron-containing amino acids and the work of Spielvogel and Hall [14] in developing boron analogues of amino acids, namely the aminecarboxyboranes is rooted in the concept that more rapidly dividing cells will require greater numbers of such biochemical building blocks of proteins (Figure Three and Figure Four).

BORONATED METHIONINE

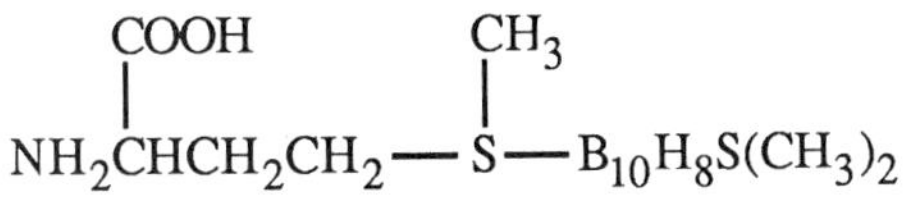

$$\begin{array}{ccc} COOH & & CH_3 \\ | & & | \\ NH_2CHCH_2CH_2\!-\!S\!-\!B_{10}H_8S(CH_3)_2 \end{array}$$

Figure 3.

$$CH_2CH(\overset{+}{N}H_3)COO^-$$

$$B(OH)_2$$

$$CH\underset{B_{10}H_{10}}{\overset{O}{-\!\!\diagup}}CCH_2CH(\overset{+}{N}H_3)COO^-$$

$$H_3\overset{+}{N}BH(CH_3)\overset{-}{C}OOH$$

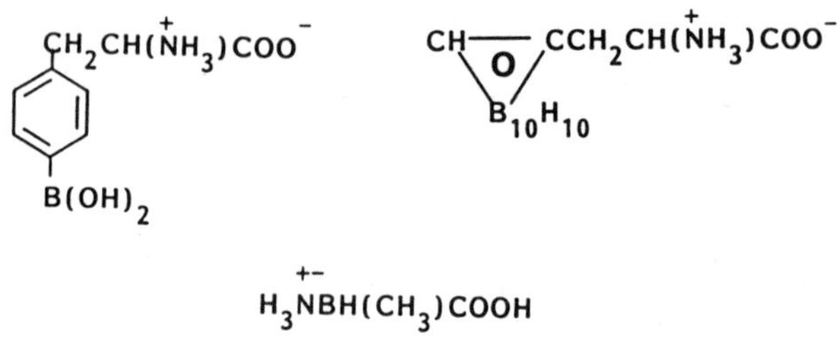

Figure 4.

Another approach with respect to melanomas has been the observation that thiouracil is uniquely and selectively incorporated into melanotic melanomas during melanogenesis. Efforts by Gabel in Germany and Wilson in Australia have centered on the synthesis of boron-containing thiouracils [15]. A third direction, first initiated by Mishima and his collaborators has been the synthesis of boron-containing promazines (Figure Five). The basis for this research has been the observation that chlorpromazine has exhibited a high degree of localization in tissue that contains pre-formed melanin [16]. The compound shown in Figure Six is an initial one which we have synthesized in which the S-N-N axis, that is involved in binding to melanin, is not interfered with [17]. This structure appears to have some promise as a tumor targeting agent but lacks water solubility. Efforts are now underway to make carboranylpromazines that are water soluble.

Two classes of compounds with a demonstrated propensity for localizing in a wide variety of malignant tumors are the porphyrins and the phthalocyanines (Figure Seven). Porphyrin accretion by neoplasms has resulted in the destruction of cancer cells in vivo through photo-activation. This result encouraged researchers to synthesize a number of boronated porphyrins (Figure Eight) and their high concentration and persistence intracellularly in tumors makes them very attractive candidates for use in BNCT [18]. Drs. Kahl and Gabel have been actively involved in the development of such boron-containing porphyrins. A key question with all of these compounds and one which ultimately must be addressed is the biochemical and physiological mechanisms by which such structures achieve elevated concentrations in neoplasms.

BORONATED PROMAZINES

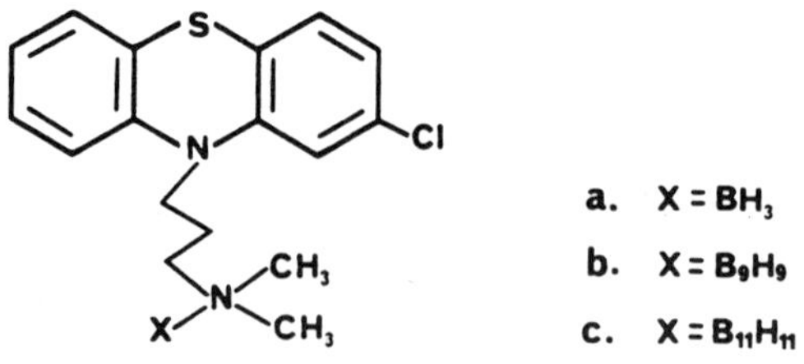

a. X = BH_3

b. X = B_9H_9

c. X = B_{11}H_{11}

Figure 5.

CARBORANYL PROMAZINE

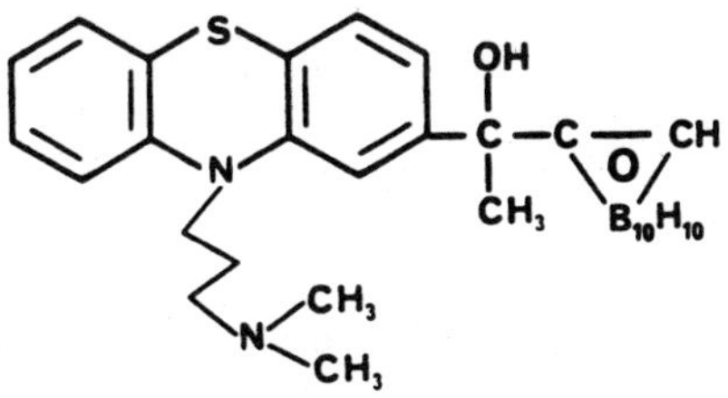

Figure 6.

One final category of low molecular weight boron compounds which should be mentioned are the boron-containing purines, pyrimidines, and their nucleosides. The rationale for their synthesis is that such structures may function as fraudulent bases and thereby be intracellularly incorporated into the tumor cell. This could occur via several pathways: (1) generation and cellular trapping of the nucleotide; (2) inhibition of enzymes involved in pyrimidine and purine nucleoside biosynthesis; and (3) incorporation into tumor cell DNA. A cytoplasmic or preferably a nuclear localization of such a boron nucleoside in tumor would result in greater relative biological effectiveness of BNCT and thereby permit the use of lower boron concentrations. Dr. Schinazi of Emory University has succeeded in synthesizing 5-dihydroxyboryl-2^1-deoxyuridine shown in Figure Nine [19]. Studies by him in collaboration with Laster, Popenoe, and Fairchild at Brookhaven National Laboratory have shown that this compound replaces thymidine to the extent of 5-13% with a boron concentration of approximately 6 μg B/g cell. This level may well be approaching a useful therapeutic range. Others are now engaged in synthesizing carboranylnucleosides.

The above compounds have been administered as low molecular weight compounds. Whether they remain as such is very unclear in many instances. It is certainly conceivable that they could interact with various biopolymers in the vascular system and be transported as conjugates into tumor. It has become apparent that there is a need to understand reactions which occur in the blood, either through ionic, hydrophobic, or covalent linkages with macromolecules and how such conjugates affect tumor concentration. To date, very little has been done in this regard. It is also possible to administer boron-containing macromolecules and in Figure Ten are presented some of the macromolecules which have been considered. The development of monoclonal antibodies against specific tumor-associated antigens offers significant potential for specifically targeting many different solid tumors for both diagnosis and therapy. The key requirement is to translate potential to actuality. In the case of BNCT, it has been estimated that 10^9 Boron-10 atoms per cell will be required to assure cellular destruction. Thus, of the order of one thousand boron atoms per antibody will be

MACROCYCLIC TUMOR LOCALIZING COMPOUNDS

1. Boron Containing Porphyrins

2. Boron Containing Phthalocyamines

Figure 7.

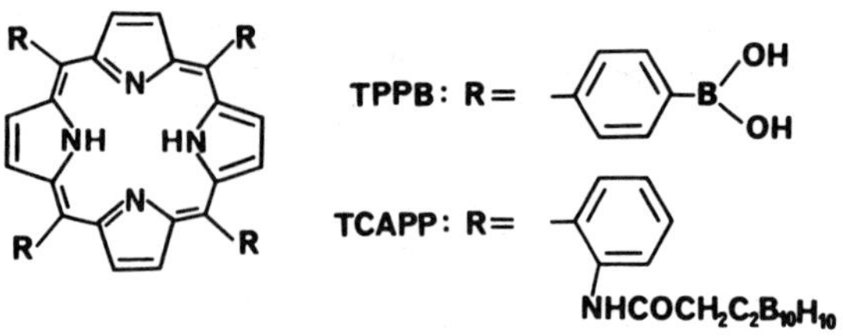

Figure 8.

necessary with an antigen site density of approximately 10^6. Since recent data indicates a degree of internalization of antibodies [20], such a high surface site density may not be required. A number of groups have been actively engaged in attaching boron compounds to antibodies. These include Hawthorne and his associates in the United States, as well as Hersey, Moore, and Tamat in Australia, Gabel in Germany, and our own efforts. Initially, the concept was to attach small molecular weight compounds, containing 10 to 12 boron atoms, to antibodies. An example of one such structure is shown in Figure Eleven [21]. However, in order to obtain 1000 boron atoms on the antibody one would require 100 discrete chemical reactions between the boronating reagent and the antibody. As a consequence, the monoclonal antibody becomes sufficiently modified so that it is no longer capable of targeting antigens on the tumor cell membrane. In order to achieve the requisite number of boron atoms while minimizing conformational alterations of the antibody, small boron polymers have been prepared which could be incorporated at only several sites on the antibody molecule. Hawthorne has referred to these structures as the "^{10}B-trailer reagents." Two approaches have been undertaken: (1) those derived from readily obtainable nonuniform macromolecules such as polylysine to which small reactive boron cages are conjugated and (2) the preparation of carborane-containing amino acid from which boron-containing polypeptides can be prepared using solid-state synthetic techniques (the Merrifield method).

BORON-CONTAINING NUCLEIC ACID PRECURSORS

1. Dihydroxyboronodeoxyuridine

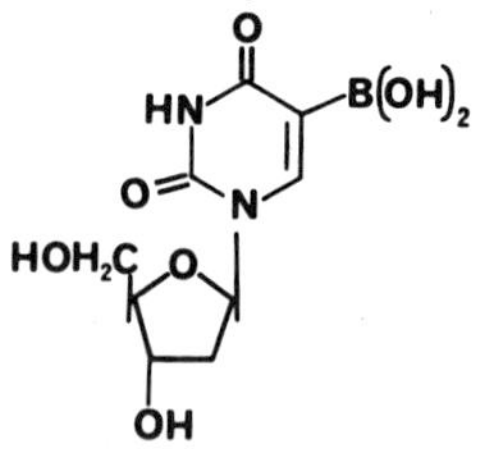

2. Boronothiouracil

3. Other Boron-Containing Bases

Figure 9.

1.   Monoclonal Antibodies and Antibody Fragments

2.   Boronated Peptides

3.   Boron-Containing Compounds that Bind to Surface Receptors

Figure 10.

The disadvantage of the former is the nonuniformity of the polymeric backbone. However, boron polymers can be prepared with several thousand boron atoms. The disadvantage of the latter technique is the difficulty in preparing high molecular weight species in sufficient amount for biological testing. However, the compounds which can be prepared are single species of uniform composition. Further research is needed in order to develop boron polymers which will combine the best features of both procedures.

One of the techniques for incorporating such structures in antibodies has been to prepare the boronated polymer with one heterobifunctional group attached to it by means of the active ester functionality and similarly to modify the Mab with a second functional group. The appropriate two functional entities will react selectively with each other to form a stable covalent linkage. This procedure developed by Tsukada and his associates in Japan [22] has now been adapted to link 10^3 boron atoms to Mabs with the retention of 40-90% of the native Ab's immunoreactivity under in vitro conditions. The incorporation technique used is shown in Figure Twelve [23].

Promising as these in vitro studies have been, when these boronated antibodies were injected into tumor-bearing animals, the antibodies concentrated in the liver with low levels in tumor. Thus it is essential that the antibody following boronation preserve its in vivo tumor targeting capability. It is necessary to determine what are the physiochemical parameters of the boron polymer that will allow the antibody to retain its localizing properties.

One final category of macromolecular structures are what may be termed encapsulating complexes (Figure Thirteen). Structures such as liposomes, containing boron compounds, fall into this category. Liposomes of appropriate size and proper composition accumulate 10-15 times more in tumors than in adjacent normal tissue. Schmidt and Hawthorne are attempting to apply these encapsulating structures to BNCT. Another related approach by Kahl is to deplete cholesterol from low density lipoproteins (LDLs) and

HETEROBIFUNCTIONAL BORONATING AGENT (SBDP)

N-Succinimidyl-3-(undecahydro-closo-dodecaboranyldithio) propionate

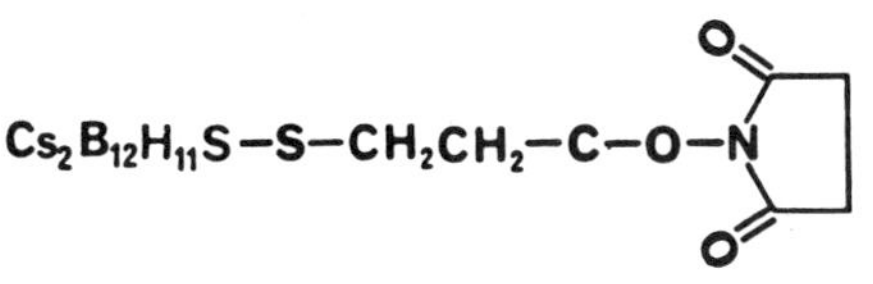

$$Cs_2B_{12}H_{11}S-S-CH_2CH_2-C-O-N$$

Figure 11.

CONJUGATION OF BORONATED POLYLYSINE TO ANTIBODIES

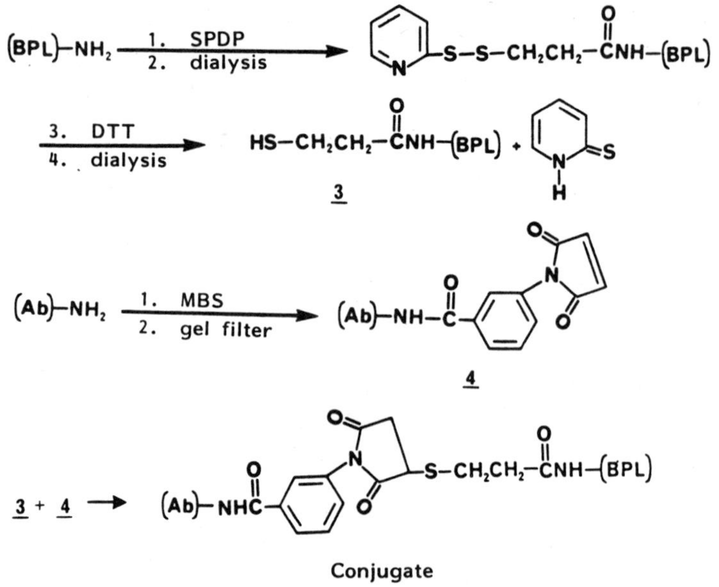

Figure 12.

to replace part of it with carboranyl esters of long chains of alcohols. This approach is in the developing stages and is based upon the fact that LDLs are targeted to cell surface receptors, especially those on tumors. Because the core carries approximately 50% cholesterol, replacement of a suitable percentage with lipophilic boron esters may permit such structures to be used in BNCT [24].

Up to this point, only ^{10}BNCT has been discussed. Other nuclides with high cross section capture for thermal neutrons may also be considered as shown in Table Two. Initially, all nuclides which were either radioactive and/or did not yield high LET particles were discounted. However, more recently, Martin, Allen, and their collaborators in Australia have raised an important issue which requires a reexamination of this tenet. ^{157}Gd, with a capture cross section more than 50 fold greater than ^{10}B, is associated with electron internal conversion in the capture reaction and this produces Auger electron cascade. Their initial studies [25], examining DNA damage as a result of GdNCT are promising and certainly warrant much greater study of Auger electron effects as a result of NCT.

ENCAPSULATING STRUCTURES

1. Boronated Liposomes

2. Low Density Lipoproteins Containing Boron

Figure 13.

Table Two

Thermal Neutron Capture Cross Section Values in Barns
of Potential Nuclides for Neutron Capture Therapy

Nuclide	Cross Section (σ) (barns)	Nuclide	Cross Section (σ) (barns)
^{3}He	5,500	^{155}Gd	58,000
^{6}Li	953	^{157}Gd	240,000
^{10}B	3,837	^{174}Hf	400
^{113}Cd	20,000	^{199}Hg	2,000
^{135}Xe*	2,720,000	^{235}U*	678
^{149}Sm	41,500	^{241}Pu*	1,375
^{151}Eu	5,900	^{242}Am*	8,000

*Nuclides are radioactive.

ACKNOWLEDGMENTS

This work has been supported by Grant No. 5R01CA41288 from the National Institute of Health, Contract No. DE-AC02-82ER60040 from the U.S. Department of Energy, and the Ohio State University Research Grant Program.

REFERENCES

1.   W. H. Knoth, J. C. Sauer, D. C. England, W. R. Hertler, and E. L. Muetterties, J. Am. Chem. Soc., 86:3973 (1964).

2.   A. H. Soloway, H. Hatanaka, and M. A. Davis, "Penetration of Brain and Brain Tumor. VII. Tumor-Binding Sulfhydryl Boron Compounds," J. Med. Chem., 10:714 (1967).

3.   Y. Hayashi, H. Furukawa, Y. Kawano, M. Matsuwa, A. Harihara, S. Araki, Y. Muraoka, Y. Maeda, T. Yamashita, and M. Muto, "Toxicity of Sodium Monomercaptoundecahydro-closo-dodecaborate by Single Intravenous Inspection," Research Report from the Shionogi Research Laboratory (Privileged Communication).

4.   H. Hatanaka, K. Amano, H. Kanemitsu, I. Ikeuchi, and T. Yoshizki, "Boron Uptake by Human Brain Tumors and Quality Control of Boron Compounds," in Boron-Neutron Capture Therapy for Tumors, H. Hatanaka, ed., Chapter V, Nishimura Co., Ltd., Niigata, Japan, p. 77 (1986).

5.   G. C. Finkel, C. E. Polotti, R. G. Fairchild, D. N. Slatkin, and W. H. Sweet, J. Neurosurg., 24:6 (1989).

6.   H. Hatanaka and Y. Urano, "Eighteen Autopsy Cases of Malignant Brain Tumors Treated by Boron-Neutron Capture Therapy between 1968 and 1985," in <u>Boron-Neutron Capture Therapy for Tumors</u>, H. Hatanaka, ed., Chapter XXVI, Nishimura Co., Ltd., Niigata, Japan, p. 381 (1986).

7.   F. Alam, A. H. Soloway, and R. F. Barth, "Boronation of Antibodies with Mercaptoundecahydro-closo-dodecaborate (2–) Anion for Potential Use in Boron Neutron Capture Therapy," <u>Intl. J. Rad. Appl. and Instrumentation</u>, Part A, 38:503 (1987).

8.   O. W. Press, J. A. Hansen, A. Farr, and P. J. Martin, <u>Cancer Res.</u>, 48:2249 (1988).

9.   D. Joel, D. Slatkin, R. Fairchild, P. Micca, and M. Nawrocky, "Pharmacokinetics and Tissue Distribution of the Sulfhydryl Boranes (Monomer and Dimer) in Glioma-Bearing Rats," <u>Strahlenther. Onkol.</u>, 165(2/3):167 (1989).

10.   G. R. Wellum, E. I. Tolpin, A. H. Soloway, and A. Kaczmarczyk, "Synthesis of μ-Disulfido-bis(undecahydro-closo-dodecaborate) (4–) and of a Derived Free Radical," <u>Inorg. Chem.</u>, 16:2120 (1977).

11.   Y. Mishima, <u>Pigment Cell</u>, 1:215 (1973).

12.   J. A. Coderre, J. D. Glass, R. G. Fairchild, P. L. Micca, and J. Kalef-Ezra, "Boronophenylalanine for Neutron Capture Therapy," <u>Strahlenther. Onkol.</u>, 165(2/3):215 (1989).

13.   H. R. Snyder, A. J. Reedy, and W. J. Lennarz, <u>J. Am. Chem. Soc.</u>, 80:835 (1958).

14.   B. F. Spielvogel, A. Sood, I. H. Hall, R. G. Fairchild, and P. L. Micca, "Synthesis, Antineoplastic Activity and Bio-Distribution of Boron Analogues of Amino Acids and Related Compounds," <u>Strahlenther. Onkol.</u>, 165(2/3):123 (1989).

15.   W. Tjarks and D. Gabel, "Synthesis of Boronated Thioureas for Neutron Capture Therapy," <u>Third Int. Symp. on Neutron Capture Therapy</u>, Bremen, FRG, 31 May - 3 June 1988.

16.   R. G. Fairchild, S. Packer, D. Greenberg, P. Som, A. B. Brill, I. Frank, and W. P. McNally, <u>Cancer Res.</u>, 42:5126 (1982).

17.   F. Alam, B. V. Bapat, A. H. Soloway, R. F. Barth, N. Mafune, and D. M. Adams, "Boronated Compounds for Neutron Capture Therapy," <u>Strahlenther. Onkol.</u>, 165(2/3):121 (1989).

18.   S. B. Kahl, "Present Developments in the Synthesis and Isolation of Atropisomers of Porphyrinic Carburanyl Anilides," <u>Second Int. Symp. on Neutron Capture Therapy</u>, Tokyo, 1985, H. Hatanaka, ed., Nishimura Co., Ltd., Niigata, Japan (1986).

19.   R. F. Schinazi, B. H. Laster, R. G. Fairchild, and W. H. Prusoff, "Synthesis and Biological Activity for a Boron-Containing Pyrimidine Nucleoside," <u>First Int. Symp. on Neutron Capture Therapy</u>, Cambridge, MA, 1983, R. G. Fairchild and G. L. Brownell, eds., Brookhaven National Laboratory, <u>BNL-51730</u>, p.260 (1984).

20.   R. F. Barth, N. Mafune, F. Alam, D. M. Adams, A. H. Soloway, G. E. Makroglu, O. A. Oredipe, T. E. Blue, and Z. Steplewski, "Conjugation, Purification and Characterization of Boronated Monoclonal Antibodies for Use in Neutron Capture Therapy," <u>Strahlenther. Onkol.</u>, 165(2/3):142 (1989).

21.   F. Alam, A. H. Soloway, J. E. McGuire, R. E. Barth, W. E. Carey, and D. Adams, "Dicesium N-Succinimidyl 3-(Undecahydro-closo-dodecaboranyldithio) Propionate, a Novel Heterobifunctional Boronating Agent," <u>J. Med. Chem.</u>, 28:522 (1985).

22. Y. Tsukada, Y. Kato, N. Umemoto, Y. Takeda, T. Hara, and H. Hirai, <u>J. Natl. Cancer Inst.</u>, 73:721 (1984).

23. F. Alam, A. H. Soloway, R. F. Barth, and D. M. Adams, "Chemoradiotherapy of Cancer: Boronated Antibodies and Boron Containing Derivatives of Promazine for Neutron Capture Therapy," in <u>Neutron Capture Therapy</u>, H. Hatanaka, ed., Nishimura Co., Ltd., Niigata, Japan, p. 8 (1986).

24. S. B. Kahl, <u>First Boron USA Workshop</u>, Dallas, Texas, 10-13 April 1988.

25. B. J. Allen, B. J. McGregor, and R. F. Martin, "Neutron Capture Therapy with Gadolinium-157," <u>Strahlenther. Onkol.</u>, 165(2/3):156 (1989).

NEUTRON BEAM DESIGN

MONTE CARLO METHODS OF NEUTRON BEAM DESIGN FOR NEUTRON CAPTURE THERAPY AT THE MIT RESEARCH REACTOR (MITR-II)

S. D. Clement,[†] J. R. Choi,[†] R. G. Zamenhof,[*] J. C. Yanch,[†]
and O. K. Harling[†]

[†] Nuclear Reactor Laboratory, Massachusetts Institute of Technology
Cambridge, MA

[*] Department of Radiation Oncology, Tufts - New England Medical Center
Boston, MA

ABSTRACT

Monte Carlo methods of coupled neutron/photon transport are being used in the design of filtered beams for Neutron Capture Therapy (NCT). This method of beam analysis provides segregation of each individual dose component, and thereby facilitates beam optimization. The Monte Carlo method is discussed in some detail in relation to NCT epithermal beam design. Ideal neutron beams (i.e., plane-wave monoenergetic neutron beams with no primary gamma-ray contamination) have been modeled both for comparison and to establish target conditions for a practical NCT epithermal beam design. Detailed models of the 5 MWt Massachusetts Institute of Technology Research Reactor (MITR-II) together with a polyethylene head phantom have been used to characterize approximately 100 beam filter and moderator configurations. Using the Monte Carlo methodology of beam design and benchmarking/calibrating our computations with measurements, has resulted in an epithermal beam design which is useful for therapy of deep-seated brain tumors. This beam is predicted to be capable of delivering a dose of 2000 RBE-cGy (cJ/kg) to a therapeutic advantage depth of 5.7 cm in polyethylene assuming 30 μg/g ^{10}B in tumor with a ten-to-one tumor-to-blood ratio, and a beam diameter of 18.4 cm. The advantage ratio (AR) is predicted to be 2.2 with a total irradiation time of approximately 80 minutes. Further optimization work on the MITR-II epithermal beams is expected to improve the available beams.

INTRODUCTION

Neutron capture therapy (NCT) is being developed in a joint effort by the Massachusetts Institute of Technology's Nuclear Reactor Laboratory and the Tufts - New England Medical Center. A major component of these studies is the use and development of epithermal neutron beams. Characterization of beams for NCT use requires extensive information on the neutron and photon fluxes, both energy and intensity, as well as the resulting depth-dose distributions in an experimentally measurable head phantom and an actual human head. This paper focuses on epithermal neutron beam design using probabilistic (i.e., Monte Carlo) radiation transport techniques.

Neutron Beam Design, Development, and Performance for Neutron Capture Therapy
Edited by O. K. Harling *et al.*
Plenum Press, New York, 1990

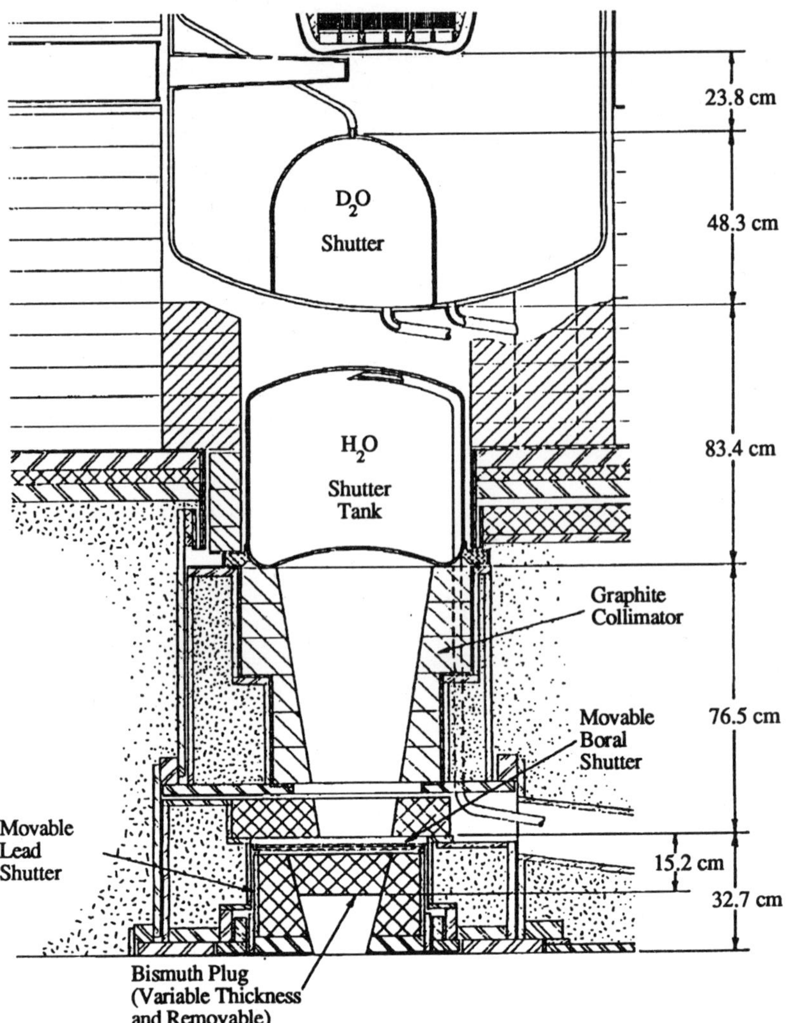

Figure 1. Cross-sectional view of the medical therapy beam port of MITR-II.

MONTE CARLO METHODS OF NEUTRON AND PHOTON TRANSPORT

The primary computational design tool in the development of filtered neutron beams is a three dimensional, pointwise-continuous energy cross section Monte Carlo code, MCNP [1], capable of performing neutron, photon, or coupled neutron/photon transport. The medical therapy facility at the MITR-II, illustrated in Figure 1, consists of a vertical beam port leading to a surgical operating room directly beneath the compact undermoderated reactor core. The beam, in its original configuration, is controlled by a series of four shutters: (1) a 95-liter semielliptical D_2O blister tank used to vary the fraction of epithermal neutrons entering the beam port, (2) a 51-cm thick H_2O shutter used to attenuate the beam, (3) a 0.64-cm thick boral plate, and (4) a 30-cm thick lead shutter which contains a 15.2-cm thick bismuth plug insert. Adequate modeling of the medical beam requires a complex three dimensional geometry scheme. MCNP provides this by defining geometric regions from first, second, and some special fourth degree surfaces and then combining these surfaces with Boolean operators. An extensive list of surface equations is recognized by MCNP, resulting in models whose complexity is limited only by user dedication. An internal geometric plotting capability as well as extensive internal error checking aid the user in debugging complex geometries.

MCNP uses pointwise-continuous energy cross section data from the Evaluated Nuclear Data File (ENDF/B-V) [2], the Evaluated Nuclear Data Library (ENDL) [3], the Activation Library (ACTL) [4], and the Applied Nuclear Science Group [5-7] at the Los Alamos National Laboratory (LANL). All possible neutron reactions and interactions for which data is tabulated are taken into account. Because resonance scatterers are promising filters for neutron beams, an intensive cross-section structure is required. The density of the available data is such that linear interpolation between points reproduces evaluated data within a tolerance of 1% [1]. Cross-section data tables are also available to correct for chemical binding and crystalline effects for neutron energies below 4 eV. Available tables include hydrogen-corrected neutron cross-section data for light water, heavy water, and polyethylene as well as tables for graphite. Photon cross-section tables are available for materials with Z=1 to Z=93 and include the following interactions: elastic and inelastic scattering, photoelectric absorption with the possibility of fluorescent emission, and pair production with local emission of annihilation quanta.

The tally structure of MCNP provides the user with either neutron or photon information, anywhere throughout the problem geometry, in any of the following forms: number of particles crossing a surface, particle flux averaged across a surface, the average flux throughout a cell volume, or energy deposition throughout a cell. All tallies may be segmented to a desired surface area and/or to a specific direction. Tallies may also be tabulated as a function of energy or modified by a particular dose-energy function. All MCNP tallies are normalized to per starting particle and are printed with an estimate of the relative error corresponding to one standard deviation. To convert the relative results to absolute fluxes, and therefore doses, a calibration constant must be used. This constant is equal to the source strength in number of particles produced per second. If this data is not available, the MCNP output can be normalized to a single experimentally measured data point from which all remaining tallies can be inferred. One of the most powerful features of the MCNP tally structure is its ability to modify a tally which is a function of energy by an energy dependent function. This dose-energy function allows MCNP to calculate reaction rates for materials which have tabulated or known energy-dependent cross sections, dose rates for materials with known flux-to-dose conversion factors, absorbed dose rates for materials with known energy-dependent KERMA factors, or any reaction of the form:

$$\int_E \Phi(E)\,f(E)\,dE \tag{1}$$

NEUTRON BEAM DESIGN USING MONTE CARLO METHODS

Beam analysis for NCT applications is based on the resulting depth-dose distributions of neutrons and photons in a head or head phantom. Dose rates as a function of depth can be calculated by tallying neutron and gamma-ray fluxes as a function of energy, then modifying those fluxes with energy-dependent KERMA factors. The KERMA factors used in this study were calculated based on the elemental composition of brain. This composition was calculated from the data presented by Brooks et al. [8] and the assumption of a 50 by 50 weight-percent average of grey and white matter. The photon KERMA factors were calculated from the data presented by Hubbel [9] and the neutron KERMA factors were calculated from the data presented by Caswell et al. [10]. The KERMA factors for the $^{10}B(n,\alpha)^7Li$ reaction used in this study were calculated by Zamenhof et al. [11]. The brain equivalent elemental composition and all KERMA factors used are presented and discussed in detail by Zamenhof et al. [12] in another paper in these proceedings.

We use three criteria or figures of merit to describe the performance of NCT beams. First is the advantage depth which provides a measure of the maximum useful depth for therapeutic benefit. Advantage depth (AD) is defined as the depth in a given material at which the total therapeutic dose to the material is equal to the maximum total background dose to the material. The total therapeutic dose is the sum of the total background dose and the $^{10}B(n,\alpha)^7Li$ dose. Figure 2 shows a graphical representation of AD. A maximum advantage depth (AD_{max}) occurs when the boron dose ratio between the tumor and the

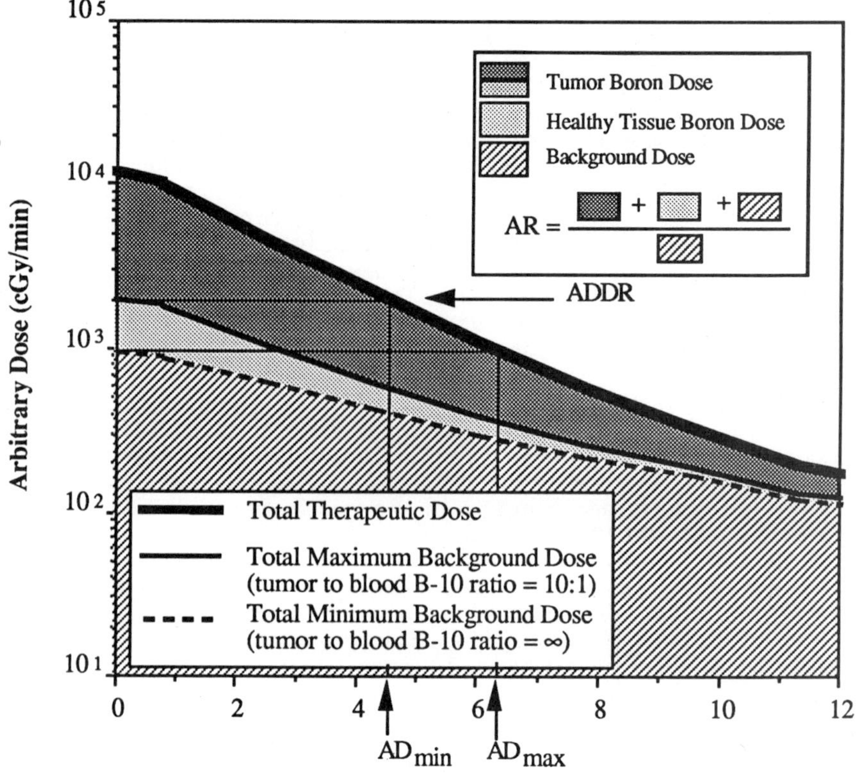

Figure 2. Typical dose versus depth plot illustrating the concept of advantage depth (AD), advantage ratio (AR), and advantage depth dose rate (ADDR). An effective tumor to blood B-10 concentration of 10 to 1 has been assumed.

healthy tissue/blood is infinite. However, more realistically this ratio is set at ten-to-one. This arises from the evidence that a physical boron concentration ratio between the tumor and blood of approximately three-to-one seems achievable, and that there exists a calculated geometrical dose absorption factor of about three-to-one for those boron reactions that are initiated in the microvasculature of the brain [13-14]. By using an effective tumor-to-blood ratio of ten-to-one, we can represent a reasonable boron dose partition between tumor and normal tissue. This results in the concept of a minimum advantage depth (AD_{min}), also shown in Figure 2.

The second figure of merit in characterizing the various neutron beams, also illustrated in Figure 2, is the concept of advantage ratio (AR). Simply stated, the advantage ratio indicates the relative ratio of the integral of the total therapeutic dose versus the integral of the total background dose over a given depth in the material. For this paper, that depth is chosen to be the maximum advantage depth.

While the AD and AR can be considered as a measure of the neutronic quality of a beam, the third important concept in characterizing the NCT beam, advantage depth dose rate (ADDR), has to do with its intensity. Because a useful neutron beam has to be able to deliver sufficient dose to the tumor within a "reasonable time," the intensity of the neutron beam becomes an important quality factor. "Reasonable time" depends on a number of factors such as patient throughput, a requirement which would initially be very low, whether or not the patient must be given general anesthesia, and so on. These factors will not be discussed in a comprehensive way in this paper. However, as a rough

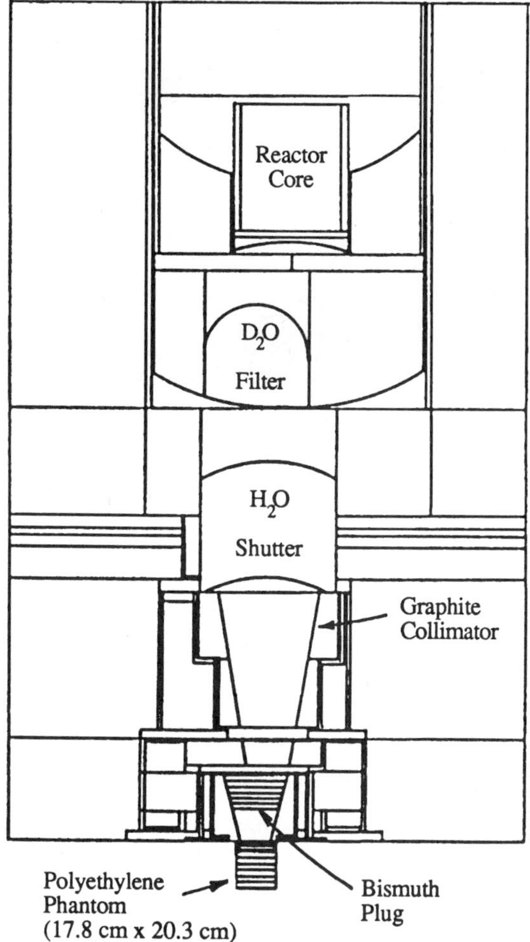

Figure 3. MCNP model of MITR-II medical therapy beam port.

guideline for useful NCT beams we suggest the following: (1) the beam be able to deliver a therapeutic dose of approximately 2000 RBE-cGy (cJ/kg) at the minimum advantage depth, in less than one hour if the irradiation is performed through an opened skull, and (2) for irradiation through intact skulls, where fractionation is beneficial, that each fraction be deliverable in about one hour or less. Clearly no hard boundaries on useful beam intensities are necessary and we offer the above as suggestions only in judging beam performance, especially for initial clinical trials when patient loads will be light.

Our design philosophy emphasizes producing a beam with the highest possible AD while also maximizing the ADDR and the AR. To accomplish this, it is necessary to acquire depth versus dose data for all dose components, including the following: ^{10}B therapeutic dose, fast neutron dose, thermal neutron dose (including ^{14}N(n,p)), dose from gamma rays induced in the head phantom, dose from gamma rays induced in structural components, and core gamma-ray dose. Three MCNP runs are required to produce all of the information described. The first run is a coupled neutron/photon run with photon production occurring only in the head phantom. The neutron and photon fluxes are tallied versus depth in the phantom and modified by the appropriate KERMA factors to produce the ^{10}B therapeutic dose, the fast neutron dose, the thermal neutron dose, and the dose resulting from gamma rays produced in the head phantom. The second MCNP run is also a coupled neutron/photon run but with photon production occurring everywhere except in the head phantom. This run provides the dose versus depth as a result of gamma rays

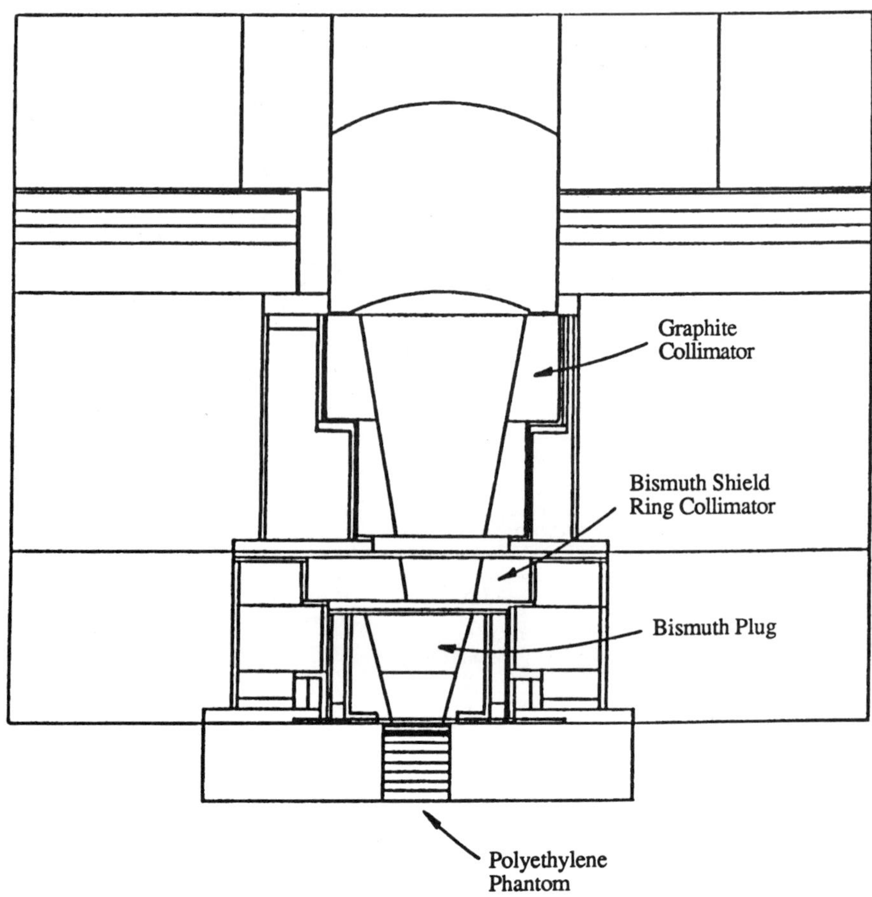

Figure 4. Scaled-down MCNP model of the lower section of MITR-II medical therapy beam line.

produced in structural components including filter or moderator materials. The final run is a photon run which provides the dose versus depth resulting from core gamma rays. The background dose components are added together to provide the total background dose versus depth. This dose corresponds to the dose received by healthy tissue in the brain. The total background dose is then added to the ^{10}B therapeutic dose to provide the tumor dose versus depth. These data are then used to determine the AD, ADDR, and AR of the particular beam.

Detailed models of the MITR-II and a human skull have been developed, exercising care to ensure that all important reactor components and structures were included. Figures 3 and 4 illustrate the MITR-II model and demonstrate the powerful internal plotting capability of MCNP. A right circular cylinder of polyethylene was used in the model to facilitate experimental confirmation of the code's performance. In addition, a more realistic human head (i.e., skull and brain) model was developed to quantify the effects of changes in material composition, macroscopic density, and geometry on the depth-dose calculations. The skull model and cortical bone composition were originally presented by Snyder et al. [15]. This model has been used by Zamenhof et al. [16] in the development of Monte Carlo treatment planning techniques and by Deutsch and Murray [14] for Monte Carlo dosimetry calculations. The skull model in both a radial and a rectangular grid is shown in Figure 5. It is assumed that correcting the hydrogen cross section for molecular binding, as if it were bound in water, is sufficient for brain material.

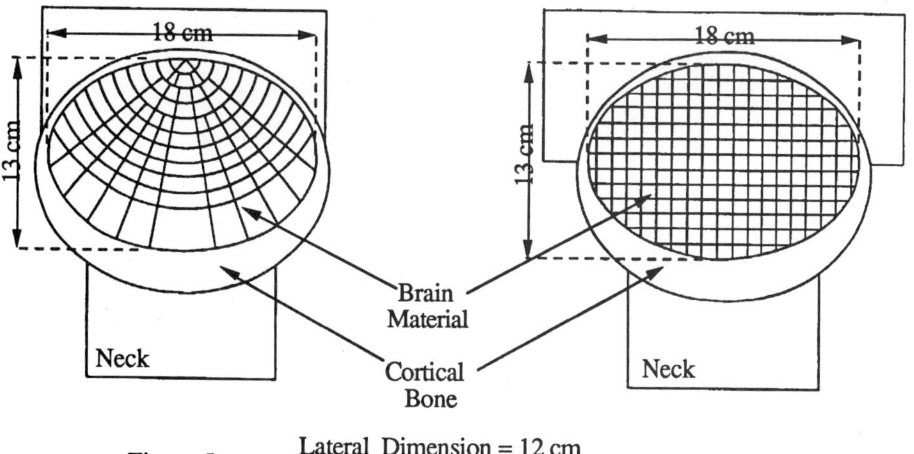

Figure 5a.   Snyder model of the human skull and brain. Brain-equivalent material is within the inner ellipsoid. Cortical bone (the skull) is within the two non-concentric ellipsoids. A lateral view is shown.

Figure 5b.   Snyder model used for Monte Carlo calculations in Cartesian coordinates. Each cube is 1 cc of brain-equivalent material.

IDEAL BEAM STUDY

Several series of Monte Carlo calculations were performed on ideal neutron beams. Ideal beams are those which are modeled as plane-wave currents of monoenergetic neutrons with no primary gamma-ray contamination. The first series of runs was performed to establish the relationship between neutron energy and the resulting AD, AR, and ADDR. The geometry, beam diameter, and incident neutron current were all held constant while the incident neutron energy was varied from thermal energies up to 1 MeV. The modeled phantom geometry is shown in Figure 6.

Because these runs were designed for a relative comparison of various beam energies, the effects of radiation fractionation were not considered. The relative biological effectiveness (RBE) was accounted for using values of 1.0 for all gamma rays, 1.6 for fast neutrons and $^{14}N(n,p)$ recoil protons, and 2.3 for the $^{10}B(n,\alpha)$ reaction. The ADs were calculated assuming 30 μg/g of ^{10}B in tumor and both an infinite tumor-to-blood ratio and a ten-to-one tumor-to-blood ratio. Due to the changing radial flux distribution in phantom, all dose components were calculated along the beam centerline averaged over a 3-cm radius. The results are listed in Table One. These data indicate that as neutron energies increase above 20 keV, both the resulting AD_{min} and AR decrease. At neutron energies below 20 keV, an inverse relationship appears to exist between the AD_{min} and the AR. These results are graphically summarized in Figure 7. A maximum in the AD_{min} versus AR curve occurs at around 20 keV. It is not obvious that 20 keV is the best beam energy for NCT; however, it can be concluded that beam energies above 20 keV, while possibly useful, are not as good as lower energy beams. At 100 keV, the AD_{min} is greater than that of a pure thermal beam, but the AR is only 60% of that of the pure thermal beam. It is apparent from Figure 7 that good AD and good AR are achievable with a wide range of epithermal neutron energies (i.e., ~5 eV to >20 keV). This makes the problem of implementing a practical therapy beam on a fission reactor considerably easier than would be the case if only a narrow neutron energy band width were found to be suitable.

Figure 8 shows the comprehensive depth-dose distribution of all background and therapeutic dose components for an ideal 2-keV neutron beam. The beam is 12.7 cm in diameter and the phantom is polyethylene ($\rho = 0.913$ g/cc). The maximum RBE AD of

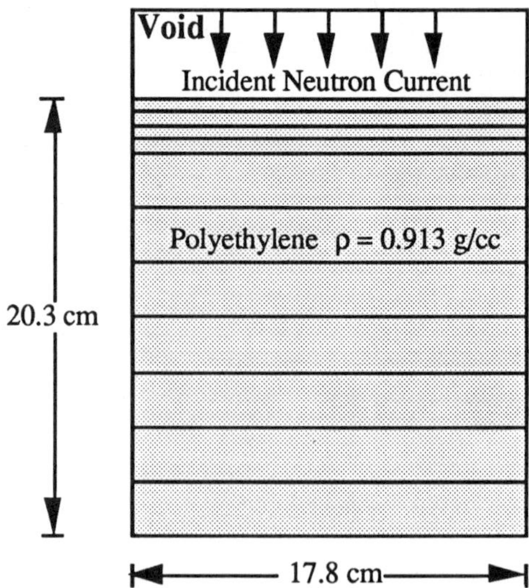

Figure 6. MCNP model of right-circular cylinder of polyethylene head phantom. Diameter = 17.8 cm, height = 20.3 cm, density = 0.913 g/cc, hydrogen weight fraction = 14.38%.

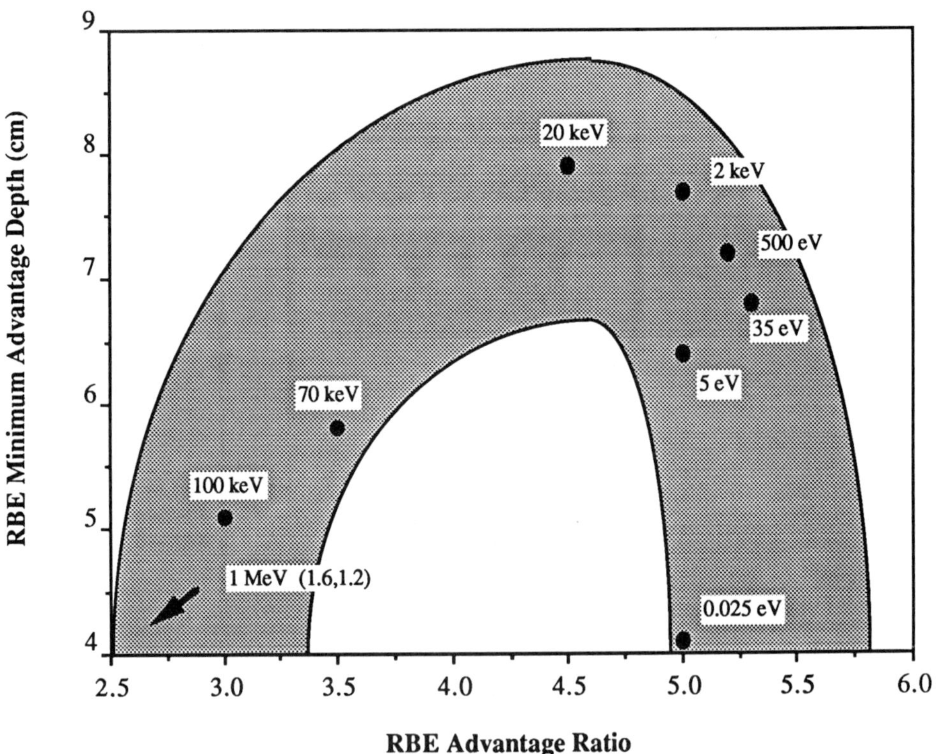

Figure 7. Advantage ratio versus advantage depth for ideal beams of various energies. Phantom is polyethylene and the beam diameter = 17.8 cm. The gray zone in the figure represents the "general region" in which all ideal beam results are found for the given phantom and beam diameter.

Table One.   Results for ideal neutron beams. Advantage depths (AD) with and without RBE, advantage ratios (AR), and advantage depth dose rates (ADDR) for ideal neutron beams of various energies. Beam diameter = 17.8 cm and the head phantom is polyethylene.

Energy	RBE AD (cm) Max/Min	AD (cm) Max/Min	RBE AR	RBE ADDR[a] (RBE-cGy/min)
0.025 eV	5.4/4.1	4.4/3.7	5.0	26
5 eV	7.4/6.4	6.3/5.7	5.0	38
35 eV	7.8/6.8	6.6/6.1	5.3	36
500 eV	8.2/7.2	7.0/6.5	5.2	31
2 keV	8.9/7.7	7.6/6.9	5.0	30
20 keV	8.3/7.9	7.5/7.3	4.5	30
70 keV	5.8/5.8	5.1/5.1	3.5	63
100 keV	5.1/5.1	3.9/3.9	3.0	80
1 MeV	1.6/1.6	1.3/1.3	1.2	300

[a] Data calculated assuming an incident neutron current $J = 10^9$ n/cm^2-s.

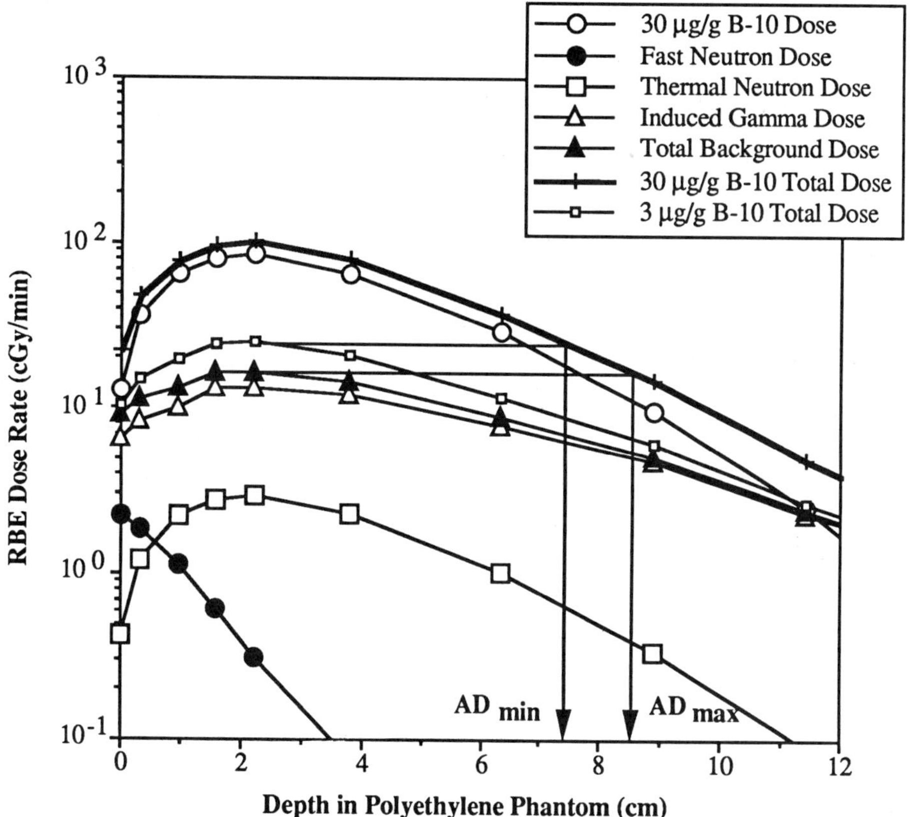

Figure 8.   Calculated comprehensive depth-dose distribution of a 2-keV ideal beam (diameter = 17.8 cm) through a polyethylene head phantom. RBE dose rate calculated assuming an incident neutron current $J = 10^9$ n/cm^2-s.

Table Two. Advantage depths (AD), advantage ratios (AR), and advantage depth dose rates (ADDR) for an ideal beam in various geometries. Beam energy is 2 keV. The phantom is 17.8 cm in diameter by 20.3 cm in height.

Head Phantom and Geometry	Beam Diameter (cm)	RBE AD (cm) Max/Min	AD (cm) Max/Min	RBE AR	RBE ADDR[a] (RBE-cGy/min)
Cylindrical Polyethylene	12.7	8.6/7.4	7.3/6.7	5.4	25
Cylindrical Polyethylene	17.8	8.9/7.7	7.6/6.9	5.0	30
Cylindrical Brain	17.8	9.4/8.3	8.2/7.6	4.6	23
Elliptical[b] Skull & Brain	6.0	8.8/7.8	7.6/6.8	6.7	5.5
Elliptical[b] Skull & Brain	12.7	9.6/9.0	8.6/8.0	5.3	15
Elliptical[b] Skull & Brain	17.8	10.2/9.6	9.6/9.0	3.2	14

[a] Data calculated assuming an incident neutron current $J = 10^9$ n/cm^2-s.

[b] The beam is transported through the skull and the AD_{min} is given for the depth in brain tissue.

the beam is 8.6 cm and the minimum AD is 7.4 cm in polyethylene, slightly smaller than that for the 17.8-cm beam of Table One. The primary background dose component is from gamma rays induced in the phantom and is inherent in the NCT approach for hydrogenous materials. The magnitude of the induced gamma-ray dose is on the order of that produced by the $^{10}B(n,\alpha)$ reaction with 5 $\mu g/g$ ^{10}B in tumor based on RBE-rads. The therapeutic dose component peaks at a depth of approximately 2.5 cm for the ideal 2-keV beam.

A second series of MCNP runs was performed to identify the effects of material and geometrical changes between a cylindrical polyethylene head phantom and a human head. The results of the study for an ideal neutron beam in various geometries are summarized in Table Two. First, a right-circular cylinder of brain-equivalent material was modeled to quantify the depth-dose differences between it and polyethylene. Because brain KERMA factors are used even with the polyethylene phantom, the resulting differences are due exclusively to the radiation transport and $^1H(n,\gamma)$ production changes resulting from lowering the hydrogen weight fraction from 14.38% in polyethylene to 10.57% in brain tissue. An increase of approximately 0.5 cm in AD_{min} was observed while the AR and ADDR decreased by 8% and 23% respectively. Another MCNP computation consisted of changing the cylindrical polyethylene model into the elliptical brain and skull model illustrated in Figure 5 while maintaining the beam energy and diameter constant. A further increase of 1.3 cm in AD_{min} was observed but a decrease in AR and ADDR of 30% and 40% respectively accompanied the AD_{min} increase. It should also be noted that the beam was transported through the skull and the AD_{min} was calculated inward from the surface of the brain. Monte Carlo runs in the final ideal beam series were designed to establish the effects of beam diameter on the depth-dose distribution. A cylindrical polyethylene phantom was modeled with both a 12.7-cm and a 17.8-cm diameter ideal beam of monoenergetic 2-keV neutrons. The resulting AD_{min} of the wide beam was 4% greater than that of the narrow beam, while the AR decreased by approximately 7%. An ideal 2-keV neutron beam was also modeled with the elliptical brain using 17.8-cm, 12.7-cm, and 6.0-cm diameter beams. In this geometry, the influence of beam diameter on the AD_{min} and the AR was much more dramatic. With increased beam diameters, the penetration (AD_{min}) of the beam increased by up to 25% but the AR monotonically decreased by up to 50%. These results indicate that a 2 keV ideal beam which is 12.7 cm in diameter will have an additional minimum 1-cm penetration into a brain through an intact skull over that of a polyethylene head phantom, while the AR will remain relatively unchanged. If wide beams (i.e., diameter = 17.8 cm) are used, an additional 2-cm penetration can be expected.

Based upon the results of this ideal beam study, we can conclude that: (1) for monoenergetic beams, an energy of about 2 keV to 20 keV is the best for the treatment of deep-seated tumors by NCT, (2) good therapeutic figures of merit can be achieved with a wide range of neutron energies (i.e., in the 5 eV to >20 keV range), (3) neutron beams of up to 70 keV, while not ideal, are still useful for NCT applications, (4) for the purposes of beam development, the deepest useful penetration of an ideal beam into an experimentally measurable polyethylene head phantom by unilateral irradiation is approximately 8.9 cm (RBE maximum advantage depth assuming 30 μg/g ^{10}B in tumor and an infinite tumor-to-blood ratio), and (5) for the realistic human head geometry, wide epithermal neutron beams provide greater penetration over narrow beams but sacrifice significantly in AR.

EPITHERMAL NEUTRON BEAM DESIGN AT THE MIT RESEARCH REACTOR

As previously stated, our design philosophy is to provide a practical reactor treatment beam with the highest possible AD_{min} while simultaneously maximizing the AR and retaining a useful dose rate at the AD. The MITR-II medical beam (Figure 1) is equipped with a 61-cm high graphite collimator located directly beneath the light water shutter. This graphite collimator and a 12-cm thick bismuth shield ring provide a 76.5-cm space within which beam filter or moderator materials can be placed.

The first step in our design process was to model the existing beam and to evaluate the code performance through experimental comparison. Measurements as described by Choi et al. [17], in a companion paper, were made on the existing D_2O, bismuth-filtered beam to provide depth-dose information for the fast and thermal neutron dose rates, the total gamma-ray dose rate, and the therapeutic ^{10}B(n,α) dose rate in a cylindrical polyethylene head phantom. Gamma-ray dose measurements were also made in air to segregate the ^{1}H(n,γ) induced gamma-ray dose from the incident core gammas and the gammas induced in structural components. Three MCNP runs were performed using the MITR-II model shown in Figure 3. It was immediately recognized that the CPU effort required to transport neutrons from the reactor down to the head phantom was beyond the capability of our computing system. The transport of neutrons was therefore separated into a two-step process. It was assumed that the source of neutrons entering the medical facility was due solely to those passing through the plane of the light water shutter. First, a Cray supercomputer was used to characterize the neutron energy spectrum directly above the light water shutter. Second, this source spectrum was used in a scaled-down model which consisted of only the beam line entering the medical therapy facility (shown in Figure 4). This permits characterization of the multiple beam filters, using the characterized source, without the redundant CPU intensive process of transporting particles from the reactor core through the D_2O reflector and down to the entrance of the medical therapy beam line.

The characterized neutron source is used for the first two MCNP runs of each beam filter design. All dose components previously discussed are calculated as a function of depth in the polyethylene phantom as well as the thermal neutron flux. A single calculated thermal flux data point was selected and normalized to the measured thermal flux. The resulting calibration constant was then used to calculate the depth-dose distribution of all dose components. A single measured fast neutron dose rate was also used to slightly adjust (conservatively) the characterized source spectrum. The calculated depth-dose data showed excellent agreement with experimentally measured values for each dose component. The calculated and measured total tumor dose and total background dose curves are presented in Figure 9. It was also desired to evaluate the code's performance with a slightly "harder" neutron beam. This was accomplished by placing 1.3 cm of boral and 30 cm of aluminum directly above the 15.2 cm of bismuth. This geometry was also modeled and again, the calculated and measured depth-dose distributions showed good agreement. Because these beams were filtered with large thicknesses of bismuth, any errors in the core gamma-ray energy distribution could not be recognized. The measured total gamma-ray dose rate in these configurations is strongly dominated by gammas

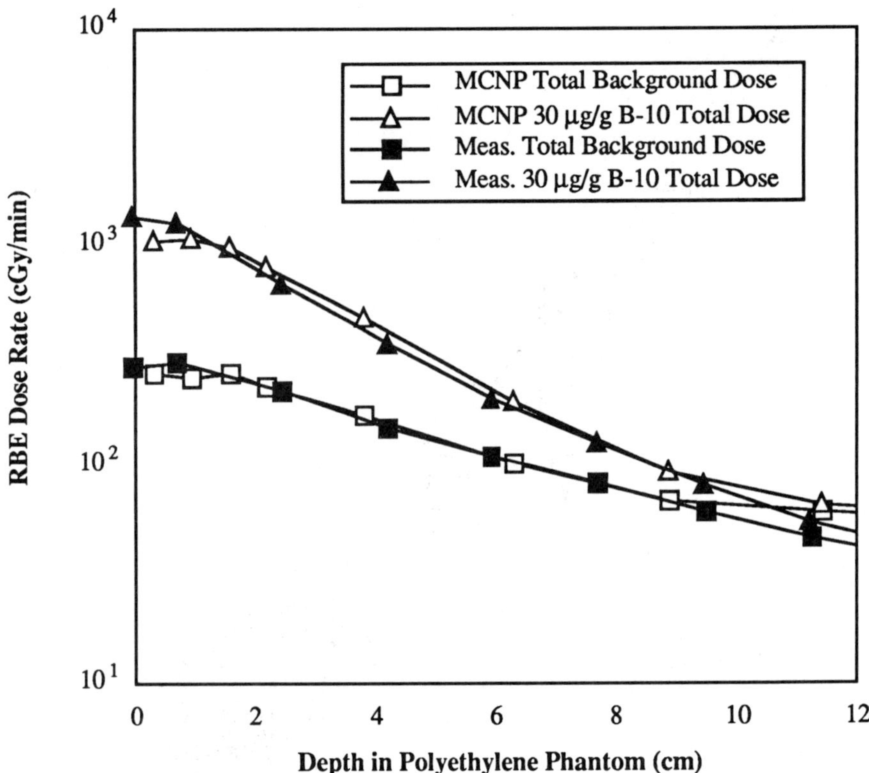

Figure 9. MCNP calculated and experimentally measured total tumor dose and total background dose curves. Filter is 24-cm D_2O and 15.2-cm bismuth. Phantom is polyethylene (a right-circular cylinder 17.8-cm dia. by 20.3-cm high) and the beam diameter is 12.7 cm. These results refer to one possible configuration of the MITR-II medical beam.

induced in the head phantom. Incorrectly predicting the core gamma dose by even an order of magnitude is not immediately recognizable in the total gamma-ray dose measurement for the D_2O and bismuth filtered beams. However, errors in the core gamma-ray source were very significant for our later epithermal beam designs and appropriate corrections were made, as discussed below.

Thus far, over one hundred beam filter and moderator designs for NCT applications have been modeled and evaluated with the MCNP code. For this discussion we will define a moderator as a beam material which is in contact or very "close" to the head phantom. Conversely, a filter is defined as a beam material which is "far" from the patient irradiation position. Typically, for the cases we have modeled, moderator configurations provide a much higher dose rate over filter configurations of similar material. However, filters in our study, provide a more desirable beam (i.e., higher advantage depth and advantage ratio) over that of moderators. Exploring such a large number of configurations, by computer calculations, has allowed us to obtain a reasonable understanding of the range of options for beam design at MITR-II. We have also thoroughly measured the in-phantom beam performance of more than 20 configurations.

Initial MCNP designs of epithermal neutron beams were based upon hand calculations by Harling [18] of various filter materials and thicknesses in simple geometries. An aluminum-sulfur mixture, 35-cm thick, was thought to have promising characteristics. Each filter design had a thermal neutron absorber at both the inlet and outlet. The inlet absorber provided a much reduced dose from capture gamma rays. The

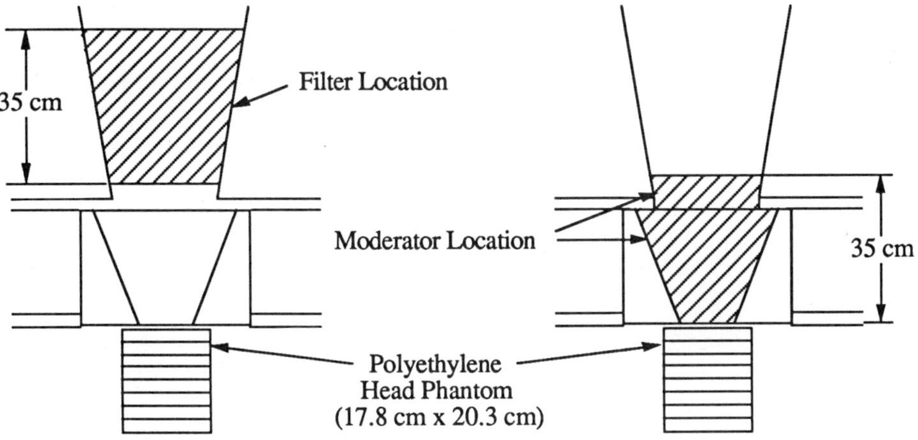

Figure 10. Simple 35-cm thick filter configuration for the MITR-II medical beam port.

Figure 11. Simple 35-cm thick moderator configuration for the MITR-II medical beam port.

materials for thermal neutron absorption were chosen based upon their neutron cross section and potential for photon emission following neutron capture. Cadmium clearly has a favorable cross section for thermal absorption and epithermal transmission; however, the high probability of high energy gamma-ray emission was thought to override its attractive neutron cross section properties. Experimental measurements confirm that use of cadmium in the beam near the patient irradiation position, with little or no shielding, produces a capture gamma dose which dominates all other dose components. We decided that a boron containing compound (boral) and/or cadmium would be used for the inlet filter and that an enriched lithium compound ($^6Li_2CO_3$), which emits few gamma rays, would be used for the final filter.

Filter materials were placed in the available space within the graphite collimator of the beam line. This region is fairly accessible and requires little in the way of reactor modification prior to introducing filter materials. A simple 35-cm thick filter configuration (arbitrary material) is illustrated in Figure 10. Moderator configurations were modeled by replacing the bismuth plug insert with the desired material. A simple 35-cm thick moderator configuration (arbitrary material) is illustrated in Figure 11.

Upon completion of the modeling using simple configurations, work was undertaken to model a practical epithermal therapeutic beam that could be built in accordance with realistic engineering criteria. An aluminum support structure, designed to support two independent regions of filter materials and facilitate simple insertion and removal of the materials, was introduced into the model. The support structure is shown in place within the graphite collimator region in Figure 12. Multiple MCNP runs were performed with various combinations and thicknesses of aluminum and sulfur in the regions labeled D and E. Regions F and G were typically modeled with materials such as aluminum shot or sulfur in aluminum cans.

Beam optimization was carried out by making small perturbations in the filter system and observing the results on individual dose components. Moderator versus filter configurations were studied for identical materials and thicknesses. The moderator configurations produced much higher dose rates but were not as effective as filters for the removal of fast neutrons. The gamma-ray dose induced from neutron capture in structural materials (including the moderator material itself) was also much higher for the moderators than with the filter configurations. Attempts at doping the moderator material with ^6Li to

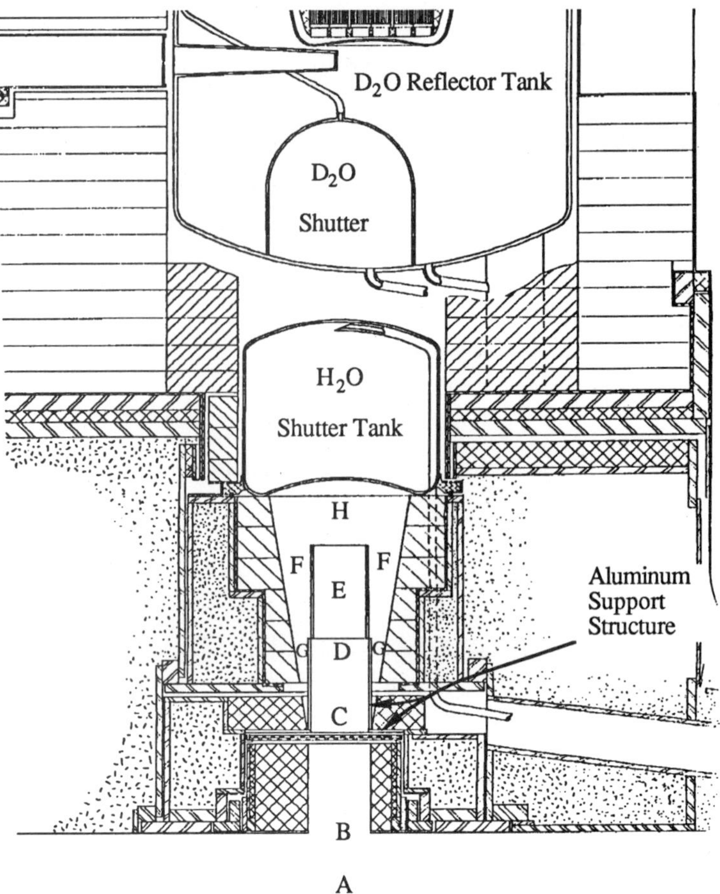

Figure 12. Aluminum support structure designed to contain various beam-filter materials and facilitate rapid insertion or withdrawal of those materials. Various filter materials were placed at the medical beam locations D - G, and beam performance was characterized at locations A, B, and C.

reduce the structure-induced gamma dose were not successful. The gamma dose did decrease, but the ^{10}B therapeutic dose also decreased resulting in a lower therapeutic to background dose ratio (AR) and poorer overall epithermal beam quality. As the beam materials were shifted further away from the patient irradiation position (i.e., toward filter configurations), the epithermal beam quality improved. Our study of beam filter materials concentrated on aluminum and sulfur in various combinations. Aluminum and sulfur mixed homogeneously in equal weight fractions produced advantage depths similar to pure sulfur and pure aluminum. However, pure sulfur provided a substantially higher dose rate. Thicker sections of pure sulfur were found to be superior at removing fast neutrons from the beam. The geometry of the filters was found to have a significant effect on epithermal beam quality. Thin filter sections placed far from the patient irradiation position were found to produce superior epithermal beams over thicker sections placed even moderately close to the beam port exit. Filters far from the irradiation position also provided the opportunity to reevaluate the inlet thermal neutron absorber.

A promising initial filter design consisted of 0.018 cm of cadmium followed by 35 cm of sulfur and a 1-mm lithium-6 carbonate final filter. The sulfur is in the central cylindrical cavity (positions E and H of Figure 12) at the furthest possible location from the patient irradiation position. The cadmium is 16.5 cm in diameter and is placed on top of the sulfur. The outer cone region (positions F and G of Figure 12) is filled with aluminum shot, homogeneously mixed with 1000 ppm of natural lithium. The aluminum shot is modeled as solid aluminum with 50% density for computational purposes. The thicknesses of both the cadmium inlet filter and the lithium-6 carbonate final filter were optimized for thermal neutron absorption with good epithermal neutron transmission. The beam opening in the bismuth plug insert is a vertical cylinder 18.4-cm in diameter. This particular design has many positive features including the following: (1) the aluminum support structure provides the option for rapid (within one day) conversion between various types of beams, e.g., thermal and epithermal beams, (2) the medical facility can be accessed while at full power because operation of the existing shutters is not inhibited by filter components, (3) the medical beam can be turned on and off very rapidly, and (4) the patient irradiation position is unchanged from the original design thereby facilitating use of the existing medical facility features (i.e., hydraulic lift, operating table, etc.).

Initial fast neutron dose and thermal neutron flux measurements as a function of depth in phantom, showed excellent agreement with calculation; however, initial total gamma dose measurements were a factor of seven greater than predicted. The high gamma dose rate was traced to high energy gamma rays originating in the reactor core. The average energy of gamma rays emanating from a ^{235}U-fueled reactor core is 0.9 MeV. The flux of gamma rays consists of essentially equal fractions of fission prompt gammas and saturated short-lived fission products, both of which have similar energy distributions [19, 20]. Our initial assumption concerning the core gamma-ray energy distribution was that modeling monoenergetic 1-MeV gammas would be a conservative approximation. Several independent experiments on our sulfur-filtered epithermal beam indicated that the effective energy of gamma rays entering our filter structure is on the order of 3-MeV. Subsequent calculations have shown a significant contribution to the core gamma dose to be from hard gammas (prompt and delayed) emanating from the aluminum in the core structure.

The core gamma-ray energy distribution was modified, and using this corrected model, the MCNP code was able to predict the total gamma-ray dose versus depth for several configurations within ~20%. Following additional calculations, it was obvious that substantial thicknesses of either bismuth or lead would have to be introduced to the beam line to attenuate the high energy core gammas. Models were developed incorporating 23 cm of bismuth shot into region G (Figure 12) and placing various thicknesses of solid bismuth directly beneath the sulfur filter in region D. A promising filter configuration, shown in Figure 13, consists of 0.018-cm cadmium, 35-cm sulfur, 2.5-cm aluminum, 13.5-cm bismuth, and 0.1-cm ^{6}Li$_2$CO$_3$. The inlet and final thermal neutron filters reduce the thermal flux incident upon the phantom thereby giving the typical epithermal depth-dose profile. The 24 cm of D$_2$O directly beneath the reactor core serves as a spectral shifter and the D$_2$O and H$_2$O shutters are voided providing a path for the neutrons to enter the beam filter region. The 35-cm sulfur filter attenuates the residual high energy neutron tail while the 0.018-cm cadmium inlet filter limits gamma-ray production from thermal neutron capture in the sulfur. The aluminum shot (1000 ppm natural Li) attenuates high energy neutrons which could otherwise stream around the sulfur filter. The lithium mixed with the shot limits the production of gamma rays from thermal neutron capture. The 13.5 cm of bismuth attenuates the core gamma rays while permitting substantial epithermal neutron transmission. The 0.1-cm ^{6}Li$_2$CO$_3$ final filter absorbs thermal neutrons arising from epithermal neutron moderation in the filter materials.

The calculated comprehensive depth-dose distribution of this beam in a cylindrical polyethylene head phantom is shown in Figure 14. The AD$_{min}$ of the beam is 5.7 cm while the AD$_{max}$ is 6.2 cm in polyethylene assuming 30 μg/g ^{10}B in tumor for both, and an effective ten-to-one tumor-to-blood ratio for the former. The AR of the beam was calculated to be 2.2 and the dose rate at the ADDR$_{min}$ was calculated to be 23 RBE-cGy/min. Some of the dose components have large error bars, and this should be considered in assessing the general trend versus depth.

Further improvements in this epithermal beam will emphasize reduction in incident fast neutron dose and changes which improve the figures of merit. Mixed field dosimetry will continue to be used as a final determinant of beam performance and to verify Monte Carlo predictions.

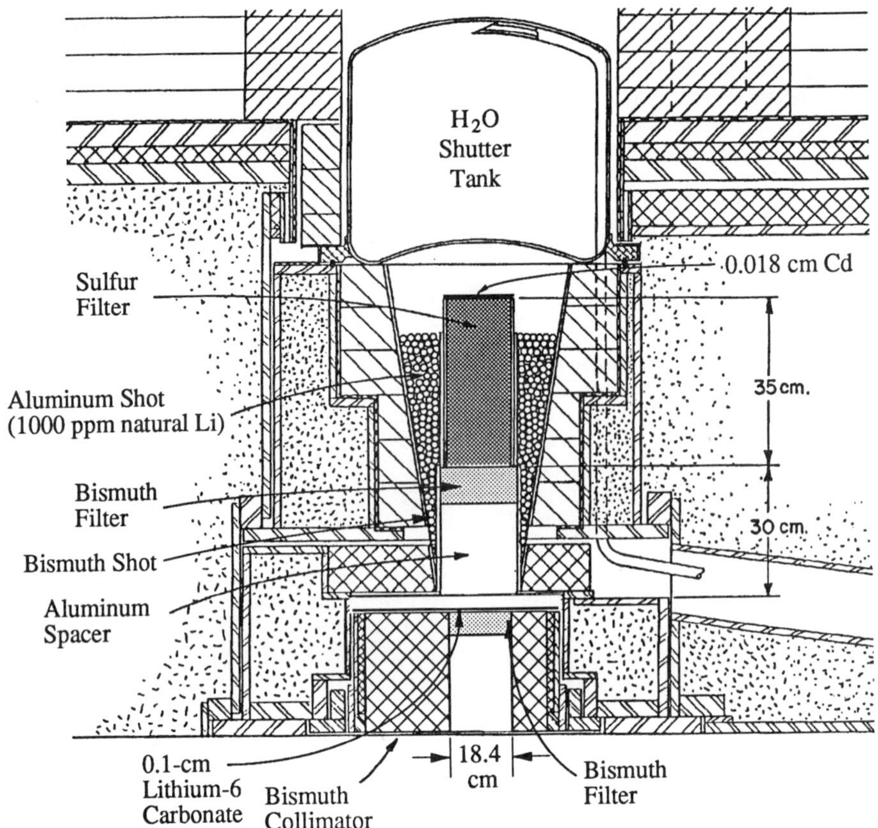

Figure 13. Epithermal neutron beam filter configuration. Filter consists of 0.018-cm cadmium, 35-cm sulfur, 2.5-cm aluminum, 13.5-cm bismuth, and 0.1-cm $^6Li_2CO_3$. The outer cone region contains 37 cm of aluminum shot (50% void fraction) homogeneously mixed with 1000-ppm natural lithium and 23 cm of bismuth shot (50% void fraction). Beam diameter = 18.4 cm.

We have discussed one promising epithermal beam design for the MITR-II. There are many options for varying epithermal beam design within the boundary conditions of the MITR-II beam line. These options will be explored and it is expected that several good designs for epithermal medical therapy beams can be achieved and will be made available at MITR-II. Furthermore, with our current engineering approaches to the installation of filters, we expect to be able to switch from one beam configuration to another with limited effort, thus offering the possibility of optimizing beam characteristics based on specific therapeutic needs.

CONCLUSIONS

Considerable neutron beam development for NCT applications has been performed at the MITR-II. Monte Carlo methods of coupled neutron/photon transport have been employed in the design of an epithermal beam filter. Beam optimization thus far has been performed by careful examination of all dose components versus depth in a polyethylene head phantom. Our efforts have been concentrated on designing a beam with the highest possible AD_{min} while also maximizing the AR and retaining an adequate ADDR.

A filter consisting of 24 cm of D_2O, 0.018 cm of cadmium, 35 cm of sulfur and 2.5 cm of aluminum (surrounded by aluminum shot doped with 1000 ppm natural lithium), 13.5 cm of bismuth, 1 mm of $^6Li_2CO_3$, and 18.4 cm in diameter has been determined to

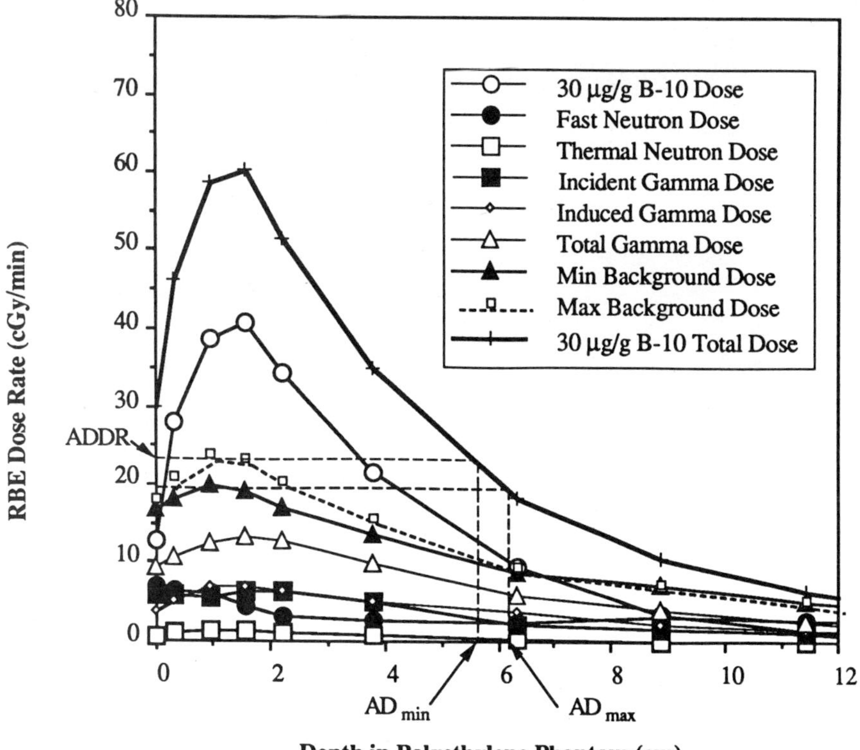

Depth in Polyethylene Phantom (cm)

Figure 14. MCNP calculated and normalized comprehensive depth-dose distribution of a sulfur and aluminum filtered beam. Filter consists of 0.018-cm cadmium, 35-cm sulfur, 2.5-cm aluminum, 13.5-cm bismuth, and 0.1-cm $^6Li_2CO_3$. The beam diameter is 18.4 cm and the phantom is polyethylene.

have good characteristics for the MITR-II medical irradiation facility. This filter is predicted to produce an RBE dose rate of 23 RBE-cGy/min which is capable of delivering a therapeutic dose of 2000 RBE-cGy (cJ/kg) to the AD_{max} of 6.2 cm in polyethylene with an AR of 2.2. These results neglect the beneficial effects of bilateral irradiation and radiation fractionation which are discussed by Zamenhof et al. [12]. The AD_{min} of 5.7 cm is, in polyethylene, within 2 cm of that achievable with a wide monoenergetic ideal neutron beam. With this beam, a therapeutic dose can be delivered to a deep-seated glioma with unilateral irradiation in approximately 80 minutes of total beam time. With bilateral irradiation, a tumor anywhere within the brain can be treated with good therapeutic advantage as discussed by Choi et al. [17] in another paper in these proceedings.

ACKNOWLEDGMENTS

The authors wish to thank Dr. John Bernard for his assistance. The authors also wish to thank Georgia Woodsworth, Leonard Andexler, and Ara Sanentz for their assistance in preparation of the manuscript. This research was funded by the U.S. Department of Energy under Grant No. DE-FG02-87ER-6060.

REFERENCES

1.   J. F. Briesmeister, ed., "MCNP – A General Monte Carlo Code for Neutron and Photon Transport, Version 3A," Los Alamos National Laboratory, <u>LA-7396-M</u>, Rev. 2 (1986).

2. R. Kinsey, "Data Formats and Procedures for the Evaluated Nuclear Data File, ENDF," Brookhaven National Laboratory, BNL-NCS-50496 (ENDF 102), 2nd Ed. (ENDF/B-V) (Oct. 1979).

3. R. J. Howerton, D. E. Cullen, R. C. Haight, M. H. MacGregor, S. T. Perkins, and E. F. Plechaty, "The LLL Evaluated Nuclear Data Library (ENDL): Evaluation Techniques, Reaction Index, and Descriptions of Individual Reactions," Lawrence Livermore National Laboratory, UCRL-50400, Vol. 15, Part A (Sept. 1975).

4. M. A. Gardner and R. J. Howerton, "ACTL: Evaluated Neutron Activation Cross Section Library - Evaluation Techniques and Reaction Index," Lawrence Livermore National Laboratory, UCRL-50400, Vol. 18 (Oct. 1978).

5. E. D. Arthur and P. G. Young, "Evaluated Neutron-Induced Cross-Sections for ^{54,56}Fe to 40 MeV," Los Alamos Scientific Laboratory, LA-8626-MS (ENDF-304) (Dec. 1980).

6. D. G. Foster, Jr. and E. D. Arthur, "Average Neutronic Properties of 'Prompt' Fission Products," Los Alamos National Laboratory, LA-9168-MS (Feb. 1982).

7. E. D. Arthur, P. G. Young, A. B. Smith, and C. A. Philis, "New Tungsten Isotope Evaluations for Neutron Energies Between 0.1 and 20 MeV," Trans. Am. Nucl. Soc., 39:793 (1981).

8. R. A. Brooks, G. DiChiro, and M. R. Keller, "Explanation of Cerebral White-Gray Contrast in Computed Tomography," J. Comp. Assist. Tomog., 4(4):489 (1980).

9. J. H. Hubbel, "Photon Mass Attenuation and Energy-Absorption Coefficients from 1 keV to 20 MeV," Int. J. Appl. Radiat. Isot., 33:1269 (1982).

10. R. S. Caswell, J. J. Coyne, and M. L. Randolph, "KERMA Factors of Elements and Compounds for Neutron Energies Below 30 MeV," Int. J. Appl. Radiat. Isot., 33:1227 (1982).

11. R. G. Zamenhof, B. W. Murray, G. L. Brownell, G. R. Wellum, and E. I. Tolpin, "Boron Neutron Capture Therapy for the Treatment of Cerebral Gliomas: I. Theoretical Evaluation of the Efficacy of Various Neutron Beams," Med. Phys., 2(2):47 (1975).

12. R. G. Zamenhof, S. D. Clement, O. K. Harling, J. F. Brenner, D. E. Wazer, H. Madoc-Jones, and J. C. Yanch, "Monte Carlo Based Dosimetry and Treatment Planning for Neutron Capture Therapy of Brain Tumors." (These Proceedings.)

13. R. A. Rydin, O. L. Deutsch, and B. W. Murray, "The Effect of Geometry on Capillary Wall Dose for Boron Neutron Capture Therapy," Phys. Med. Biol., 21(1):134 (1976).

14. O. L. Deutsch and B. W. Murray, "Monte Carlo Dosimetry Calculations for Boron Neutron Capture Therapy in the Treatment of Brain Tumors, Nucl. Technol., 26:320 (1975).

15. W. S. Snyder, M. R. Ford, G. G. Warner, and H. L. Fisher, Jr., "Estimates of Absorbed Fractions for Monoenergetic Photon Sources Uniformly Distributed in Various Organs of a Heterogeneous Phantom," MIRD, J. Nucl. Med., Suppl. No. 3, Pamphlet 5, 47 (1969).

16. R. G. Zamenhof, S. D. Clement, K. Lin, C. Lui, D. Ziegelmiller, and O. K. Harling, "Monte Carlo Treatment Planning and High-Resolution Alpha-Track Autoradiography for Neutron Capture Therapy," Strahlenther. Onkol., 165(2/3):188 (1989).

17. J. R. Choi, S. D. Clement, O. K. Harling, and R. G. Zamenhof, "Neutron Capture Therapy Beams at MITR-II." (These Proceedings.)

18. O. K. Harling, "Preliminary Design of Epithermal Beam for MITR-II," Internal Memo (July 13, 1987).

19. H. Soodak,, ed., <u>Reactor Handbook</u>, 2nd Ed., Interscience Publishers, New York (1962).

20. A. B. Chilton, J. K. Shultis, and R. E. Faw, <u>Principles of Radiation Shielding</u>, Prentice-Hall, Inc., New Jersey, pp. 90-92 (1984).

NEUTRON CAPTURE THERAPY BEAM DESIGN AT HARWELL

G. Constantine

Materials Physics and Metallurgy Division
Harwell Laboratory
Harwell, Oxon, United Kingdom

ABSTRACT

At Harwell, we have progressed from designing, building, and using small-diameter beams of epithermal neutrons for radiobiology studies to designing a radiotherapy facility for the 25-MW research reactor DIDO. The program is well into the survey phase, where the main emphasis is on tailoring the neutron spectrum. The incorporation of titanium and vanadium in an aluminium spectrum shaper in the D_2O reflector has been shown to yield a significant reduction in the mean energy of neutrons incident on the patient by suppression of streaming through the cross-section window in aluminium at 25 keV.

INTRODUCTION

A major part of the UK effort in boron neutron capture therapy (BNCT) has been the work at Harwell on neutron beam design. During the past four years, we have progressed from an iron-filtered beam, reported at the Second International Symposium on Neutron Capture Therapy, in Tokyo [1] to an aluminium/sulphur/liquid argon-filtered beam, described at the Third International Symposium on Neutron Capture Therapy, in Bremen [2]. Both beams were of only 50-mm diameter, dictated by their generation within the hollow fuel elements of DIDO and PLUTO, Harwell's 25-MW research reactors. The first gave a 93% pure beam of 24-keV neutrons, utilizing the DENIS principle [3] to enhance the beam current by taking advantage of the buildup in intensity of neutrons at the energy of the anti-resonance cross-section window in iron. The second beam, giving a broad spectrum with a mean energy of ~10 keV, demonstrated a markedly superior performance in reduced proton recoil damage per neutron delivered in an extensive series of vitro cell and chromosome aberration measurements [2, 4]. Our subsequent steps in designing a therapy beam have therefore been aimed at broad spectrum filtering.

At this juncture, we have not fully updated the specification of the therapy facility described at Bremen for the DIDO 10 H beam tube, preferring to build up a bank of experience in modelling the performance of various designs, effects of modifications to dimensions and materials, and "getting a feel" for the physics of the combined moderation and transport processes. While other reactors are of markedly different geometry from DIDO, we offer our results in the hope that they may find application elsewhere.

Generation of a high intensity field of epithermal neutrons for the treatment of glioblastoma multiforme presents many intriguing problems. In parallel with high intensity, a suitable spectrum is a prime requirement. Proton-recoil damage must be minimised and adequate penetration through the scalp and skull assured. Skin sparing is achieved by filtering out thermal and low energy epithermal neutrons with ^{10}B. The depth of penetration into tissue to generate the thermal neutron flux profile needed for the $^{10}B(n,\alpha)^7Li$ reaction central to BNCT depends upon the incident neutron energy in a relatively insensitive manner. The reduction of proton-recoil damage by tailoring the spectrum is thus the more pressing feature and the one which warrants our concentration.

THE PROPOSED HARWELL BNCT FACILITY

The 10 H beam in DIDO is a 27-cm diameter horizontal tube which penetrates the reactor tank and extends to within 9 cm of the core. Sufficient space is available outside the reactor for a shielded patient enclosure, incorporating a turntable for the couch so that bilateral irradiations can be accommodated. The neutron beam is generated within the spectrum shaper at the core-end of the beam tube. The spectrum produced there is further softened by filters within the biological shield. Figure 1 shows a quadrant of the reactor at the core midplane level, with the 10 H facility and spectrum shaper included. A part of the shaper may take the form of aluminium blocks deployed in the heavy water moderator (though more on that subject later), while further components are loaded inside the beam tube. Water cooling will be necessary to dissipate nuclear heat generated in these components.

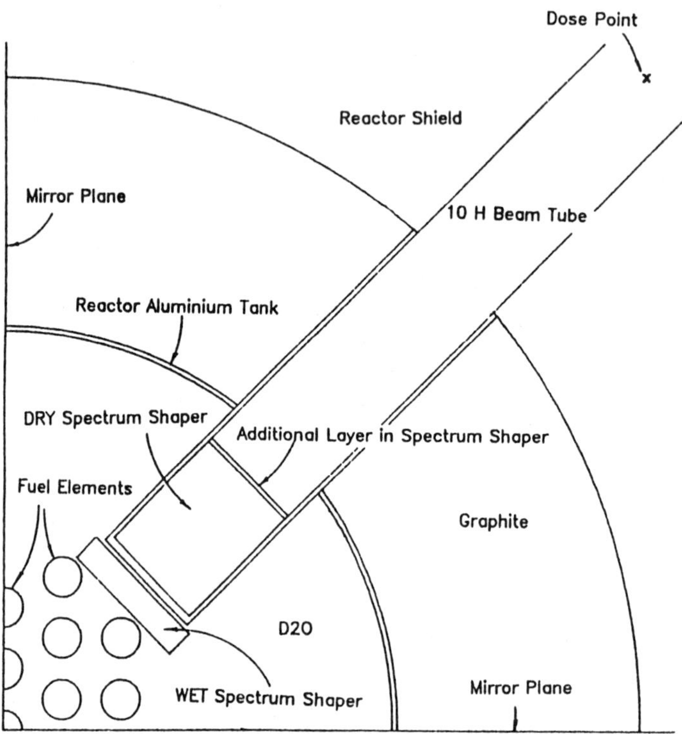

Figure 1. Plan View of One Quarter of the DIDO Reactor.

CALCULATION AND COMPUTING METHODS

Since the work reported in Bremen was carried out, we have turned to the Monte Carlo code MCNP [5]. This code has the advantage over the previously used MORSE of adopting a point energy treatment which gives far superior representation in the region of resonances in total cross section. This is an important consideration because resonances are very relevant to much of the physics of spectrum shapers and filters. In common with other Monte Carlo codes, MCNP has point detector capability, which is ideal for dealing with beam geometry. In this, a subsidiary calculation is performed for every neutron

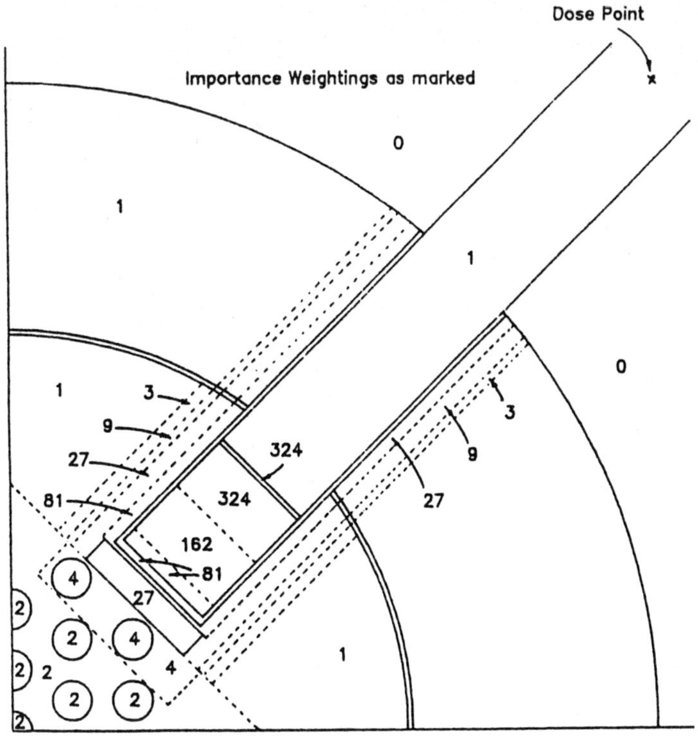

Figure 2. Plan View with Importance Function.

collision to compute the probability of that neutron reaching the dose point from the point of collision. This takes into account the probability of scattering in the right direction, attenuation by intervening material, and the effect of the inverse square law (distance).

A valuable additional asset lies in the line-of-sight attenuator option, in which filter materials can be specified and their effect on the beam arriving at the dose point can be taken into account. No allowance is made for multiple scattering so this option is only appropriate for filters that are well-separated from the dose point. Several combinations of line-of-sight attenuating materials can be dealt with, taking very little extra computing time in addition to that expended for the neutron tracking.

A recently acquired Mistral Computer Systems Hitech-10 workstation has revolutionised our calculation turnaround. To achieve high accuracy with Monte Carlo calculations in complex geometries with significantly wide flux variation across the model requires the accumulation of large numbers of particle histories. This is economically achievable with the 10 mips capability of this workstation coupled with the ability to leave it computing over nights and weekends. The wide variation in fluxes from one part of the model to another leads to poor statistics in the low flux regions. Means of ameliorating this disadvantage include splitting, wherein neutrons crossing "splitting" boundaries specified in the problem are subdivided into several sub-particles of appropriately lower weight as they cross in one direction. By tracking a larger number of lower weight particles the statistical uncertainty in the results can be reduced for a given CPU time, much less time being spent tracking neutrons in regions well away from the spectrum shaper. Several consecutive crossings of splitting boundaries leads to further subdivision into even lower weight particles, while conversely, motion in the opposite direction entails randomised killing of neutrons with increased weight for survivors. Figure 2 shows a typical importance distribution, the weights of individual particles being inversely proportional to the figure quoted in each region. Material boundaries are shown as solid lines, while splitting boundaries, which may coincide with the material ones, are drawn as dashed lines.

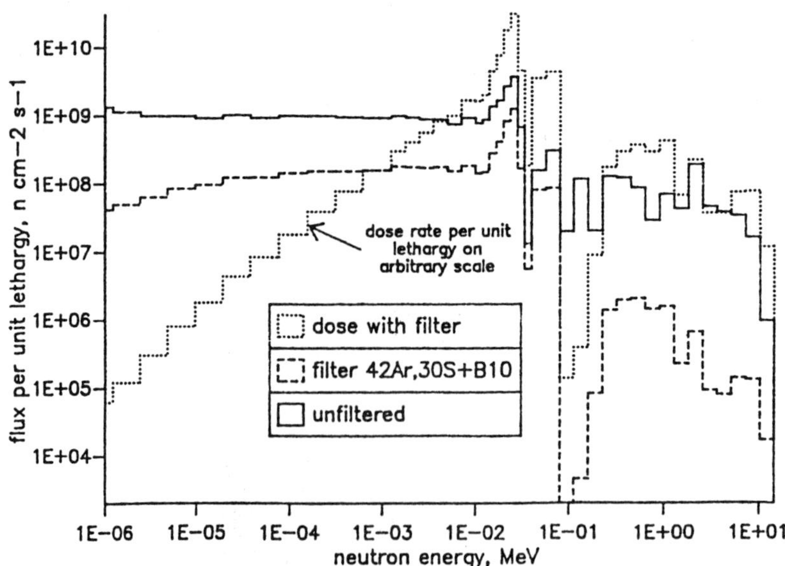

Figure 3.   Standard Spectrum Shaper with and without Filter-Dose Point Spectra.

The MCNP code was run in the fixed source mode for a quarter-core geometry. The source distributions were in proportion to the measured fission powers of the individual fuel elements in the quarter core, with division between the four concentric fuel tubes of each element and axial profiles according to measured data. (Note: The four concentric fuel tubes are not shown in Figure 1.) The specified neutron production rate was 2.45 per fission. Modelling only one quarter of the core in the point detector mode was done by employing reflecting planes. This entailed some risk in that contributions coming directly from the sources could not arrive by reflection from these planes, thus excluding the legitimate part played by the fuel elements outside the quadrant. However, scrutiny of the tallies for the uncollided contributions from the eight elements included showed that they represented a very small fraction of both total dose and flux. Hence, the effect of fuel elements in the more distant parts of the core could be taken as negligible.

OPTIMISING THE SPECTRUM SHAPER CONFIGURATION

The task of maximising the current of neutrons at the dose point while minimising their dose-weighted mean energy is an underspecified one. Information on an acceptable tradeoff is currently lacking. However, the interdependence of these two parameters is of considerable interest in enabling judgements on therapy protocol to be made.

Initially, a standard benchmark case was chosen that consisted of two components. These were a 7-cm thick aluminium slab located in the D_2O between the end of the 10 H facility and the core, and a 50-cm long cylinder of aluminium positioned inside the beam tube. These components are referred to as "WET" and "DRY" respectively. The resulting output spectrum is shown in Figure 3. This is actually the spectrum that would arrive at the dose point just outside the biological shield in the absence of any further filtering. A "standard" filter of 42 cm of liquid argon, 30 cm of sulphur, and 0.0456 g-cm^{-2} of ^{10}B was interposed and then analysed using the tally facilities described above. With the exception of the aluminium component, this filter is identical to that in use in the broad spectrum beam in DIDO. This modified the spectrum that was shown in Figure 4, giving a mean neutron energy of 9.6 keV. This is somewhat higher than that quoted for the same geometry at Bremen. This discrepancy is not the result of differences between MORSE and MCNP, but rather is due to the fact that we have used recently published total cross-section data for argon [6] in preference to that available in the data libraries of both codes.

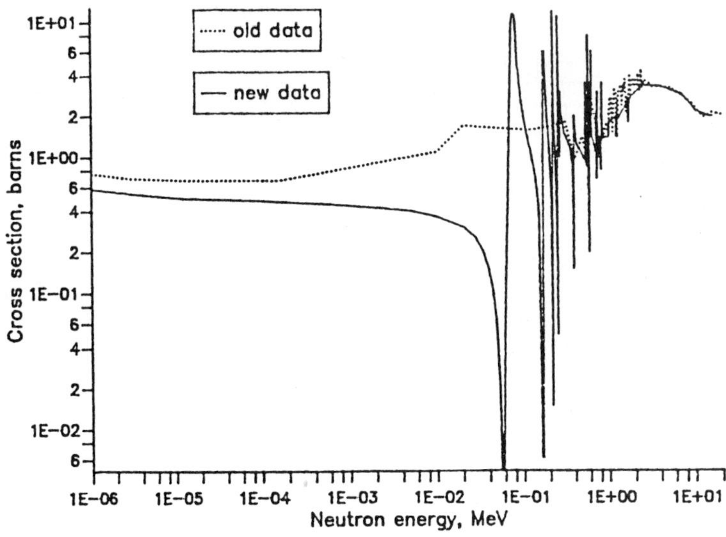

Figure 4. Comparison of Old and New Argon Data.

The considerable revision of the cross section that has taken place is shown in Figure 4. Our concern at this state of affairs has prompted us to undertake a program of measurements on the HELIOS electron linear accelerator at Harwell, results from which are expected shortly. The "sample" for these measurements, a 100-cm long, liquid argon-filled chamber, is ideally suited for measurements in the region of low cross section. This is the region where the gas-filled tubes used by other experimentalists give results of poor precision. The chamber was designed to be used later in filtered beam experiments. The improved data is employed by incorporating the energy-dependent transmission through 42-cm liquid argon in the point detector scoring tallies, both those weighted with the neutron energy dependent dose function and without. This permits an assessment of the dose-weighted mean energy of individual spectra to be made. Similar transmission functions were prepared for 100 cm of liquid argon.

INVESTIGATIONS

A broad range of Monte Carlo "experiments" have been performed with the Hitech-10. Because of the wide variation of parameters, complete matrix coverage of all the possibilities is not feasible. Hence, progress toward an optimum solution is not direct. Rather, it appears as a series of excursions in different directions, some of which may initially appear profitable but may be countered later by improvements in another parameter or introduction of different materials. Mathematically, optimising the spectrum shaper/filter configuration is equivalent to seeking a minimum in multi-dimensional space.

A comment on the precision of the Monte Carlo calculations is appropriate at this stage. Because of acceleration procedures such as the splitting technique and the typical 3-5 hours of computing time that we are able to devote to each case, uncertainties can be reduced to approximately 2-3% in flux values and 4-6% in dose (greater because a smaller and thus relatively less precise number of higher energy neutrons contribute disproportionately). The ratio of these quantities, the dose per neutron delivered, will be in error by perhaps 7%. Hence, it is unwise to ascribe too much significance to small differences between calculated mean energies. The option of computing for longer times to improve accuracy can be adopted later when we move on from the survey phase.

Fission neutrons emitted from the core are moderated by the aluminium spectrum shaper combined with its surrounding environment of D_2O. In optimising the spectrum shaper, each reactor configuration presents a different problem and each has its own unique

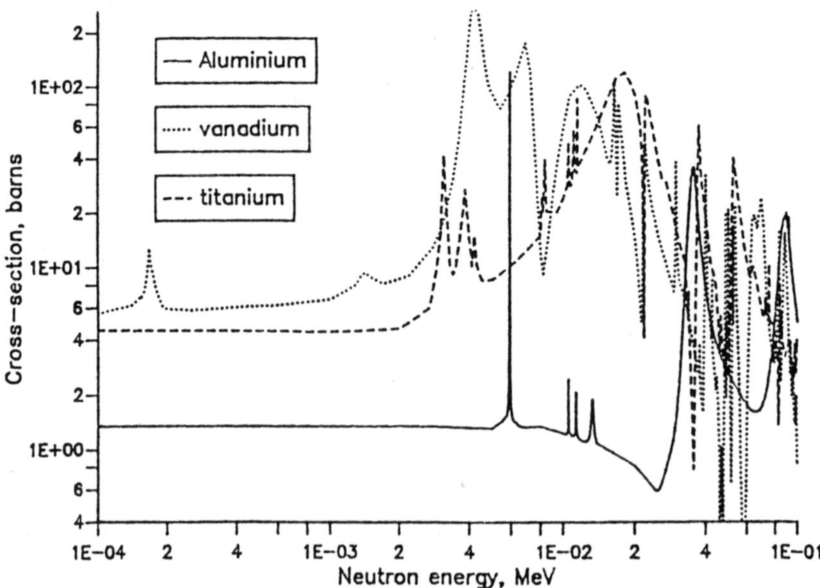

Figure 5. Comparison of Al, Ti, and V Cross Sections.

solution. However, the common trend is for the high energy loss per collision in D_2O, in conjunction with the decreasing cross section of aluminium as moderation proceeds, to combine to give a soft spectrum. Streaming through the aluminium at the energy of the 25-keV window, as shown in Figure 5, where the mean free path is in excess of 30 cm, and to a lesser extent at the 70-keV window, is responsible for undesirable high spots in the dose point spectrum. The 70-keV minimum cross section is of a comparable size to the continuum value of ~1.4 barns at 1-keV and below. However, the 25-keV window lies well below this and passage through ever more aluminium only enhances the spectrum relatively at that energy. Other components such as sulphur cannot help. The earlier argon data erroneously showed an increase in cross section between 1 keV and 25 keV, so that argon appeared to suppress this problem. However, recent changes in cross section contradict this. Fortunately, titanium and vanadium have cross sections which can be used to fill this gap, although their values below 1 keV are such that they must be used sparingly.

Figure 6 shows diagrammatically the dependence of the neutron current and mean energy at the dose point for "DRY" spectrum shapers containing different configurations of titanium and vanadium, in each case with a 7-cm "WET" aluminium block in the D_2O, and a filter within the shield of 30-cm sulphur, 0.0456 g-cm^{-2} of ^{10}B, and either 42 cm or 100 cm of liquid argon. In each case, apart from the first 50 cm of pure aluminium, the same number of both titanium and vanadium atoms are present in the spectrum shifters, which were either 35 cm or 45 cm in length. The beneficial effect of Ti and V is clear. It is also apparent that they are better when placed to the rear of the shaper, presumably because the spectrum has been enhanced relatively in neutrons around 25 keV, by passage through the full length of aluminium. The short mean free paths in Ti and V over the range 4-25 keV lead not only to considerable outscattering of such neutrons into the D_2O (with a chance of returning to the beam at a lower energy after moderation), but also to significant local moderation in the Ti and V, although their common decrease in total cross section at ~8 keV does somewhat hinder this latter effect. These processes give considerable extra weight to the argument for incorporating Ti and V in the spectrum shaper rather than in the filter, where differential attenuation by scattering is their sole mechanism for improving the spectrum. Arguing against such incorporation is the fact that thermal neutron absorption in Ti and V will give rise to increased gamma production, a problem as yet unaddressed. One possible solution would be to sandwich the Ti and V between layers of lithium/ aluminium alloy to absorb the thermal neutrons without generating gamma rays.

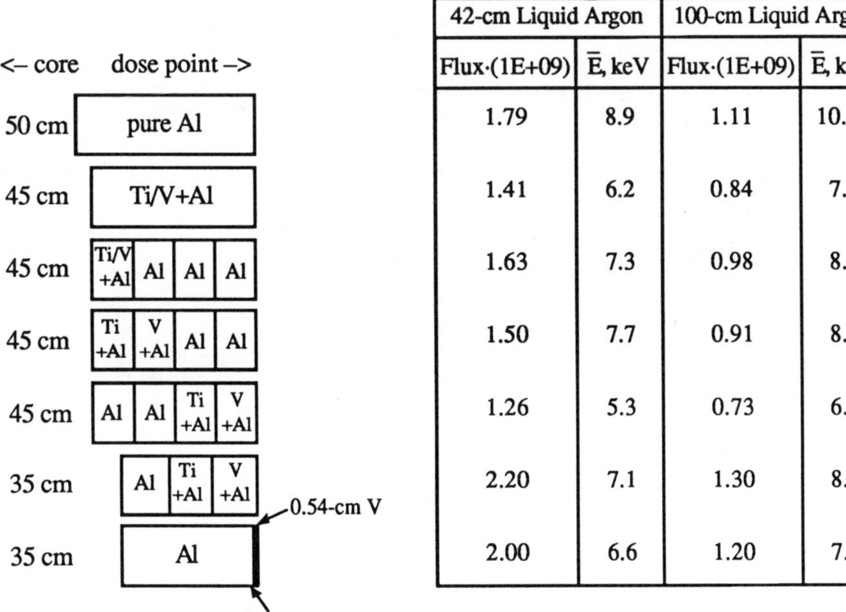

	42-cm Liquid Argon		100-cm Liquid Argon	
	Flux·(1E+09)	$\bar{E}$, keV	Flux·(1E+09)	$\bar{E}$, keV
50 cm — pure Al	1.79	8.9	1.11	10.0
45 cm — Ti/V+Al	1.41	6.2	0.84	7.2
45 cm	1.63	7.3	0.98	8.1
45 cm	1.50	7.7	0.91	8.6
45 cm	1.26	5.3	0.73	6.0
35 cm	2.20	7.1	1.30	8.1
35 cm — Al	2.00	6.6	1.20	7.2

Figure 6. Positional Effects of Ti and V in the Spectrum Shaper. (Note: Spectrum shaper includes a 7-cm aluminium block in the D_2O as well as the configurations above. The filter comprises 30-cm S, 0.0456 g-cm^{-2} of boron-10, in addition to 42-cm or 100-cm liquid argon.)

A belief held since our earlier calculations on a BNCT facility for PLUTO [7] was that the aluminium of the spectrum shifter should be extended to be as close to the core as possible. This was re-examined and the results are shown in Figures 7 and 8, where the relationship between neutron flux and mean energy at the dose point, filtered by 42-cm or 100-cm liquid argon, 0.0456 g-cm^{-2} of ^{10}B, and various combinations of sulphur and aluminium are compared for "WET" aluminium thicknesses of 7, 3.5, and 0 cm, with D_2O of course replacing any aluminium removed. It is evident that a tradeoff is involved. In each case, a reduction in aluminium thickness leads to both a decrease in mean energy and intensity. In view of the fact that deploying the "WET" part of the spectrum shifter in the D_2O would be a difficult engineering task and would undoubtedly introduce an as yet unquantified reactivity penalty, it has been provisionally omitted.

Having decided that the aluminium component of the spectrum shaper between the end of the 10 H beam tube and the core could be dispensed with, further calculations with only D_2O in that region have been carried out both with and without Ti and V in the "DRY" part of the shaper. To achieve roughly comparable fluxes at the dose point with the arbitrarily adopted equal-atom quantities of Ti and V, we compared a 50-cm pure aluminium shaper with a 45-cm total length shaper incorporating 1.18 cm of Ti/V mixture at the rear. Neutron spectra are compared in Figure 9, where it is evident that considerable benefits are available from Ti and V with a ^{10}B + 42-cm liquid filter. Increasing the argon component of the filter to 100 cm gives a further reduction in beam energy, much of it due to additional differential attenuation of the >100-keV neutrons, as seen in Figure 10. This change is supported by the "new" data on argon cross section that was shown in Figure 4. Not only the relative concentrations and spatial dispositions of the Ti and V, but also their absolute values, are matters that must be left to a more thorough investigation. A preliminary calculation, in which we doubled the quantities of both Ti and V in the spectrum shaper, led to decreases in both mean neutron energy and flux at the dose point, with no obvious net benefits.

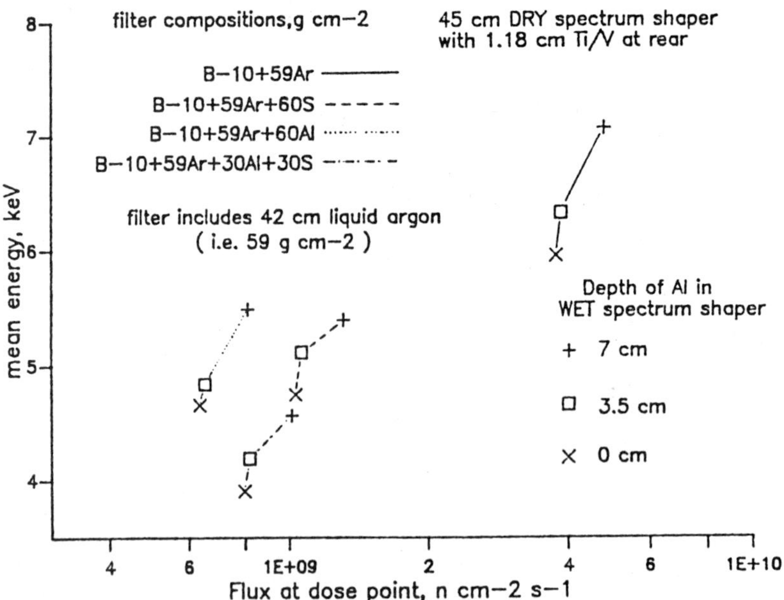

Figure 7. Effects of "WET" Aluminium Thickness on Flux/Mean Energy Relationship (42-cm Long Argon Filter).

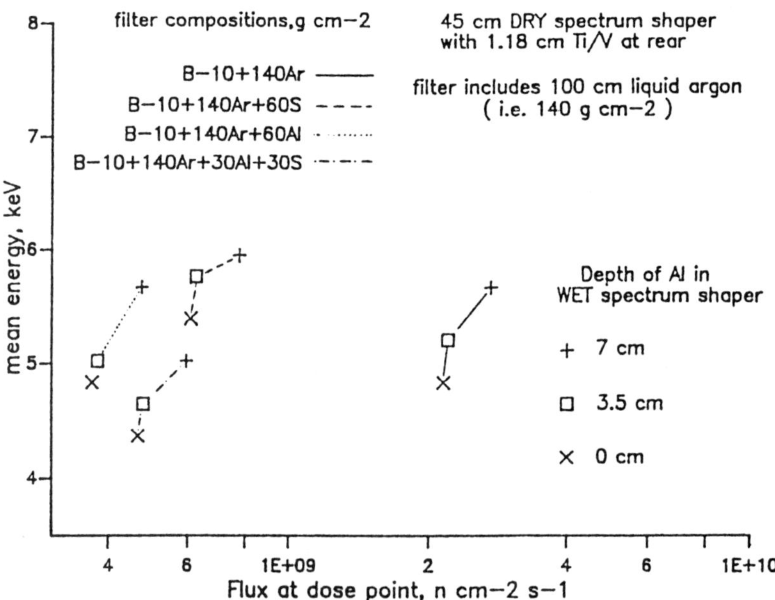

Figure 8. Effects of "WET" Aluminium Thickness on Flux/Mean Energy Relationship (100-cm Long Argon Filter).

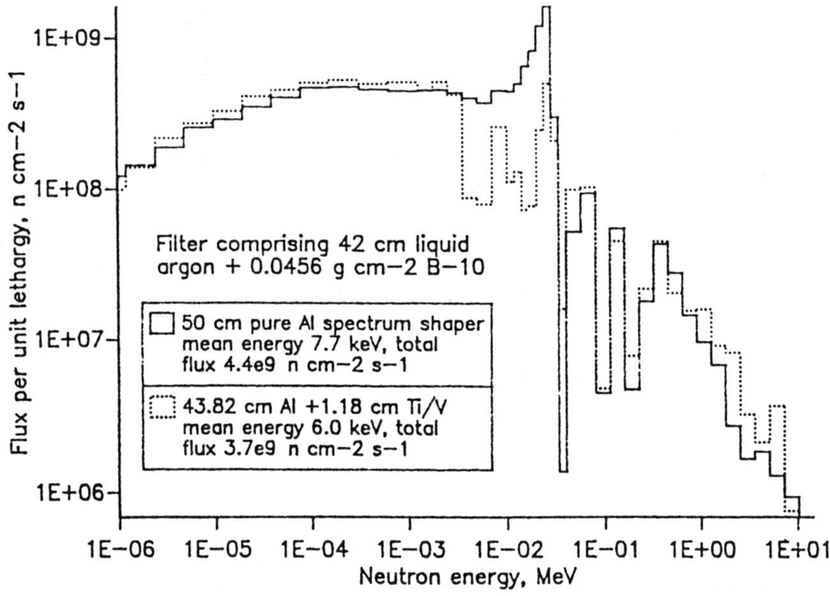

Figure 9. Neutron Spectra: Effect of Ti and V in the Spectrum Shaper.

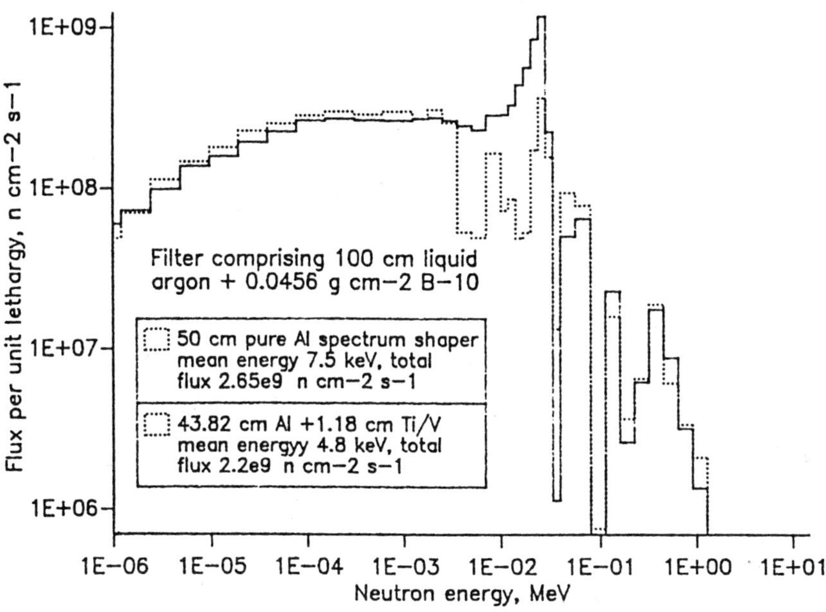

Figure 10. Neutron Spectra: Effect of Ti and V in the Spectrum Shaper.

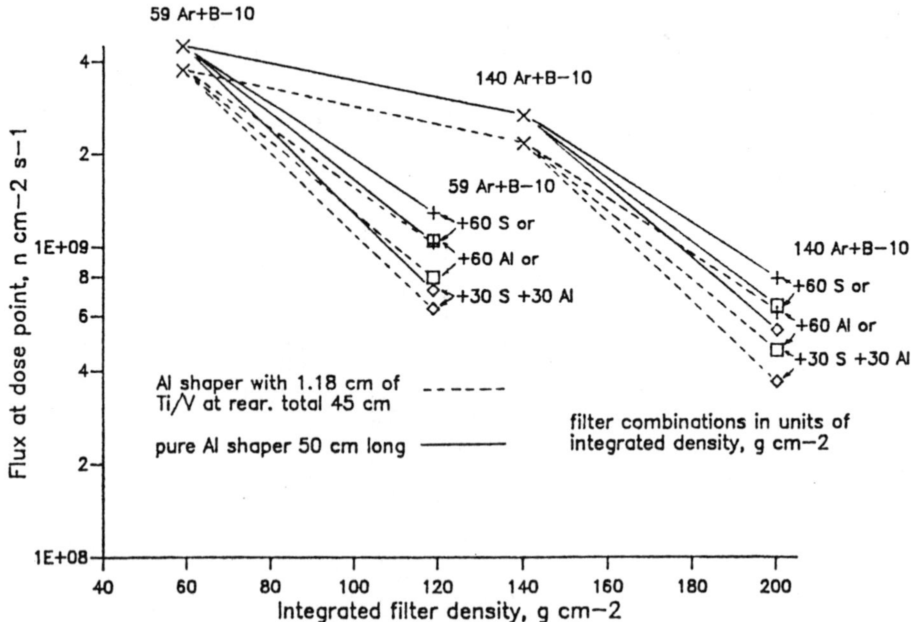

Figure 11.   Effects on Flux of Ti/V in Spectrum Shaper and of Integrated Filter Density.

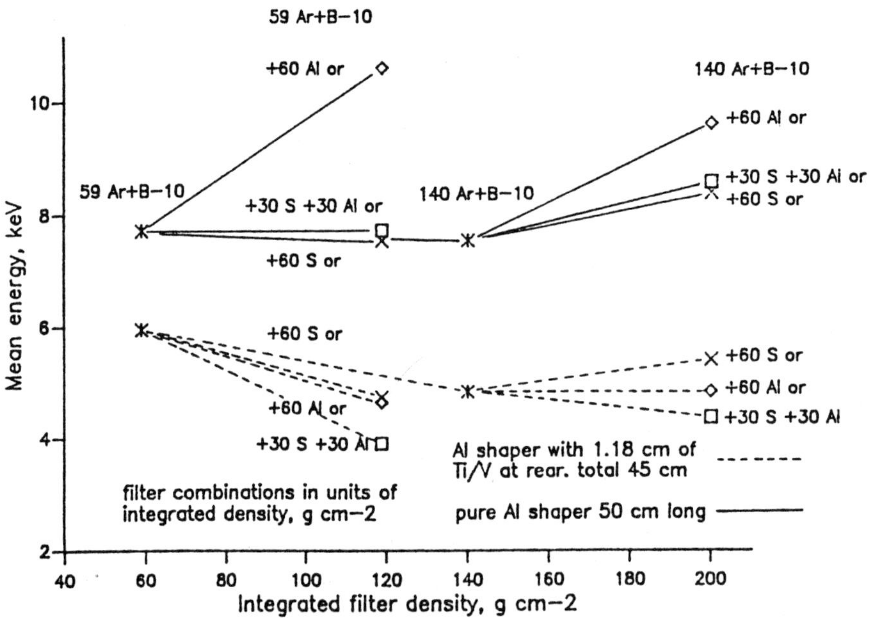

Figure 12.   Effects on Mean Energy of Ti/V in Spectrum Shaper and of
Integrated Filter Density.

The question of the gamma-ray background in the beam is a complex one, assuming increasing importance as that background approaches the level of the gamma component of the dose produced by neutron capture in the patient, principally from the H(n,γ) reaction. It was considered superficially during earlier work [2] using MORSE but has yet to be investigated in the current computing campaign. The contributions at the dose point from core gammas that have been attenuated by the spectrum shifter and filter, and from gammas arising from neutron capture in the spectrum shaper that have been subject to rather less attenuation are roughly comparable. A useful parameter for comparing the effectiveness of filters against gamma rays is the integrated density. Figures 11 and 12 illustrate the effects on dose-point flux and mean energy respectively of different ways of assembling filters of a given integrated density from various component materials. The clear-cut advantages of liquid argon are immediately apparent, while the beneficial influence of incorporating Ti and V in the spectrum shaper is re-emphasised.

Other areas investigated covered the incorporation of light water not only as a coolant in the spectrum shaper but also as a moderator. The coolant function was included as a water-filled annulus between the shaper and the beam tube (not shown in Figure 1), equivalent in area to four 15-mm diameter pipes. Further quantities of water, whether included as cylindrical shells or as transverse discs across the shaper, led only to a diminution in performance. The possibility of using D_2O as both coolant and moderator will be explored shortly, in spite of misgivings in respect to the complications involved in heavy water circuitry with attendant tritium activity.

CONCLUSIONS AND FURTHER WORK

The work carried out since the Third International Symposium on Neutron Capture Therapy in Bremen has reinforced the view that a therapy facility that generates a flux in excess of 10^9 n-cm^{-2}-s^{-1} can be built, with a mean neutron energy reduced to ~5 keV by including titanium and vanadium in the spectrum shaper. As part of our future efforts in furtherance of our design study, it is proposed to carry out spectrum measurements on another beam in DIDO. This is of smaller diameter and terminates at a greater distance from the core. A cylindrical aluminium spectrum shaper ~15 cm in diameter, with or without titanium and vanadium, will be positioned between the end of the beam tube and the core, so that spectrum measurements at the outer end of the beam tube can be compared with calculations. A variety of filters, including ^{10}B, various lengths of sulphur and aluminium, and the 100-cm long liquid argon-filled chamber can be incorporated for validation of our calculations.

ACKNOWLEDGMENTS

The support of the UK Department of Health for the work described in this paper is gratefully acknowledged.

REFERENCES

1.   G. Constantine, L. J. Baker, P. G. F. Moore, and N. P. Taylor, "The Depth Enhanced Neutron Intense Source (DENIS)," in Proc. Second Int. Symp. on Neutron Capture Therapy, Tokyo, 1985, H. Hatanaka, ed., Nishimura Co., Ltd., Niigata, Japan, p. 208 (1986).

2.   G. Constantine, J. A. B. Gibson, K. G. Harrison, and P. Schofield, "Harwell Research on Beams for Neutron Capture Therapy," Strahlenther. Onkol., 165(2/3):92 (1989).

3.   G. Constantine, L. J. Baker, and N. P. Taylor, "Improved Methods for the Generation of 24.5 keV Neutron Beams with Possible Application to Boron Neutron Capture Therapy," Nucl. Inst. Meth., A250:565 (1986).

4.   G. R. Morgan, A. J. Mill, C. J. Roberts, S. Newman, and P. D. Holt, "The Radiobiology of 24 keV Neutrons," _Brit. J. Radiol._, 61:1127 (1988).

5.   J. F. Briesmeister, ed., "MCNP – A General Monte Carlo Code for Neutron and Photon Transport, Version 3A," Los Alamos National Laboratory, _LA-7396-M_, Rev. 2 (1986).

6.   V. McLean, C. L. Dunford, and P. F. Rose, "Neutron Cross-Section Curves for Z=1-100," in _Neutron Cross Sections_, Vol. 2, Academic Press, Boston (1988).

7.   G. Constantine, G. R. Morgan, and N. P. Taylor, "Progress toward Boron Neutron Capture Therapy at Harwell," in _Proc. Int. Symp. Utilisation of Research Reactors_, Grenoble, IAEA-SM-300/74 (1987).

PHYSICS DESIGN FOR THE BROOKHAVEN MEDICAL RESEARCH REACTOR EPITHERMAL NEUTRON SOURCE

F. J. Wheeler,[1] D. K. Parsons,[1] D. W. Nigg,[1] D. E. Wessol,[1]
L. G. Miller,[1] and R. G. Fairchild[2]

[1] Idaho National Engineering Laboratory
EG&G Idaho, Inc.
Idaho Falls, ID

[2] Brookhaven National Laboratory
Medical Department
Upton, Long Island, NY

ABSTRACT

A collaborative effort by researchers at the Idaho National Engineering Laboratory and the Brookhaven National Laboratory has resulted in the design and implementation of an epithermal-neutron source at the Brookhaven Medical Research Reactor (BMRR). Large aluminum containers, filled with aluminum oxide tiles and aluminum spacers, were tailored to pre-existing compartments on the animal side of the reactor facility. A layer of cadmium was used to minimize the thermal-neutron component. Additional bismuth was added to the pre-existing bismuth shield to minimize the gamma component of the beam. Lead was also added to reduce gamma streaming around the bismuth. The physics design methods are outlined in this paper. Information available to date shows close agreement between calculated and measured beam parameters. The neutron spectrum is predominantly in the intermediate energy range (0.5 eV - 10 keV). The peak flux intensity is 6.4E+12 $n/(m^2 \cdot s \cdot MW)$ at the center of the beam on the outer surface of the final gamma shield. The corresponding neutron current is 3.8E+12 $n/(m^2 \cdot s \cdot MW)$. Presently, the core operates at a maximum of 3 MW. The fast-neutron KERMA is 3.6E-15 $cGy/(n/m^2)$ and the gamma KERMA is 5.0E-16 $cGy/(n/m^2)$ for the unperturbed beam. The neutron intensity falls off rapidly with distance from the outer shield and the thermal flux realized in phantom or tissue is strongly dependent on the beam-delimiter and target geometry.

INTRODUCTION

The Power Burst Facility/Boron Neutron Capture Therapy (PBF/BNCT) Program for Cancer Treatment is a comprehensive plan for the development of BNCT. The program is centered at the Idaho National Engineering Laboratory (INEL) and incorporates collaborators at several institutions nationwide.

The neutron-source development plan is to realize a neutron beam adequate for therapy. This will be done in three stages. The first stage is the development of an intermediate-intensity epithermal-neutron beam at the Brookhaven National laboratory (BNL). The second stage is the development of a high-intensity epithermal-neutron source at the PBF reactor located at INEL. The third stage, currently supported by nonprogrammatic funding, is the design of a new medical reactor [1] that would be

Neutron Beam Design, Development, and Performance for Neutron Capture Therapy
Edited by O. K. Harling *et al.*
Plenum Press, New York, 1990

deployable at a medical facility in a metropolitan area. The first stage of the program, the BNL Medical Research Reactor (BMRR) epithermal source, has been realized by a joint BNL/INEL effort.

The basic principles and history of neutron capture therapy (NCT) are well-documented [2,3]. An epithermal-neutron* beam is thought to offer an advantage compared to a thermal-neutron beam for capture therapy because the neutrons will penetrate a few centimeters into tissue before forming a thermal flux peak. Therefore, with an epithermal beam it may be possible to treat tumors at greater depths and with greater protection for the outer layers of healthy tissue compared to treatment by a thermal-neutron beam.

In 1965, Fairchild [4] performed experiments with an "epithermal" beam at the BMRR. This beam was obtained by filtering out the thermal neutrons from a thermal beam with a Cd filter. This beam had sufficient intensity, but therapeutic application was not possible because the fast-neutron component was at least one order of magnitude too high. Funding limitations restricted further research until 1986, when the PBF/BNCT program was initiated. A collaborative effort among researchers at BNL and INEL involved both a design effort and an experimental program which has resulted in modifications to the BMRR. This effort has produced an epithermal-neutron source suitable for medical experiments. That source forms the subject matter of this paper.

This paper describes the physics design methods used in the BMRR study. Other papers at this Workshop describe dosimetry performed for the new beam.

REACTOR FACILITY DESCRIPTION

The BMRR facility is described in [5] and in other papers at this Workshop. The core is cooled and moderated by light water and is reflected by graphite. The fuel elements are standard curved-plate design with high-enriched uranium fuel in an aluminum matrix with aluminum cladding. Each element contains 140 grams of ^{235}U. The fueled portion of the fuel element is 0.6-m high. The control system consists of three safety rods and one regulating rod. These core components and the light-water-cooled graphite reflector fill the inside of a 0.6-m (inside diameter) aluminum vessel. The present BMRR core loading consists of 23 fuel elements. An air-cooled graphite reflector, shielding, beam ports, and treatment facility are located outside the aluminum vessel. The modifications to provide an epithermal beam were made on the animal side of the reactor facility.

Figure 1 shows a sketch of the reactor, reflector, thermal shield, and three compartments (designated A, B, and C) where materials may be placed to provide a filter for tailoring the neutron (and gamma) flux arriving at the irradiation target. (Note: The material layers used to tailor the spectrum are referred to as a filter even though these layers contain moderating as well as filtering materials.) Two of the compartments (A and B) are accessible for modification. Compartment C is inaccessible except for piping. Compartment C contains two aluminum tanks which can be filled with D_2O (or perhaps other liquids) if desired. Compartments A and B are sections of a massive column which acts as a shutter. When the column is lowered, A and B are dropped to a lower level and are replaced by heavy concrete at the irradiation level. The beam is significantly quenched when the column is in the down position.

DESIGN STUDIES

Design Goals

A neutron source for NCT should have the following attributes:

---

* The U.S. National Bureau of Standards, Handbook 63, defines fast neutrons as those with energy E>10 keV, epithermal neutrons as those with energy 10 keV>E>0.5 eV, and thermal neutrons as those with E<0.5 eV.

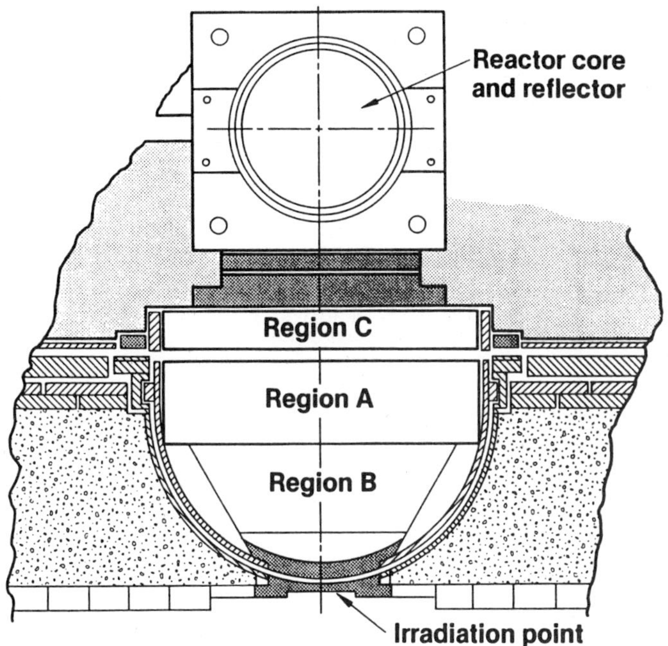

Figure 1. Reactor, Reflector, Thermal Shield, and
Compartments A, B, and C.

1.   High intensity – So that measurements and irradiations can be performed in reasonable periods of time.

2.   Low fast-neutron KERMA – This component is nonselective and should be minimized.

3.   Low incident-gamma KERMA – This component is also nonselective.

4.   Good collimation – The neutron beam should have a high forward component to maximize tissue penetration and minimize intensity falloff with depth.

5.   Flexibility – It will be desirable to change beam openings and to tailor the neutron spectrum for different applications. This will be difficult to do if the source is marginal for any of the other attributes.

These factors were the guidelines for the BMRR design. However, the type of reactor core and the pre-existing geometry dictated much of the design.

For the BMRR design, several constraints exist because major modifications could not be made. The core is not ideal for producing an epithermal source because (as for all plate-type research-reactor cores) the water fraction is high, thus depleting the higher-energy neutron current entering the filter. This limits the achievable intensity of the final beam.

It is possible to reduce the fast KERMA to a very low value. However, the cost is a greatly reduced intensity. Therefore, for the BMRR, the fast KERMA may be higher than the desirable level in order to achieve usable intensity. This KERMA can still be reduced further by adding material to Compartment C. Again, this would result in lowered intensity. The gamma KERMA can be easily reduced to a low value, as pointed out by Fairchild [6].

The collimation of the beam is poor for the BMRR because there is no space for a collimator and, even if there were space, a collimator would reduce the viable flux intensity. Neutrons emanating from the outer shield will have an angular distribution close to isotropic because the last scatter event was an isotropic elastic scatter occurring a short distance from the beam exit.

<u>One-Dimensional Studies</u>

One-dimensional calculations were performed to compare various materials and material thicknesses. Two-dimensional calculations were performed for the final design studies.

The one-dimensional model utilized the SCAMP [7] discrete ordinates code in cylindrical geometry. A 30-energy group P_1 ENDF/B-V cross-section library was derived using the COMBINE code [8]. This model, with appropriate axial-leakage corrections, yields results in reasonable agreement both with earlier BNL measurements [4] for 0–0.18 m D_2O in Region A and with more recent results from the D_2O-Al sensitivity studies [9]. This model was used to investigate several materials and material thicknesses. Material configurations, based on transmission windows at resonance interference minima such as Fe, Al, or S were investigated and, although significant intensity can be provided by these materials, they were not considered for final design because the fast-neutron component is an order of magnitude higher than desired.

The best materials for the BMRR were found to be the aluminum-heavy water combination previously identified by Oka et al. [10] and alumina, previously identified by Brugger et al. [11]. Figure 2 shows the predicted intensities for homogeneous Al-D_2O mixtures and for alumina versus thickness of these materials. Figure 3 shows the fast-neutron KERMA versus thickness of these two materials. The gamma KERMA is easily reduced by adding an appropriate thickness of bismuth or lead shielding.

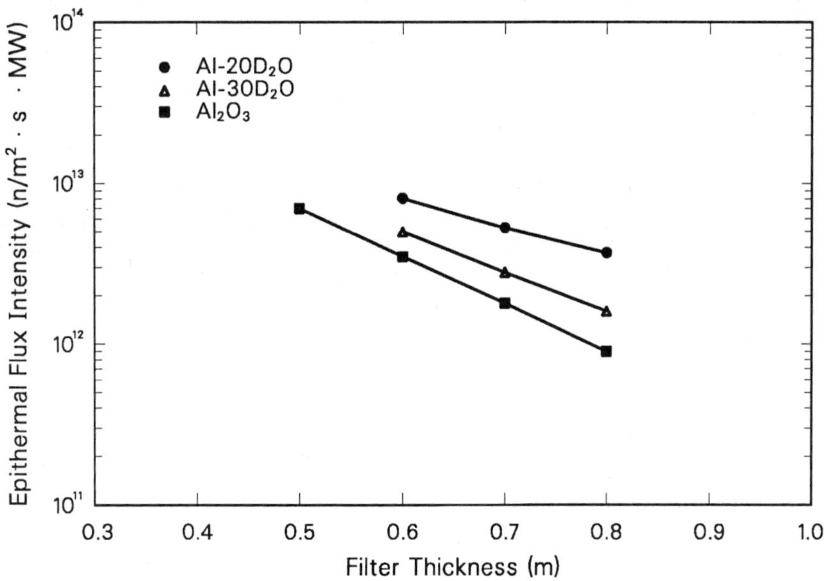

Figure 2.   Calculated Epithermal Flux at Center of Outer Shield Surface for BMRR vs. Filter Thickness.

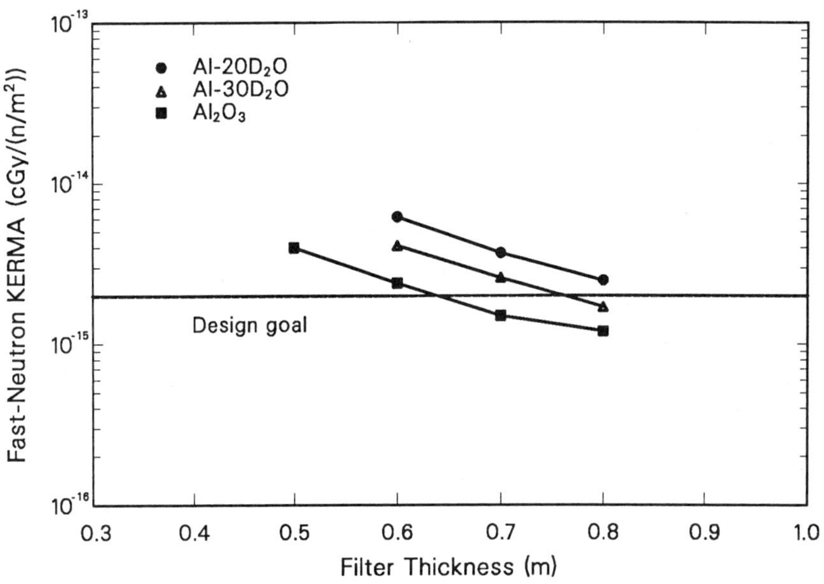

Figure 3.   Calculated Fast-Neutron KERMA for BMRR vs. Filter Thickness in Centimeters.

<u>Two-Dimensional Studies</u>

Final design and as-built analyses were performed with a two-dimensional model using the DOT 4.3 code [12] and the BUGLE-80 [13] ENDF/B-IV cross-section library. Cylindrical (r-z) geometry was employed with the z-axis coinciding with the beam centerline. The core was modeled as a homogeneous cylinder. Thus, the actual geometry of the BMRR is distorted due to the r-z model, but this is not a severe approximation if key volumes and material thicknesses are preserved.

It is possible to solve the transport problem for the BMRR r-z model in one step. However, the problem was divided into two segments to be more tractable. The first segment, or reactor model, was run as an eigenvalue problem and the second segment, or filter model, was run as a fixed-boundary flux problem with boundary angular fluxes taken from the output of the first segment. This approach results in the same solution as would be obtained if the entire model were run as one problem provided that the reactor and filter problem have sufficient overlap. It was necessary to run the reactor problem only twice, once for the design studies and once for the as-built configuration.

The BMRR DOT models did not require the detailed biased quadrature set used for the PBF design [14] because the final configuration did not have the large voided collimator employed in PBF. The BMRR model made use of symmetric S_8 angular quadrature for the 67-group calculation with diamond differencing spatial discretization. The reactor segment was overlayed as a 62 x 79 mesh and took four hours of CPU time on the INEL CRAY-XMP/24. The calculated eigenvalue was 1.021. The final model of the filter segment was divided into a 62 x 224 mesh and typically required four equivalent CRAY CPU hours per run.

Based on calculated results, we selected a filter composed of 75-volume percent aluminum with 25-volume percent D_2O. This meant that the filter would have to be 0.8 meters in length and be fairly homogeneous, employing alternating Al and D_2O layers where the Al was no thicker than 25 mm. This was not practical because the filter would have to occupy Compartment C and, as previously stated, this compartment could not be modified.

Instead, the Al_2O_3 filter was selected because Al_2O_3 is a more effective material when space is limited such as in the case of the BMRR. Al_2O_3 is a refractory material with low thermal conductivity. Hence, heating was a concern. However, analyses showed that this would not be a severe problem in Compartments A and B where the neutron fluxes are low relative to regions near the core.

The as-built BMRR configuration is depicted in Figure 4. Material thicknesses on a line at the beam axis are: 0.4572-m Al_2O_3, 0.1965-m Al, 0.51-mm Cd, and 0.15-m Bi in Compartments A and B. The filter is contained in a steel shell, with an external layer of B_4C in aluminum. This B_4C layer, not shown in Figure 4, extends around the filter except for a 0.25-m x 0.25-m opening at the beam port.

Table One provides integral results from a few of the DOT filter runs. Case 15G was the design configuration and Case 17 was the as-built configuration. The difference in Case 17 being a somewhat less Al_2O_3-thickness because of loading constraints. This reduced thickness resulted in a higher flux intensity and a higher fast KERMA than did the design. Also shown in Table One are initial measured results [15]. The agreement of the unadjusted calculations and the measured results is excellent for this complex transport problem.

The calculated neutron energy spectrum is presented in Figure 5. The fast-neutron KERMA has two primary components: a high-energy component from neutron energies in the MeV range and a component from neutron energies in the 10 - 30 keV range. The high-energy component is more apparent for Al_2O_3 than for Al-D_2O because of the 2.3-MeV oxygen interference minimum. This component is also more penetrating in tissue with a mean free path of about 0.1 m compared to a mean free path of about 0.01 m for the keV component.

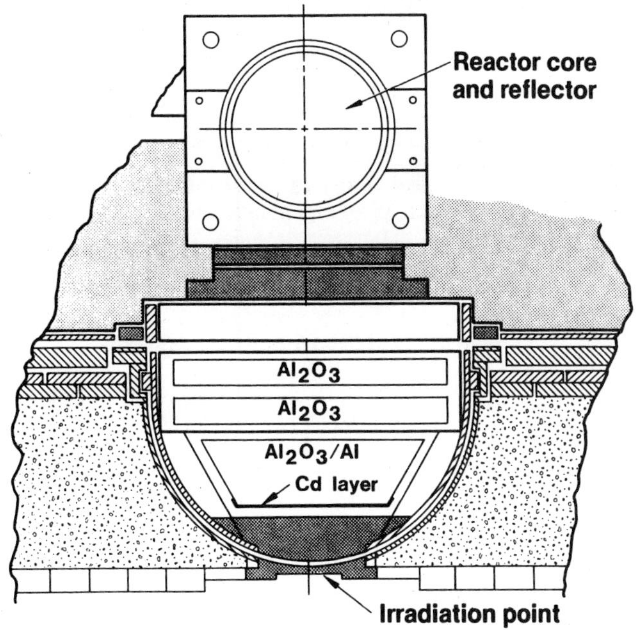

Figure 4.   As-Built Al$_2$O$_3$ Filter Installed in the BMRR.

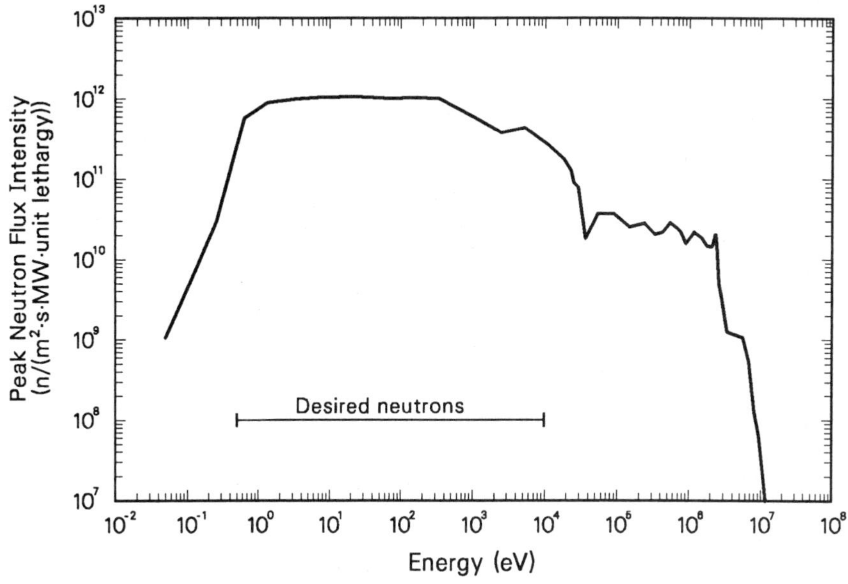

Figure 5.   Beam Port Energy Spectrum for BMRR, Case 17.

Table One

Calculated and Measured Integral Characteristics for the
Unperturbed BMRR Epithermal-Neutron Beam.[a]

Quantity	DOT Calculations		Measurements
	Case 15G[b]	Case 17[c]	(as-built)
Total Neutron Flux Intensity $[n/(m^2{\cdot}s{\cdot}MW)]$	4.3E+12	6.3E+12	6.2E+12
Fast-Neutron KERMA $[cGy/(n/m^2)]$	2.3E-15	3.1E-15	4.5E-15
Gamma KERMA $[cGy/(n/m^2)]$	4.7E-16	4.6E-16	
Fast-Neutron Dose Rate $[cGy/(MW{\cdot}min)]$	0.59	1.2	1.7
Gamma Dose Rate $[cGy/(MW{\cdot}min)]$	0.12	0.17	
Neutron Flux-to-Current Ratio	1.69	1.71	

[a] Taken at the beam centerline, on the outer surface of the final bismuth shield.
[b] Design configuration, containing 0.5132-m Al_2O_3, 0.1537-m Al (along the beam axis) in Compartments A and B.
[c] As-built configuration, containing 0.4572-m Al_2O_3, 0.1965-m Al (along the beam axis) in Compartments A and B.

APPLICATION OF THE BEAM FOR PHANTOM MEASUREMENT AND ANIMAL IRRADIATIONS

Measurements of the unperturbed beam characteristics [15] show that the unadjusted calculations are in good agreement with measurement. However, they underestimate the fast-neutron flux by about 30% in the energy range 0.1 - 3 MeV. This is likely attributable to an underestimate of neutron transmission through the oxygen window and scatter of that component to lower energies. The model will be improved to better represent the oxygen window even though the calculated results with the present model are providing excellent overall agreement with measurements.

Calculations for Cylindrical Phantoms

The DOT program was used to calculate flux and dose profiles for a cylindrical phantom with brain-equivalent composition. Figure 6 shows the positioning of the phantom relative to the filter and shields. Beam diameter is defined as the cylindrical opening (d) in a lithiated-polyethylene beam delimiter. This beam diameter is not sharply defined because of neutron scatter and penetration, but the delimiter is quite effective for the predominantly epithermal neutrons. Flux-to-dose KERMA factors for the calculations were precomputed by the MACK-IV program [16] using ENDF/B-IV data. Plots of centerline dose rates as a function of depth in the phantom are provided in Figures 7-9 for beam diameters (d) equal to 46, 100, and 200 mm. These dose rates are for a power of 1 MW. Presently the BMRR can operate at 3 MW so the data should be multiplied by 3 to obtain the maximum available dose rates. For the three beam diameters, the maximum dose rates are 12 to 50 times lower than dose rates achievable for the PBF design [17].

Table Two shows the calculated values of quantities specified for this Workshop.

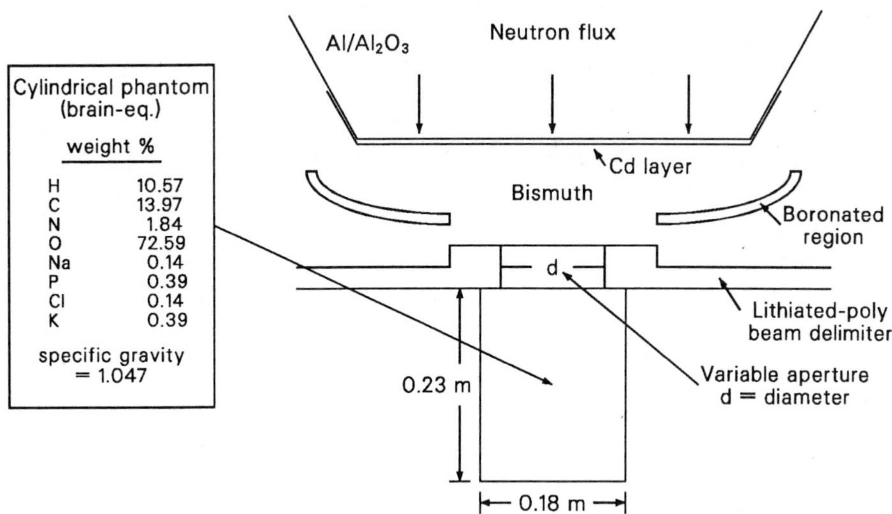

Figure 6. Arrangement of Beam, Filters, and Shields for BMRR Phantom Calculations.

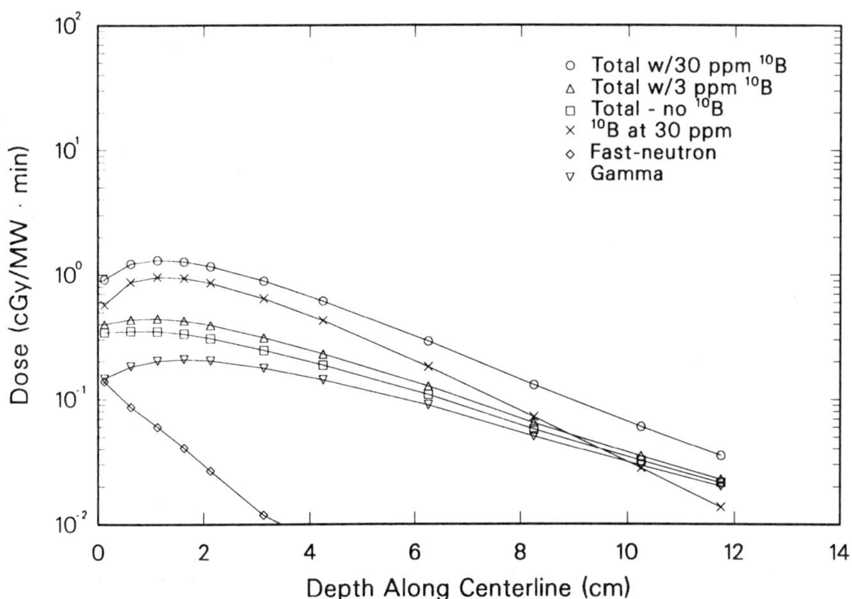

Figure 7. BMRR 4.6-cm Beam with 18x23 Phantom-Axial Distribution.

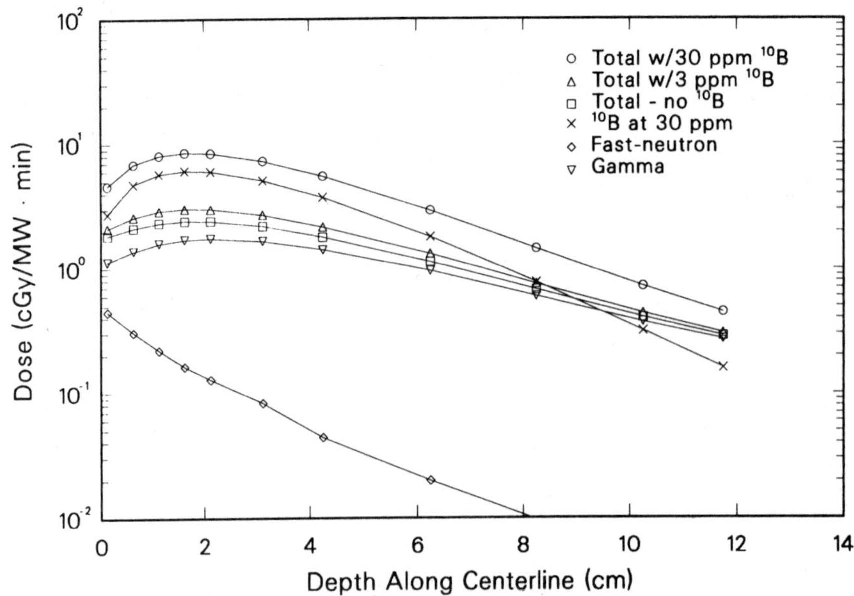

Figure 8.   BMRR 10-cm Beam with 18x23 Phantom-
Axial Distribution.

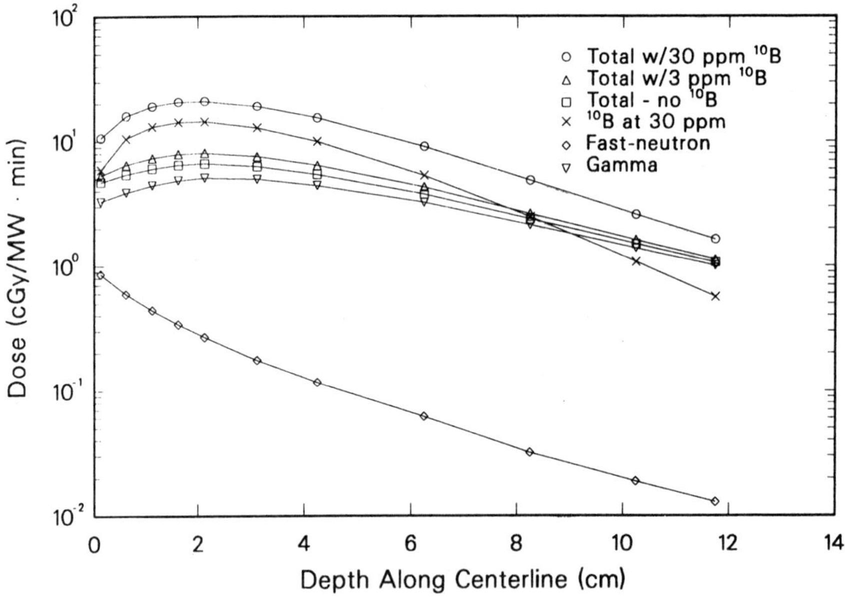

Figure 9.   BMRR 20-cm Beam with 18x23 Phantom-
Axial Distribution.

Table Two

Calculated Advantage Factors for Cylindrical Phantom – BMRR Epithermal Beam.

Quantity	No ^{10}B in Healthy Tissue			3μg/g ^{10}B in Healthy Tissue		
	d*=4.6	d=10	d=20	d=4.6	d=10	d=20
Advantage Depth:						
No RBE (cm)	5.8	6.8	7.3	5.1	6.1	6.6
With RBE[†] (cm)	6.7	8.0	8.6	5.8	6.8	7.5
Advantage Dose Rate:						
No RBE (cGy/MW·min)	0.35	2.4	6.7	0.44	3.0	8.1
Advantage Ratio:						
No RBE	3.5	3.2	2.9	2.8	2.7	2.5
With RBE	5.9	5.5	5.0	4.0	3.8	3.6
% Low-LET Dose	18.2	22.7	26.7	17.7	22.3	26.2
% High-LET Dose	10.4	8.3	7.5	10.5	8.3	7.6
% ^{10}B Dose (30μg/g)	71.4	69.0	65.8	71.8	69.4	66.2

* d = diameter of beam delimiter (cm).

† RBE = 1.0 for all gamma rays, 1.6 for fast neutrons and ^{14}N(n,p) protons, and 2.3 for the ^{10}B(n,α) reaction. (These values were set for this Workshop to provide consistency in intercomparisons.)

NOTE: The values quoted here for calculations differ substantially from those quoted by Fairchild et al., in these proceedings, for measured results. For example, the measured advantage depths were significantly higher. The results reported in both papers are each correct. The reason for the difference is beam diameter. This work assumed a beam delimiter, made of lithiated polyethylene, defining beam diameters of 4.6 cm, 10 cm, and 20 cm. The Fairchild measurement used no such delimiter. With no beam delimiter, there is a very substantial supply of neutrons beyond the 20-cm maximum diameter that this work assumed and even beyond the diameter of the phantom. Thus the neutron field falls off substantially less than with a delimiter and one obtains a greater value for advantage depth. One cannot extrapolate these results to obtain Fairchild's because there is a discontinuity when beam diameter exceeds phantom diameter. This effect was experimentally confirmed during the dog phantom measurements performed at the BMRR. There is no discrepancy in measurement or in the calculations. For clinical application, one would want a beam delimiter so as not to overirradiate healthy tissue.

These quantities are: (1) the advantage depth, defined as the depth at which the total therapeutic dose (to tumor with 30 μg/g of ^{10}B) equals the peak healthy tissue dose, (2) the advantage dose rate, or total dose rate at the advantage depth, and (3) the advantage ratio, defined as the quotient of the centerline integral of the total therapeutic dose divided by the integral of the total healthy tissue dose over depth from zero to the advantage depth.

<u>Canine Program</u>

For the INEL PBF/BNCT program, the immediate application of the beam was for canine irradiations. Prior to the canine program, phantom measurements were made with a lucite dog head phantom prepared at Washington State University.

To minimize the number of dogs used in the study, it was desirable to perform hemispherical irradiations, or to irradiate one hemisphere to a dose roughly twice that of the other hemisphere. A study was made to determine the best available material to be used as a beam delimiter, which would effectively mask the beam except for the area to be irradiated. The best material identified for this was a composite of hydrated lithium hydroxide mixed into polyethylene. The lithium was 95% enriched in the ^{6}Li isotope. This material provides a high hydrogen content to effectively moderate the epithermal neutrons and an adequate ^{6}Li concentration to capture the neutrons without significant production of secondary gammas. Two inches of this material as a shield reduces the thermal-neutrons generated in tissue by two orders of magnitude with very little added contribution to the gamma dose. Fast neutrons and incident gamma rays in the beam are attenuated only slightly.

The transport model for the dog phantom employed the Monte Carlo module of the developing radiation transport system for the INEL PBF/BNCT program. The detailed angular fluxes from the DOT output for the as-built filter (with a cylindrical phantom in place) was the source used in the Monte Carlo model. The results were compared to measurements [18] using the lucite phantom. The unadjusted calculations predicted a peak thermal flux of 2.99E+12 n/(m$^2\cdot$s$\cdot$MW) and the measurements (with Au and Cu flux wires) gave 3.1E+12 n/(m$^2\cdot$s$\cdot$MW). The calculated peak gamma dose in the phantom was 1.22 cGy/(MW$\cdot$min) and the measured value at the peak was 1.25 cGy/(MW$\cdot$min). Detailed results for the dog phantom measurements will be published elsewhere.

ACKNOWLEDGMENT

This work was performed under the auspices of the U.S. Department of Energy, DOE Contract No. DE-AC07-761D01570.

COPYRIGHT

REFERENCES

1. W. A. Neuman, "Neutron Beam Studies for a Medical Therapy Reactor." (These Proceedings.)

2. <u>Proc. First Int. Symp. on Neutron Capture Therapy</u>, Cambridge, MA, 1983, R. G. Fairchild and G. L. Brownell, eds., Brookhaven National Laboratory, <u>BNL-51730</u> (1984).

3. <u>Proc. Second Int. Symp. on Neutron Capture Therapy</u>, Tokyo, 1985, H. Hatanaka, ed., Nishimura Co., Ltd., Niigata, Japan (1986).

4. R. G. Fairchild, "Development and Dosimetry of an 'Epithermal' Neutron Beam for Possible Use in Neutron Capture Therapy," <u>Phys. Med. Biol.</u>, 10(4):491 (1965).

5.   J. B. Godel, "Description of Facilities and Mechanical Components, Medical Research Reactor (MRR)," Brookhaven National Laboratory, BNL-600 (T-173) (1960).

6.   R. G. Fairchild, Personal Communication.

7.   G. E. Putnam, "TOPIC-A FORTRAN Program for the Calculation of Transport of Particles in Cylinders," Idaho National Engineering Laboratory, IDO-16968 (1964). (SCAMP is a multigroup version of the TOPIC Program.)

8.   R. A. Grimesey and R. L. Curtis, "COMBINE: Combined Fast and Thermal B-3 Spectrum Code to Produce Fast and Thermal Multigroup Neutron Cross Sections." (To be Published.)

9.   R. G. Fairchild, J. Kalef-Ezra, S. Fiarman, I. Wielopolski, J. Hanz, S. Mussolino, and F. Wheeler, "Optimization on an Epithermal Neutron Beam for NCT at the Brookhaven Medical Research Reactor (BMRR)," Strahlenther. Onkol., 165(2/3):84 (1989).

10.   Y. Oka, I. Yanagisawa, and S. An, "A Design Study of the Neutron Irradiation Facility for Boron Neutron Capture Therapy," Nucl. Technol., 55(3):642 (1981).

11.   R. M. Brugger, T. J. Less, and G. G. Passmore, "An Intermediate-Energy Neutron Beam for NCT at MURR," in Proc. U.S. Dept. of Energy 1986 Workshop on Neutron Capture Therapy, R. G. Fairchild and V. P. Bond, eds., Brookhaven National Laboratory, BNL-51994, p. 83 (1987).

12.   W. A. Rhoades and R. L. Childs, "Updated Version of the DOT 4 One- and Two-Dimensional Neutron/Photon Transport Code," Oak Ridge National Laboratory, ORNL-5851 (1982).

13.   R. W. Roussin, "BUGLE-80: Coupled 47-Neutron, 20-Gamma-Ray, P$_3$, Cross-Section Library for LWR Shielding Calculations," Radiation Shielding Information Center, DLC-75 (1980).

14.   D. K. Parsons, F. J. Wheeler, B. I. Rushton, and D. W. Nigg, "Neutronics Design of the INEL Facility for Boron Neutron Capture Therapy Clinical Trials," in Proc. ANS 1988 Int. Reactor Physics Conf., Jackson Hole, WY, Vol. II, p. 2-443 (Sept. 1988).

15.   G. K. Becker, Y. D. Harker, R. A. Anderl, and F. J. Wheeler, "Neutron Spectrum Measurements in the Aluminum Oxide Filtered Beam Facility at the Brookhaven Medical Research Reactor." (These Proceedings.)

16.   M. A. Abdou, Y. Gohar, and R. Q. Wright, "MACK-IV: A New Version of MACK, A Program to Calculate Nuclear Response Functions from Data in ENDF/B Format," Argonne National Laboratory, ANL/FPP-75-5 (July 1978).

17.   F. J. Wheeler, "The Power Burst Reactor Facility as an Epithermal Neutron Source for Brain Cancer Therapy," in Proc. U.S. Dept. of Energy 1986 Workshop on Neutron Capture Therapy, R. G. Fairchild and V. P. Bond, eds., Brookhaven National Laboratory, BNL-51994, p. 92 (1987).

18.   P. D. Randolph, Personal Communication.

A CALCULATIONAL STUDY OF TANGENTIAL AND RADIAL BEAMS

IN HIFAR FOR NEUTRON CAPTURE THERAPY

B. V. Harrington

Australian Nuclear Science and Technology Organization
Lucas Heights, NSW, Australia

ABSTRACT

It is generally accepted that for biological purposes a tangential neutron beam is preferable to a radial beam because of its lower gamma and fast neutron contamination. Nevertheless radial broad spectrum epithermal neutron beams are currently being considered for boron neutron capture therapy of deep-seated tumours since they have the potential to deliver a more intense dose.

A calculational study of a conceptual tangential beam and a filtered radial beam in the DIDO type reactor HIFAR was undertaken. A two-dimensional transport code was used. The tangential beam was found to be superior in therapeutic gain at depth in tissue to an aluminium fluoride (AlF_3) filtered radial beam, while the dose rates of the beams were comparable.

INTRODUCTION

An epithermal neutron beam in the eV to low keV range has been shown by Zamenhof et al. (1975) and McGregor and Allen (1983) to have close to optimal characteristics for the boron neutron capture therapy (NCT) of deep-seated tumours. The intensity of monoenergetic neutron beams using a resonance window filter such as Sc (2 keV), falls some orders of magnitude short of that required for NCT. This is due to the narrow band of neutrons selected and some attenuation even of the neutrons in that band. To achieve a suitable reactor beam of sufficient intensity the focus of research over recent years has been on developing a broad spectrum epithermal neutron beam with minimal fast neutron and gamma contamination. Oka et al. (1981) showed that a combination of heavy water and aluminium could be used as a spectrum shifter to attenuate fast neutron flux relative to epithermal flux. Since then aluminium based spectrum shifters and filter assemblies have been studied by a number of groups and were reported in the Third International NCT Symposium Proceedings (Gabel 1989). Results, both calculated and measured have shown significant improvements in the neutron spectrum.

Calculations were undertaken to explore the feasibility of producing a suitable epithermal neutron beam for the treatment of deep-seated metastatic melanoma from ANSTO's 10-MW research reactor (Harrington 1987). Initially Al, Al/S, and AlF_3 spectrum shifters were considered with AlF_3 being chosen for further study. Fluorine was chosen because it has a low threshold (~100 keV) for the inelastic scattering reaction. To demonstrate the importance of a suitable method of assessment, two radial beams resulting from 12 cm and 39 cm thickness of AlF_3 spectrum shifter in the above study were

Neutron Beam Design, Development, and Performance for Neutron Capture Therapy
Edited by O. K. Harling *et al.*
Plenum Press, New York, 1990

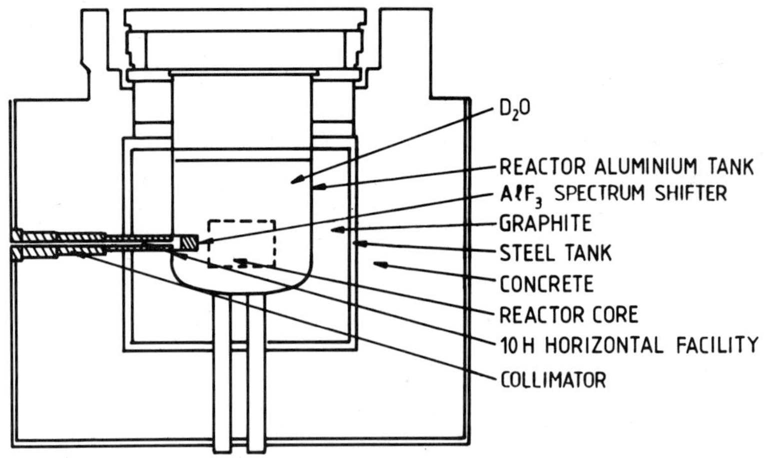

Fig. 1. Schematic Diagram of the Reactor HIFAR

compared. Initially the main consideration was the reduction of the proton recoil dose component to tissue. The results were promising. Subsequent more realistic assessments included the calculated photons, thermal, epithermal, and fast neutrons incident on a phantom. The conclusions changed, showing very little improvement in beam quality through the use of an AlF_3 spectrum shifter. Dose-depth calculations in a phantom including all components of dose were used to compare the 12 cm and 39 cm AlF_3 radial beams with an unfiltered conceptual tangential beam. These were considered to provide a suitable method of assessment.

REACTOR AND BEAM TUBE DESCRIPTION

HIFAR is a 10 MW, D_2O moderated and cooled reactor of the DIDO class (Figure 1). The core and heavy water are contained in an aluminium tank which is 2 m in diameter. The surrounding cylindrical graphite reflector is contained in a steel tank which is in turn surrounded by a concrete biological shield. The facility considered for NCT was the 25 cm diameter 10H horizontal facility which is set at core mid-plane and protrudes radially into the heavy water tank to within about 9 cm of the core. The concrete and steel collimator currently in 10H is 235 cm long and has a 5.08 x 5.08 cm square beam hole. In this study a cylindrical beam hole, 5.74 cm in diameter having an equivalent cross-sectional area, has been used. A higher beam intensity could be achieved if the beam hole were larger. The AlF_3 spectrum shifter was assumed to be positioned at the core end of 10H. A conceptual horizontal tangential beam, perpendicular to 10H, at the same distance from the core, and with an identical collimator, has also been considered.

NEUTRON AND GAMMA SPECTRA CALCULATIONS OF THE BEAM

Transport calculations were undertaken to determine the neutron and gamma spectra for varying lengths of AlF_3 spectrum shifter in the 10H horizontal facility. The ORNL two-dimensional discrete ordinates transport code DOT (Rhoades and Mynatt 1973) and a 200 group cross-section library derived from ENDF/B-IV (Garber 1975) data were used. Because of the fine structure in the aluminium cross-section data in the resonance region and the changing spectrum along the 10H horizontal facility, the calculations were done in 80 energy groups (68 neutron and 12 photon groups). A cylindrical HIFAR model was used with the axis of 10H as the axis of the cylinder (Figure 2).

The size of the 80-group transport calculation, including the reactor core, beam tube and graphite region, led to its treatment as the following three overlapping calculations:

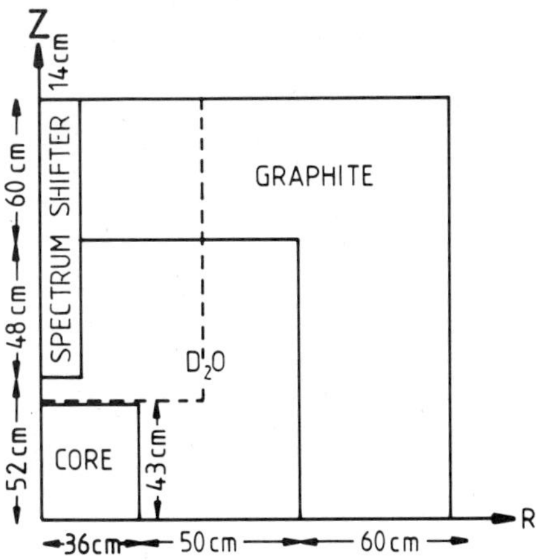

Fig. 2.   Cylindrical HIFAR Model Used for Beam Calculations.

(a)   A two-dimensional diffusion calculation of the HIFAR model (Figure 2), normalized to 10 MW thermal using the AUS (Robinson 1987) code POW (Pollard 1974), was used to generate a neutron source representation of the core. This was used as input in the transport calculation (b).

(b)   A two-dimensional source calculation, using the transport code DOT, was used to generate angular boundary fluxes for two interior boundaries (dotted lines in Figure 2). A coarse spatial mesh structure, third order Legendre polynomial P_3 scattering treatment and S_{16} angular quadrature were used. The ENDF/B cross-section data for ^{235}U include only the prompt γ-rays released in fission and not the equilibrium fission product gammas which account for approximately 50 per cent of core gammas in HIFAR. In this calculation, equilibrium fission product gammas were included as part of the input source, using the fission gamma yield data provided for the PWR shielding benchmark (Hehn 1983).

(c)   The angular boundary fluxes from (b) were used as DOT input in a two-dimensional, P_3, S_{16}, fine mesh source calculation of the area enclosed by the dotted lines of Figure 2. The calculations assumed that the AlF_3 filter material in the 10H facility extended to the edge of the graphite. The angular fluxes were read off on the axis of 10H for the required thickness of AlF_3 spectrum shifter. To approximate as closely as possible the radial and tangential beams, angular fluxes at 4° and 81° to the radial direction were used. A line-of-sight attenuation factor of $r^2/(4z^2)$, where r and z are the collimator radius and length, was used to determine the neutron and gamma current at the beam port.

PHANTOM DOSE-DEPTH CALCULATIONS

Dose as a function of depth in a phantom was used in comparative assessments of the calculated beams. Neutron and gamma fluxes resulting from beams incident on the tissue equivalent phantom head model of Zamenhof (1975) were calculated. Since this is a study of the NCT of metastatic melanoma, Rossi tissue composition (Table 1) was considered to be appropriate. A thermal neutron shield consisting of 0.05 cm ^{6}LiF and a 5 cm Pb gamma shield were also included as part of the phantom calculation (Figure 3). The ORNL one-dimensional transport code ANISN (Engle 1967) was used and a 30 cm deep tissue equivalent rectangular slab with transverse dimensions of 13.9 cm assumed. The ^{10}B was not specifically included in the model. This arbitrary and simple 1-D phantom head model, although useful for comparative beam evaluations is obviously not intended to

Table 1. Comparison of Radial and Tangential Beams.

Phantom Specification			
Shape	Rectangular slab		
Dimensions	30 cm thick, with transverse dimensions of 13.9 cm		
Composition	Weight fractions; 0.7139 O, 0.1489 C, 0.1 H, 0.0347 N, 0.001 Cl, 0.0015 Na		
Density	1.0 g/cm^3		
Hydrogen Density	5.98 E+22 atoms/cm^3		

Beam Configuration	5.74 cm diameter collimator, 0.05 cm ^{6}LiF and 5 cm Pb filters		
Description	Radial	Radial	Tangential
Spectrum Shifter	12 cm AlF$_3$	39 cm AlF$_3$	–

	Radial	Radial	Tangential
Advantage Depth (Max/Min)			
No RBE (cm)	3.6/3.3	3.3/2.9	5.0/4.5
With RBE (cm)	4.7/4.2	4.5/3.8	5.9/5.2
Advantage Dose Rate			
No RBE (cGy/min)	11	2.2	10
With RBE (cGy/min)	14	2.4	13
Advantage Ratio			
No RBE	1.6	1.6	2.2
With RBE	2.0	2.1	3.1
% Low LET Dose	42	57	24
% High LET Dose	22	7	22
% ^{10}B Dose (30 μg/g)	36	36	54

produce absolute dose values. Calculated neutron and photon fluxes were converted to absorbed doses using AUS (Robinson 1987) kerma factors which were derived from ENDFB/IV data. The RBE values of 1.6, 1.0, and 2.3 for neutron, photon, and ^{10}B components of dose, respectively, were used to obtain the RBE doses. Boron-10 loadings of 30 μg/g in tumour and 3 μg/g in tissue were assumed. The following components of dose were calculated:

a) neutron dose   H(n,n)H, i.e. proton recoil for high energy neutrons, ^{14}N(n,p)^{14}C for neutrons of low energy, other minor reactions,

b) gamma dose   induced gammas in tissue from ^{1}H(n,γ)^{2}H, reactor core gammas, induced gammas in structural materials, and

c) boron dose ^{10}B(n,α)^{7}Li.

For an epithermal neutron beam in the eV to low keV range, the minor neutron reactions can contribute up to 3% of the neutron dose near the surface of the phantom. However, because the beams in this study had significant fast neutron contamination, the minor reactions accounted for 5%-7.5% of the total neutron dose to tissue. Elastic

Fig. 3.   1-D Phantom Model with ^{6}LiF and Pb Shields.

scattering in oxygen and carbon accounted for up to 5% and 1.7% of the total neutron dose respectively. The balance was due mainly to other high energy reactions in oxygen such as the $^{16}O(n,\alpha)^{13}C$ reaction.

The dose-depth values were used to derive therapeutic gains, advantage depths, advantage dose rates, and advantage ratios. Therapeutic gain is defined as the ratio of the tumour RBE dose to the maximum tissue RBE dose. The advantage depth is defined as the depth in the phantom where the total therapeutic dose rate equals the maximum background dose rate. The advantage dose rate is defined as the total therapeutic dose rate at the advantage depth. The advantage ratio is defined as the quotient of the integral of the total therapeutic dose divided by the integral of the total background dose, assuming zero ^{10}B in normal tissue, taken from 0 cm to the advantage depth.

ASSESSMENT OF SPECTRUM SHIFTERS

Initial comparisons of AlF_3 spectrum shifters of lengths 12 cm and 39 cm were made primarily on the basis of reduced proton recoil dose considerations. These include:

a)   neutron spectrum considerations such as fast to epithermal flux ratios,

b)   mean (water kerma weighted) neutron spectrum energies, and

c)   therapeutic gain versus depth curves in a phantom for incident beams consisting only of the epithermal and fast neutron components of the beam.

Calculated neutron spectrum results showed a significant reduction in the relative fast to epithermal neutron flux as the length of AlF_3 in 10H increased. This is illustrated by the comparison of spectra in Figure 4 for 0 cm and 46.5 cm AlF_3. For 12 cm and 39 cm of AlF_3 the calculated fast to epithermal flux ratios were 0.13 and 0.06 respectively, where the epithermal flux was assumed to be between 0.4 eV and 30 keV. The mean (water kerma weighted) neutron spectrum energy was taken to be the energy corresponding to the mean water kerma value for that spectrum. For 12 cm and 39 cm of AlF_3 the values for the mean energy were calculated to be 24 keV and 12 keV respectively, implying a substantial reduction in proton recoil dose to tissue.

The initial therapeutic gain versus depth curves were calculated assuming that only fast and epithermal neutrons needed to be considered. Implicit in this approach is an ideal thermal neutron filter and an ideal gamma shield. The therapeutic gain was found to increase with length of AlF_3, for all depths in tissue (Figure 5), while the dose intensity decreased (Figure 6). These results were very promising, suggesting a trade-off between therapeutic gain and beam intensity. On the basis of all the above beam quality considerations a 39 cm AlF_3 spectrum shifter would be rated as superior to the 12 cm AlF_3 spectrum shifter.

In subsequent more realistic phantom calculations the incident thermal neutrons and gammas as well as a ^{6}LiF thermal neutron filter and a Pb gamma shield were included. The therapeutic gain curves in Figure 6 bear no resemblance to the earlier calculations. For a 12 cm AlF_3 spectrum shifter, a small increase in therapeutic gain is evident for depths in the phantom greater than about 2 cm. For longer lengths, such as 39 cm of AlF_3 the

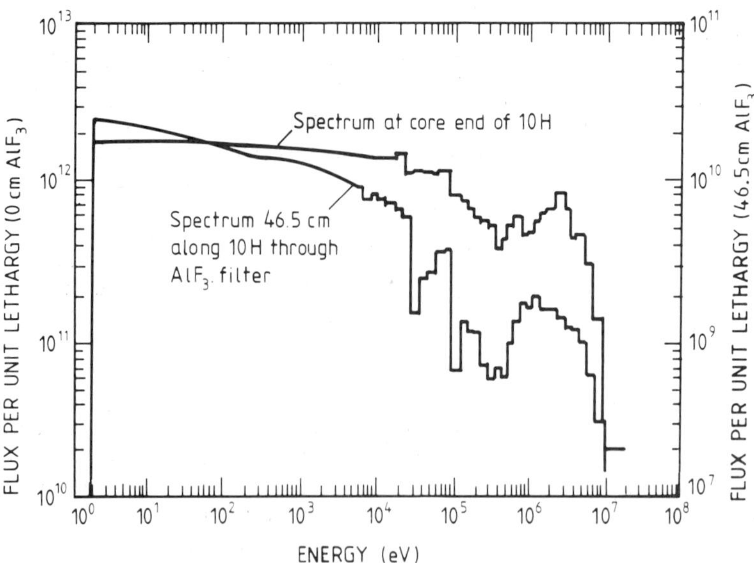

Fig. 4. Shift in Neutron Flux Spectrum through AlF_3.

therapeutic gain at depth in a phantom actually decreases. Hence using this more realistic model the conclusions were reversed and a 12 cm AlF_3 spectrum shifter was shown to be slightly superior to 39 cm AlF_3 in therapeutic gain (Figure 7) and significantly superior in intensity (Figure 8). Assessments which do not take into account the variation of all components of dose with type and geometry of filter materials proved to be inadequate in determining the optimum thickness of AlF_3 spectrum shifter.

The more realistic phantom calculations discussed above were used to determine dose as a function of depth with and without RBEs for 12 cm AlF_3, 39 cm AlF_3, and an unfiltered conceptual tangential beam. The corresponding advantage depths, advantage dose rates, and advantage ratios are given in Table 1. These values support the conclusions that the main advantage of the 12 cm AlF_3 spectrum shifter over 39 cm AlF_3 is in beam intensity. Therapeutic gain curves (Figure 7) show the 12 cm AlF_3 beam to be superior to the 39 cm AlF_3 beam for depths in tissue greater than about 2 cm. Advantage ratios, however, give no indication of the superiority at depth in tissue of the 12 cm AlF_3 beam. This is because the advantage ratio includes the tumour dose integrated from the surface to advantage depth. Hence the advantage ratio may not be a good measure of the superiority of one beam over another for a tumour at a given depth.

The dose components show a very significant reduction in the neutron (i.e. high LET) dose for 39 cm AlF_3 compared to 12 cm. There is however no improvement in the relative [10]B dose, because of the increase in the relative gamma dose. Dose-depth components (Figures 9 and 10), show a very dramatic increase in the structural gamma dose for 39 cm AlF_3 compared to 12 cm AlF_3. The beam quality improvement due to attenuation of fast neutrons through AlF_3 was counteracted by a relative increase in gamma dose because of gammas arising from neutron captures in the structural materials, particularly the aluminium. It was found that an increased thickness of Pb or Bi gamma shield was ineffective at improving the beam quality (Harrington 1987). The aluminium gammas are hard and cannot be attenuated by lead or bismuth without a detrimental effect on the neutron spectrum. In this study the AlF_3 was positioned within 9 cm of the reactor core, in a region of relatively high thermal neutron flux. Thermal neutron capture in aluminium would have accounted for most of the induced structural gamma dose. If the AlF_3 were placed further from the reactor core and if the AlF_3 were shielded from thermal neutrons, a different optimized design might result.

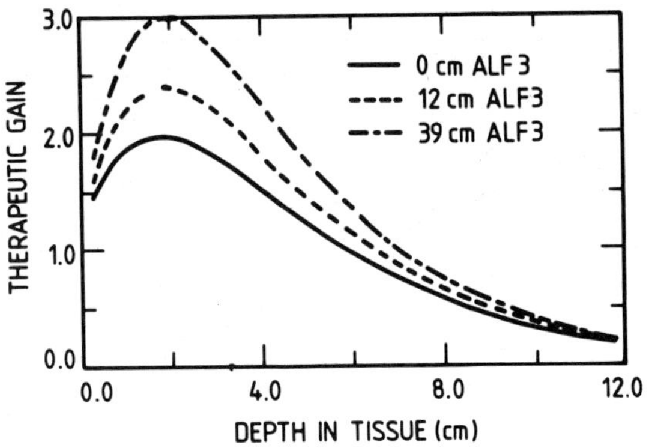

Fig. 5.   Therapeutic Gain for Beams with Fast
and Epithermal Neutron Components Only.

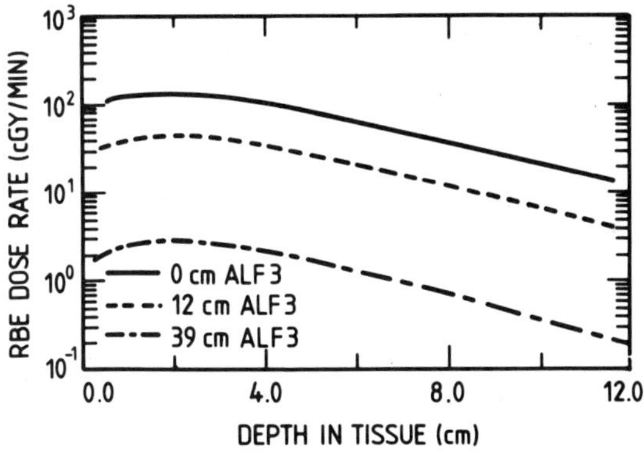

Fig. 6.   Dose for Beams with Fast and Epithermal
Neutron Components Only.

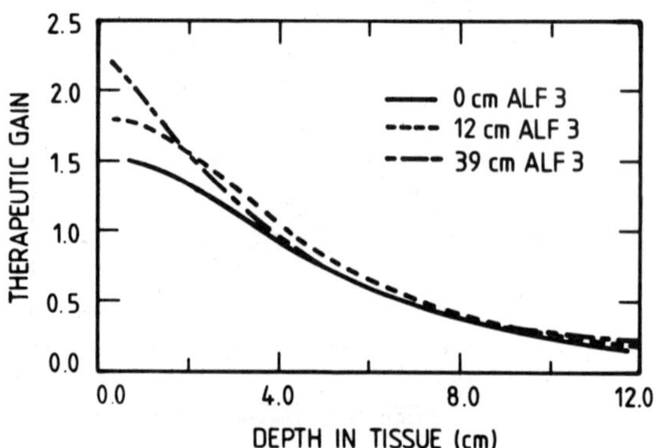

Fig. 7. Therapeutic Gain for Realistic Beams.

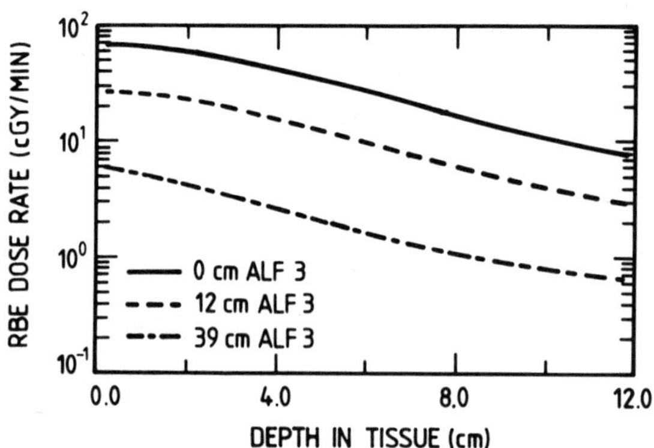

Fig. 8. Dose for Realistic Beams.

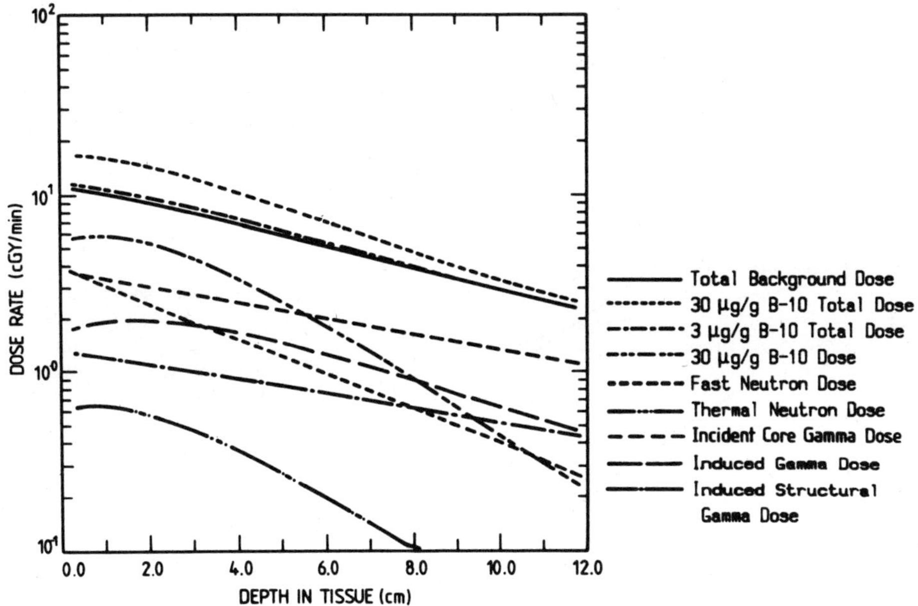

Fig. 9. Dose Components for 12 cm AlF$_3$ Radial Beam.

A conceptual tangential beam is compared to the 12 cm and 39 cm AlF$_3$ radial beams (Table 1). The dose components for the tangential beam (Figure 11) show that up to a depth of about 5 cm in the phantom the 30 µg/g ^{10}B component of dose is higher than the background dose. This is due to the relatively low incident gamma dose. The advantage depth and advantage ratio (Table 1) of the tangential beam were found to be considerably higher than that for the radial beam while the dose intensity was comparable.

CONCLUSIONS

An AlF$_3$ spectrum shifter was considered for the production of a broad spectrum epithermal neutron beam from the reactor HIFAR for the boron neutron capture therapy of metastatic melanoma. For reliable beam comparisons it was found that the variation of all the components of dose in tissue with type and geometry of filter materials needed to be included. Comparisons of an AlF$_3$ spectrum shifter in the 10H radial facility in HIFAR based on therapeutic gain at depth in a phantom showed that there is an optimum thickness of AlF$_3$, with 12 cm of AlF$_3$ being superior to 39 cm AlF$_3$. A radial beam with 12 cm AlF$_3$ spectrum shifter was compared to a conceptual tangential beam with no spectrum shifter. The dose components in a phantom for the tangential beam were lower in photon dose and comparable in fast neutron dose to a 12 cm AlF$_3$ filtered radial beam. The calculated advantage depths, advantage ratios, and dose intensities demonstrated the superiority of the tangential beam over the AlF$_3$ filtered radial beams, for the filter and shielding geometries studied here. Different placements of the filter, thermal neutron, and gamma ray shielding could significantly alter the results and should be investigated.

ACKNOWLEDGMENTS

I would like to express my appreciation of the support received from my colleagues Graham Robinson, Brian McGregor, John Connolly, and Barry Allen.

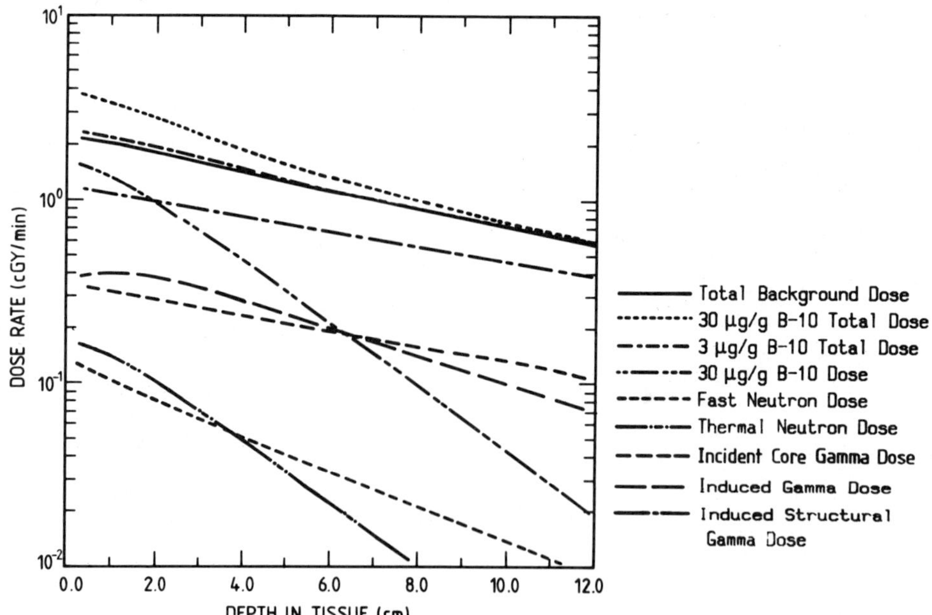

Fig. 10.   Dose Components for 39 cm AlF$_3$ Radial Beam.

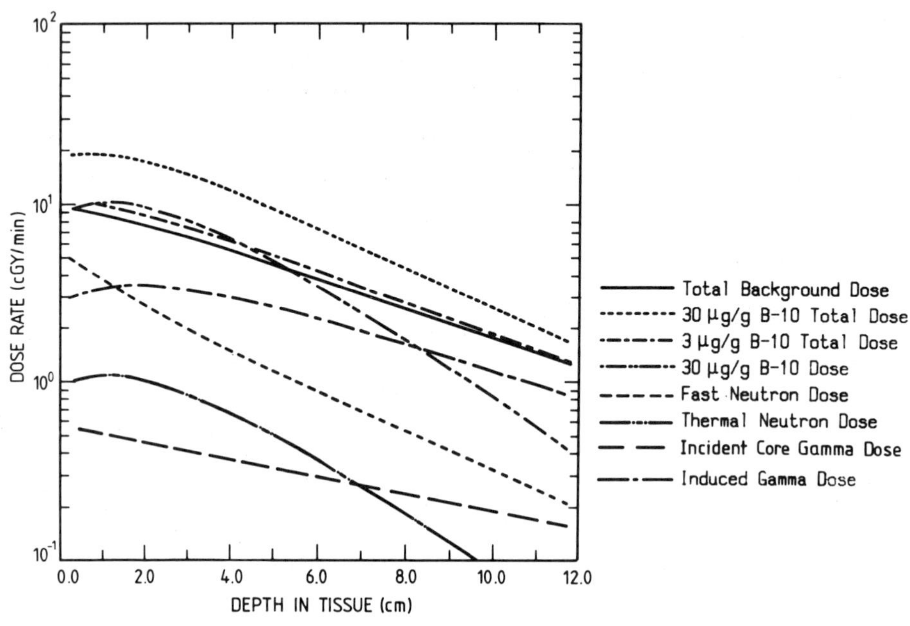

Fig. 11.   Dose Components for Tangential Beam.

REFERENCES

Engle, W. W., 1967, A User's Manual for ANISN; A One Dimensional Discrete Ordinates Transport Code with Anisotropic Scattering. K-1693.

Gabel, D., 1989, <u>Proc. Third Int. Symp. on Neutron Capture Therapy</u>, Bremen, FRG, 31 May - 3 June 1988, <u>Strahlenther. Onkol.</u>, 165(2/3):5-257.

Garber, D., 1975, ENDF/B Summary Documentation. <u>BNL-17541</u>.

Harrington, B. V., 1987, Optimization of an Epithermal Beam in HIFAR for Boron Neutron Capture Therapy. <u>ANSTO/E662</u>.

Hehn, G., 1983, PWR shielding benchmark. <u>NEACRP-L-264</u>.

McGregor, B. J. and Allen, B. J., 1983, Filtered Beam Dose Distributions for Boron Neutron Capture Therapy of Brain Tumours, in <u>Proc. First Int. Symp. Neutron Capture Therapy</u>, R. G. Fairchild and G. L. Brownell, eds., Brookhaven National Laboratory, <u>BNL-51730</u>, p.14.

Oka, Y., Nanagisawa, I., and An, S., 1981, A Design Study of the Neutron Irradiation Facility for Boron Neutron Capture Therapy, <u>Nucl. Technol.</u>, 55:642.

Pollard, J. P., 1974, AUS Module POW – A General Purpose 0, 1, and 2D Multigroup Neutron Diffusion Code Including Feedback-Free Kinetics. <u>AAEC/E269</u>.

Rhoades, W. A. and Mynatt, F. R., 1973, The DOT 3.5 Two-Dimensional Discrete Ordinates Transport Code. <u>ORNL-TM-4280</u>.

Robinson, G. S., 1987, A Guide to the AUS Modular Neutronics Code System. <u>AAEC/E645</u>.

Zamenhof, R. G., Murray, B. W., Brownell, G. L., Wellum, G. R., and Tolpin, E. I., 1975, Boron Neutron Capture Therapy for the Treatment of Cerebral Gliomas, <u>Med. Phys.</u>, 2(2):47.

RESEARCH ON NEUTRON BEAM DESIGN FOR BNCT AT THE MUSASHI REACTOR

O. Aizawa

Atomic Energy Research Laboratory
Musashi Institute of Technology
Kawasaki, Japan

INTRODUCTION

The neutron irradiation facility at the Musashi Reactor (TRIGA-Mark II, 100 kW) was modified for Boron Neutron Capture Therapy (BNCT) in 1975 [1]. A cross-sectional view of the reactor is given in Figure 1. From 1976 through the end of March 1989, ninety-four patients with brain tumors and four patients with melanoma have been treated. Although treatments have been successfully performed since 1976, it is desirable to consider improvements in the depth-dose distribution and a reduction of irradiation time.

It was recommended at the Workshop on Neutron Capture Therapy held at the Medical Department of the Brookhaven National Laboratory (BNL) in 1986 that epithermal-neutron beams be used in order to improve the depth-dose distribution. However, there remains an important role for thermal beams, especially in the treatment of melanoma and brain tumors accessible via an open craniotomy.

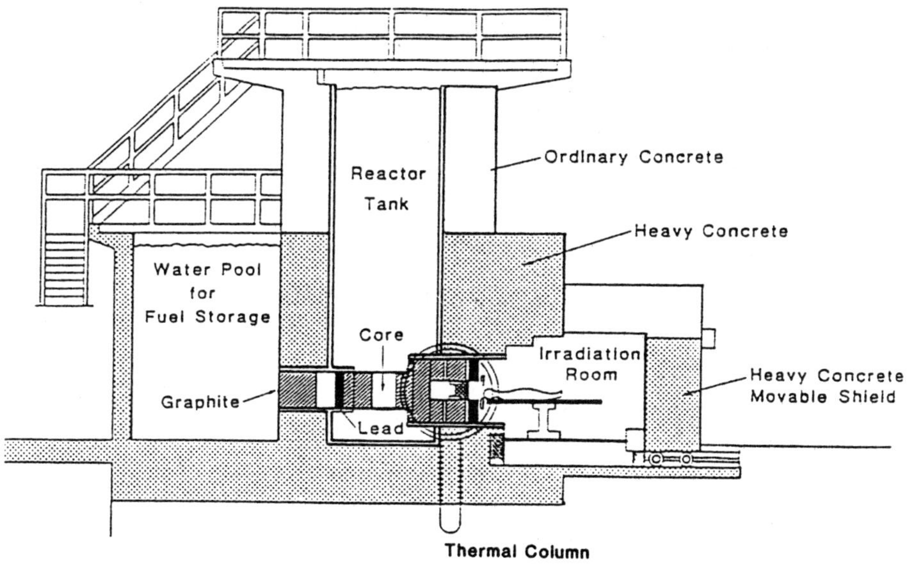

Figure 1. Cross-Sectional View of the Musashi Reactor.

Neutron Beam Design, Development, and Performance for Neutron Capture Therapy
Edited by O. K. Harling *et al.*
Plenum Press, New York, 1990

BASIC EXPERIMENTS

Enhancement of Thermal Neutrons by Using a Single-Crystal Silicon

The total cross section of single-crystal silicon was measured using a time-of-flight facility at the Musashi Reactor [2]. Given that the total cross section of this material is sufficiently small in the thermal energy region, it was expected that a thermal-neutron flux could be readily transmitted through it. In order to verify values in the cross-section library for single-crystal silicon, some benchmark experiments were performed in an irradiation port. The arrangement used is shown in Figure 2. One of the results is shown in Figure 3. It is evident that the experimentally-determined thermal-neutron flux distribution is in good agreement with the calculated one and that single-crystal silicon is very useful for enhancing thermal neutrons.

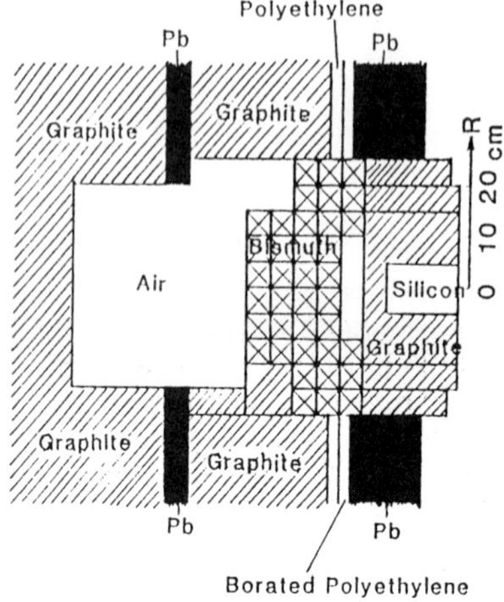

Figure 2. Arrangement of Bench-Mark Experiments at the Irradiation Port.

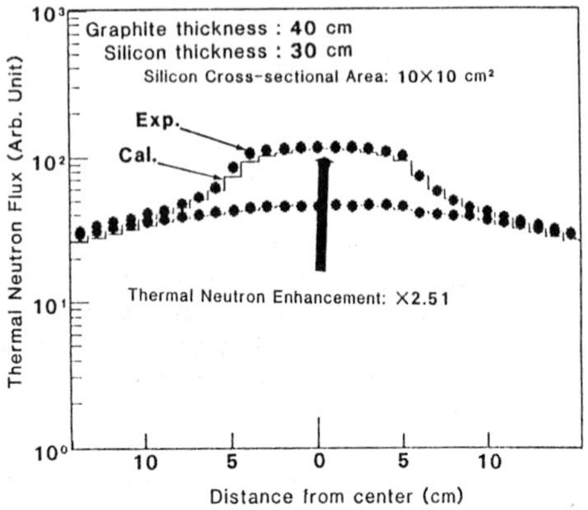

Figure 3. Thermal Neutron Enhancement at the Surface of Single-Crystal Silicon.

Measurement of Epithermal Total Neutron Flux Using Gold Foils

In this paper, we define the epithermal total neutron flux as:

$$\Phi_{epi}^{total} = \int_{0.414}^{111,000} \Phi(E)\, dE \cdot \tag{1}$$

If we assume a 1/E spectrum in this energy range, equation (1) becomes:

$$\Phi_{epi}^{total} = \int_{0.414}^{111,000} \frac{\Phi_{epi}}{E}\, dE = 12.5\, \Phi_{epi} \tag{2}$$

where Φ_{epi} is the epithermal-neutron flux at 1 eV.

The reaction rate in a cadmium-covered gold foil is:

$$R = \int_{E_{Cd}}^{\infty} f\, N_o (1-e^{-\lambda t_i})\, \sigma_{Au}(E)\, \Phi(E)\, dE$$

$$= f\, N_o (1-e^{-\lambda t_i}) \int_{E_{Cd}}^{\infty} \frac{\Phi_{epi}}{E}\, \sigma_{Au}(E)\, dE$$

$$= f\, N_o (1-e^{-\lambda t_i})\, \Phi_{epi}\, (RI)_{Au}\,. \tag{3}$$

where f is the self-shielding factor of a gold foil for epithermal neutrons, N_o is the number of atoms of gold, λ is the decay constant of gold, t_i is the irradiation time, $(RI)_{Au}$ is the resonance integral of gold, and E_{Cd} is the cadmium cut-off energy.

The self-shielding factor for a gold foil for an epithermal flux was determined from both experiments and theoretical calculations to be 0.268 for a 100-μm Au-foil. The reaction rate of a gold foil was measured to be 104 dps/mg following an irradiation of 300 minutes in the irradiation field of the Musashi Reactor. The epithermal-neutron flux at 1 eV is therefore:

$$\Phi_{epi} = \frac{104}{(0.268)(3.06 \times 10^{18})(0.0521)(1566 \times 10^{-24})}$$

$$= 1.55 \times 10^6 \ \text{n/cm}^2\text{-s at 1 eV.} \tag{4}$$

From equation (2), the epithermal total neutron flux was determined to be:

$$\Phi_{epi}^{total} = 12.5\, \Phi_{epi} = 1.94 \times 10^7 \ \text{n/cm}^2\text{-s.} \tag{5}$$

DESIGN CALCULATION

Principle and Procedure of Design Calculation

The peak flux of thermal and epithermal neutrons is usually located at the center of the core. The same is true of fast neutrons and gamma rays. The design objective is to extract a sufficiently intense beam of thermal and epithermal neutrons while suppressing the fast-neutron and gamma-ray component of the flux. It is also important to be able to calculate the neutron energy spectrum of the irradiation field because the KERMA-dose for neutrons depends on their energy. We must also consider capture gamma rays from structural materials.

A two-dimensional discrete ordinates transport code DOT 3.5 [3] was employed for the design calculations by adopting the S_{12} and P_3 approximations. The group constants used are the neutron and gamma coupled cross sections based on the DCL-23/CASK cross-section library [4]. The energy groups and the KERMA-dose factors used are listed in Table One.

The calculations were performed in three distinct steps:

1) Source Calculation (Step-1),
2) Structure-Optimization Calculation (Step-2),
3) Phantom Calculation (Step-3).

Source Calculation (Step-1)

A cross-sectional view of the Musashi Reactor was shown in Figure 1. The core

Table One

DCL-23/CASK Energy Group Structure and KERMA-Dose Factors

\multicolumn{2}{Neutron Groups}		Neutron KERMA-Dose (K-D) Factors	\multicolumn{2}{Photon Groups}		Photon KERMA-Dose (K-D) Factors

No.	Upper Group Energy	Neutron KERMA-Dose (K-D) Factors	No.	Upper Group Energy	Photon KERMA-Dose (K-D) Factors
1	15.0 MeV	6.33 E-9 rad-cm^2	23	10.0 MeV	2.33 E-9 rad-cm^2
2	12.2	5.69 E-9	24	8.0	1.96 E-9
3	10.0	5.22 E-9	25	6.5	1.76 E-9
4	8.18	4.86 E-9	26	6.0	1.53 E-9
5	6.36	4.53 E-9	27	4.0	1.20 E-9
6	4.96	4.19 E-9	28	3.0	1.06 E-9
7	4.06	3.86 E-9	29	2.6	9.17 E-10
8	3.01	3.39 E-9	30	2.0	7.89 E-10
9	2.46	3.14 E-9	31	1.66	6.81 E-10
10	2.35	3.00 E-9	32	1.32	5.60 E-10
11	1.83	2.64 E-9	33	1.00	4.51 E-10
12	1.10	2.11 E-9	34	0.80	3.63 E-10
13	650.0 keV	1.22 E-9	35	0.60	2.65 E-10
14	110.0	3.33 E-10	36	0.40	1.84 E-10
15	3.35	1.78 E-11	37	0.30	1.25 E-10
16	638.0 eV	4.17 E-12	38	0.20	6.83 E-11
17	101.0	1.25 E-12	39	0.10	3.71 E-11
18	29.0	1.25 E-12	40	0.05	4.03 E-10
19	10.7	1.94 E-12		0.01	
20	3.06	3.33 E-12			
21	1.12	5.55 E-12			
22	0.414	2.72 E-11			
	0.01				

is cylindrical in shape with both an effective radius and a height of 35.6 cm. The core is located in the lower part of the reactor tank which is 2 m in diameter and about 6.5 m in depth. Neutrons generated in the core are moderated in a graphite block and extracted at the irradiation port. The structure-optimization calculation was divided into two parts, step-1 and step-2, because of limits on the memory capacity of the available computer. The step-1 calculation was performed for the reactor core. The geometrical model is shown in Figure 4.

The source neutron spectrum was calculated from the equation:

$$N(E) = 0.453 \exp(-E/0.965) \sinh \sqrt{2.29E}. \tag{6}$$

The gamma-ray source spectrum was calculated for prompt gamma rays per fission from the equations:

$$\Gamma_p(E) = 8.0 \exp(-1.10E) \qquad (1.0 < E \leq 7.0 \text{ MeV}) \tag{7}$$

$$= 26.8 \exp(-2.30E) \qquad (E \leq 1.0 \text{ MeV}). \tag{8}$$

For delayed gamma rays per fission, it was calculated from:

$$\Gamma_d(E) = 14.0 \exp(-1.10E). \tag{9}$$

Delayed gamma rays are produced from short-lived fission products. The calculated spectra are shown in Figure 5 as a function of the mesh. The spectrum at mesh point 15 was used as the source spectrum for the step-2 calculation.

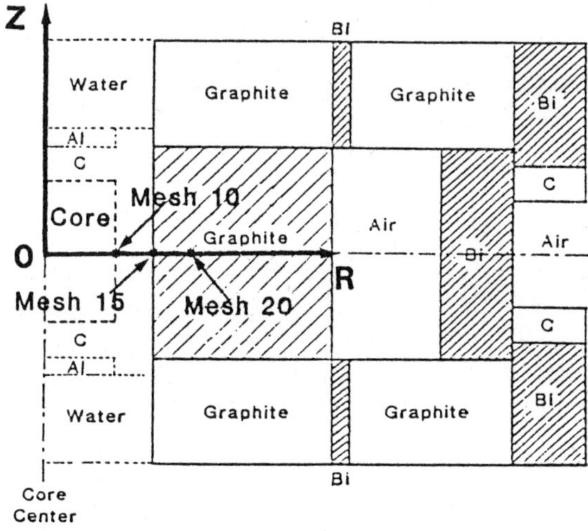

Figure 4. Geometrical Model of Source Calculation.

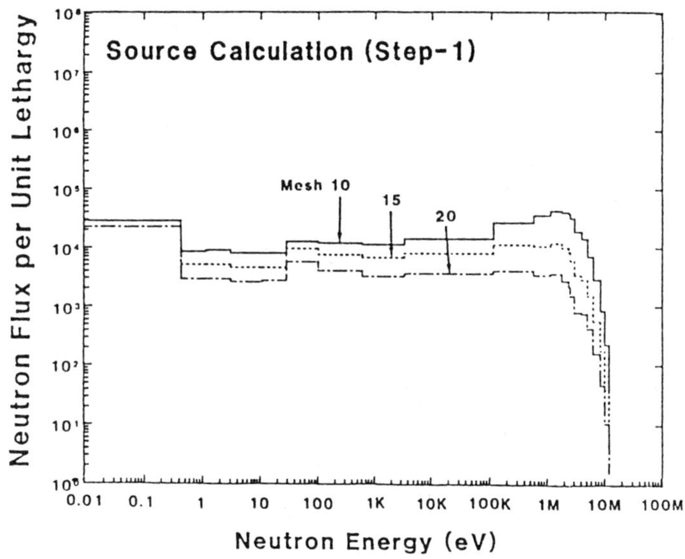

Figure 5. Calculated Neutron Energy Spectra Near the Core.

Structure Optimization Calculation (Step-2)

Source Intensity Normalization with Experiments: The structure optimization calculation was initiated for the present configuration by using a shell source at the core side as shown in Figure 6. The mesh interval was 2.5 cm and the size of the mesh was 24 x 56 in R-Z geometry. Given the availability of experimental results for the thermal, epithermal, and fast-neutron flux as well as the gamma-ray dose at the irradiation port of the reactor, the calculated results could be normalized with this experimental data. The calculated distributions are shown in Figures 7(a), (b), (c), and (d) for the thermal, epithermal, and fast-neutron flux and the gamma-ray dose rate, respectively. The values at the irradiation port are tabulated in Table Two. The source normalization factor was determined to be 1.2×10^{15} in this calculation. Although this factor is derived from the normalization of the epithermal flux, it is noted that the thermal flux, fast flux, and gamma-ray dose rate are in fairly good agreement with experiment.

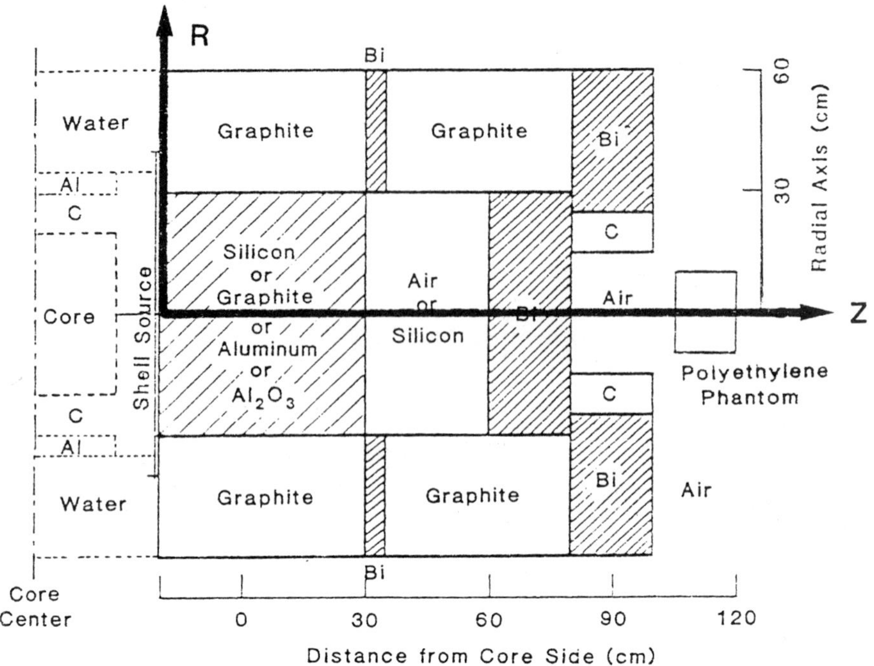

Figure 6. Geometrical Model of Structure
Optimization Calculations.

Table Two

Neutron Flux and Gamma-Ray Dose of the Present Beam

	Neutron Flux (n/cm^2-s)			Gamma-Ray Dose Rate (cGy/hr)
	Thermal	Epithermal*	Fast	
Experiments	1.3×10^9	1.9×10^7	1.0×10^6	25–30
Calculations	1.0×10^9	1.9×10^7	1.3×10^6	30.3

* Epithermal Total Flux (0.414–111,000 eV): Normalized with this Flux.

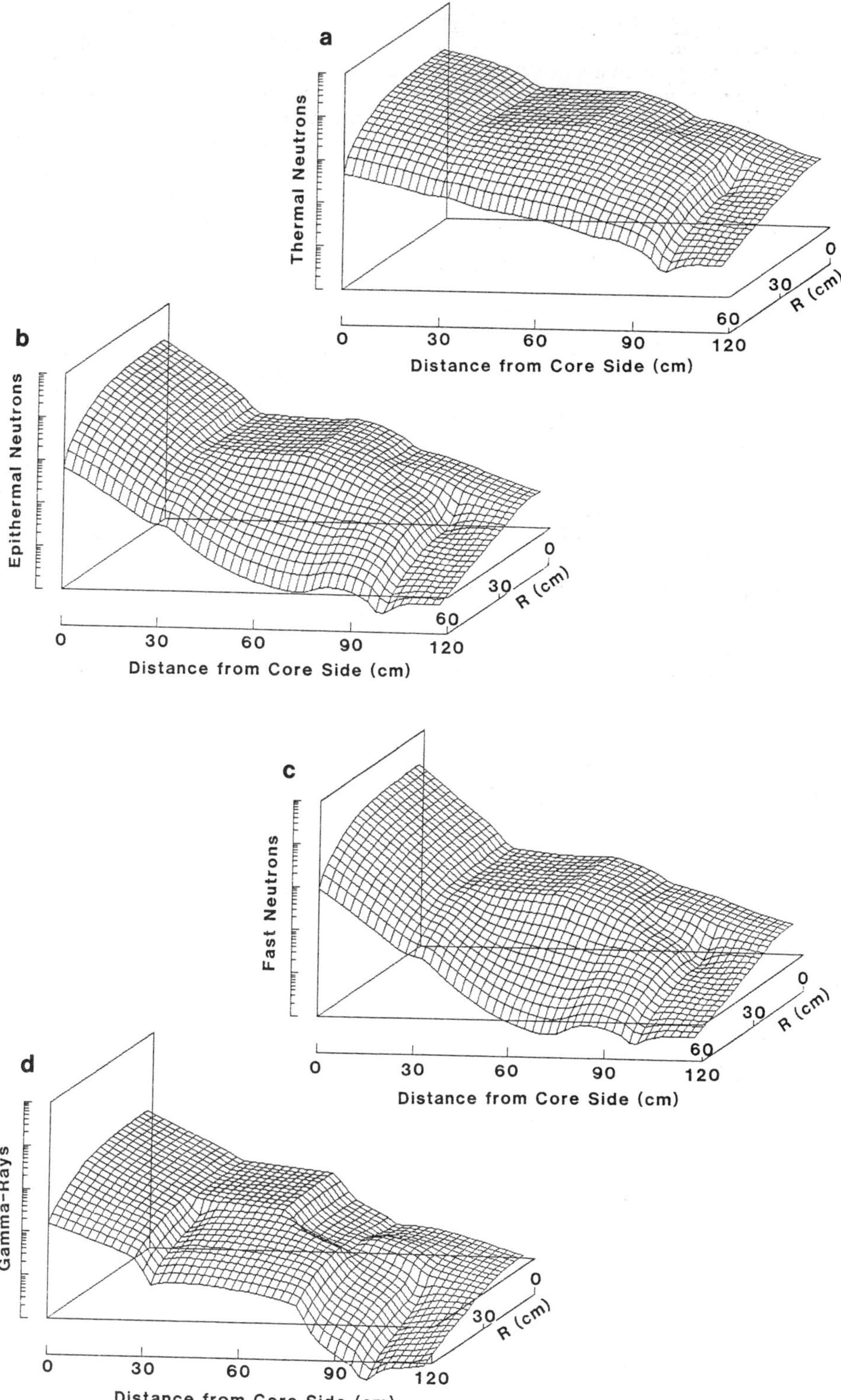

Figure 7.   Neutron Flux and Gamma-Ray Dose Distributions Calculated in the Present Configuration of the Musashi Reactor.

Trial for Enhancement of Thermal and Epithermal-Neutron Flux: In order to extract the thermal and epithermal neutrons from the core side, the graphite block (50 cm in thickness and 60 cm in diameter) shown in Figure 6 was replaced with either single-crystal silicon or aluminum. The calculated neutron spectra at the irradiation port are shown in Figure 8 and the result of the enhancement is summarized in Table Three. It is evident from this table that the epithermal flux is extremely enhanced by using either silicon or aluminum instead of graphite. However, the fast-neutron flux is also transmitted through both silicon and aluminum without scattering. It can be seen that aluminum is much better than silicon in these cases because of the difference in fast-neutron suppression.

Table Three

Enchancement of Neutron Flux for Si-50 cm and Al-50 cm Beams

	Neutron Flux $(n/cm^2\text{-}s)$			Gamma-Ray Dose Rate (cGy/hr)
	Thermal	Epithermal	Fast	
Present Configuration	1.02×10^9	1.92×10^7	1.33×10^6	30.3
Si-50 cm Beam (Enhancement Factor)	2.50×10^9 (x2.5)	6.87×10^8 (x36)	5.75×10^7 (x43)	76.9 (x2.5)
Al-50 cm Beam (Enhancement Factor)	1.14×10^9 (x1.1)	5.96×10^8 (x31)	2.24×10^7 (x17)	30.3 (x1.2)

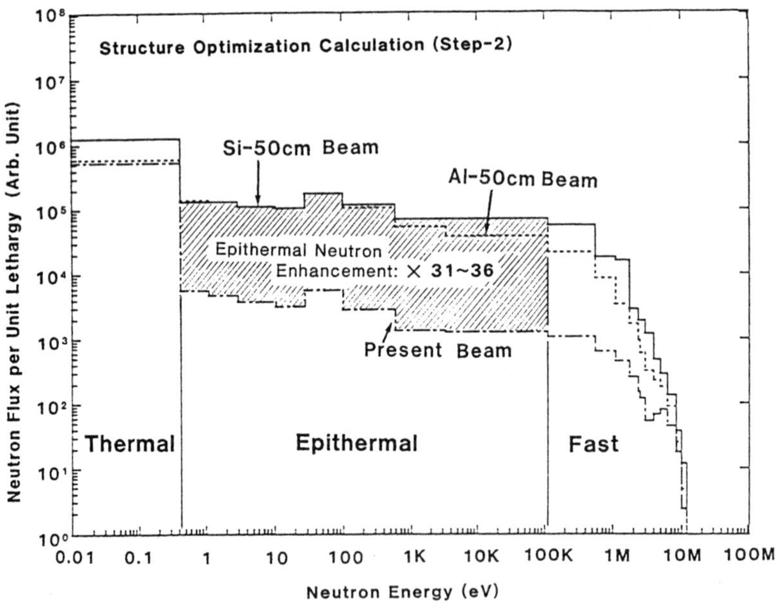

Figure 8.   Enhancement of Epithermal-Neutron Flux
for Si-50 cm and Al-50 cm Beams.

<u>Trial for Suppression of Fast Neutrons</u>: Although it is evident that silicon is a good material for enhancing the epithermal-neutron flux, the suppression of fast neutrons is not sufficient. More than 50 cm of silicon would be needed. The central region labeled air in Figure 6 is 30 cm in thickness. It was replaced with silicon. Thus, the total silicon thickness became 80 cm. The energy spectrum is shown in Figure 9 and the result of enhancement is summarized in Table Four. In this case, the epithermal enhancement factor is about 20 with a fast-neutron buildup factor of about 15. It is important to examine Figure 9 carefully because the fast-neutron component above 4 MeV is now lower than the flux in the present graphite configuration. Thus, while the fast-neutron flux above 111,000 eV has been built up by about a factor of 15, it is expected from the KERMA-dose factors that the dose rate of the fast-neutron flux will not be so large.

Table Four

Enhancement of Neutron Flux for Si-80 cm Beam

	Neutron Flux $(n/cm^2\text{-}s)$			Gamma-Ray Dose Rate (cGy/hr)
	Thermal	Epithermal	Fast	
Present Configuration	1.02×10^9	1.92×10^7	1.33×10^6	30.3
Si-80 cm Beam (Enhancement Factor)	1.35×10^9 (x1.3)	3.80×10^8 (x20)	1.94×10^7 (x15)	41.8 (x1.4)

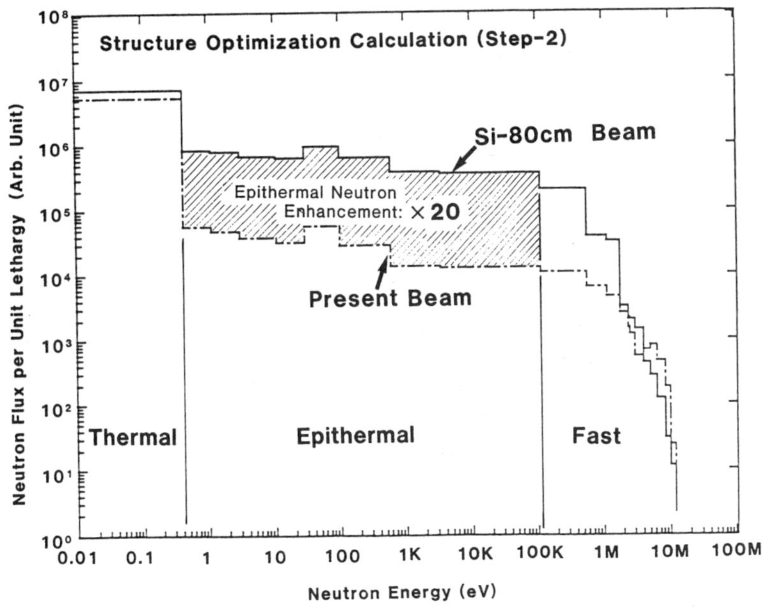

Figure 9. Enhancement of Epithermal Neutrons and Suppression of Fast Neutrons for Si-80 cm Beam.

<u>Another Trial for Suppression of Fast Neutrons</u>: Although silicon of 80 cm in thickness can produce many epithermal neutrons, the fast-neutron flux also remains large. Al_2O_3 of 50 cm in thickness was examined in order to further reduce the fast neutrons. The resulting beam characteristics are shown in Table Five. It is clear from this table that Al_2O_3 is very effective at reducing fast neutrons. Also, the epithermal neutrons are enhanced by about a factor of 5 in comparison with the present configuration.

Table Five. Al_2O_3 Beam Characteristics

	Neutron Flux (n/cm^2-s)			Gamma-Ray Dose Rate (cGy/hr)
	Thermal	Epithermal	Fast	
Present Configuration	1.02×10^9	1.92×10^7	1.33×10^6	30.3
Al_2O_3-50 cm Beam (Enhancement Factor)	3.94×10^8 (x0.4)	9.45×10^7 (x4.9)	1.68×10^6 (x1.3)	12.2 (x0.4)

<u>Gamma-Ray Buildup with Phantom</u>: The gamma-ray distribution without phantom is shown in Figure 7(d). When a polyethylene phantom is set at the irradiation port as shown in Figure 6, the gamma-ray distribution is built up by capture gamma rays from polyethylene. One such result is shown in Figure 10 and the beam characteristics both with and without phantom are listed in Table Six.

Table Six. Beam Characteristics with and without Phantom

Present Configuration*	Thermal (n/cm^2-s)	Epithermal (n/cm^2-s)	Fast (n/cm^2-s)	Gamma (cGy/hr)
Without Phantom	1.1×10^9	1.9×10^7	1.2×10^6	36
With Phantom	2.2×10^9	2.3×10^7	1.3×10^6	293

* Small Modification for Beam Aperture (from 30 cm to 20 cm in Diameter).

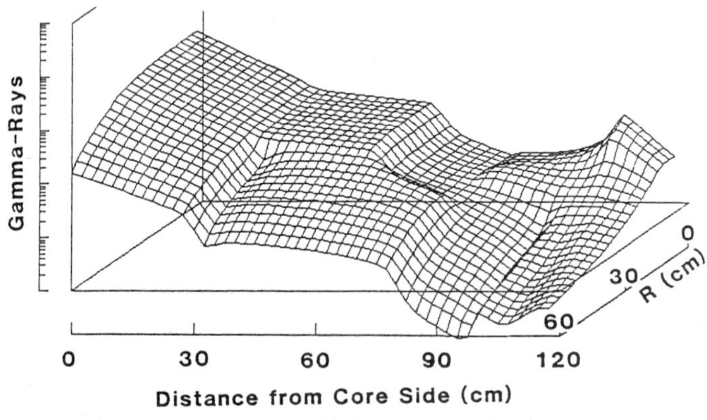

Figure 10. Gamma-Ray Dose Distribution with Phantom.

<u>Phantom Calculation (Step-3)</u>

<u>Thermal Beam from Epithermal Neutrons</u>

Five types of neutron beams were described in the previous section. These were:

(1)   Si-50 cm beam,
(2)   Al-50 cm beam,
(3)   Si-80 cm beam,
(4) Al_2O_3-50 cm beam, and
(5)   the present beam.

Neutrons from each were injected into the polyethylene phantom as a shell source of 13 cm diameter in front of the phantom as shown in Figure 11.

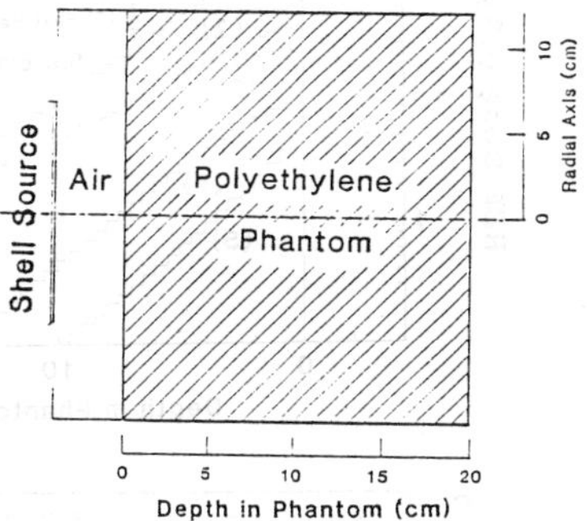

Figure 11. Geometry of Phantom Calculation.

The dose-rate distribution in the phantom was calculated assuming 30 µg/g B-10 in tumor and 3 µg/g B-10 in normal tissue. The RBE dose was calculated using RBE values of 1.0 for all gamma rays, 1.6 for fast neutrons and $^{14}N(n,p)$ protons, and 2.3 for the $^{10}B(n,\alpha)Li$ reaction products. The RBE dose distributions are shown in Figures 12(a), (b), (c), and (d) for a Si-50 cm beam, an Al-50 cm beam, a Si-80 cm beam, and the present beam, respectively. The maximum advantage depth (MAX AD) is defined here as the depth in phantom where the total dose rate with 30 µg/g B-10 equals the maximum total background dose rate. The advantage ratio is defined here as the quotient of the integral of the total dose with 30 µg/g B-10 divided by the integral of the total background dose from 0 cm to the maximum advantage depth. The advantage depth and the ratio are tabulated in Table Seven for the five types of thermal beams.

Table Seven. Dose Characteristics for Thermal Beams

Thermal / 13 cm Diameter	Beam Configuration of Thermal Beam				
	Si-50 cm	Al-50 cm	Si-80 cm	Al_2O_3-50 cm	Present
Maximum Advantage Depth					
No RBE (cm)	5.8	6.3	5.9	5.9	5.2
With RBE (cm)	6.9	7.4	7.0	7.0	6.2
Advantage Dose Rate					
No RBE (cGy/min)	23.4	12.5	12.1	3.2	6.3
With RBE (cGy/min)	29.3	15.5	14.9	3.8	7.5
Advantage Ratio					
No RBE	3.0	3.0	3.1	3.2	3.3
With RBE	4.7	4.8	4.8	5.1	5.4
% Low LET Dose	12.4	12.8	12.8	12.9	12.3
% High LET Dose	8.9	8.1	7.9	6.8	6.3
% B-10 Dose (30µg/g)	78.7	79.1	79.3	80.3	81.4

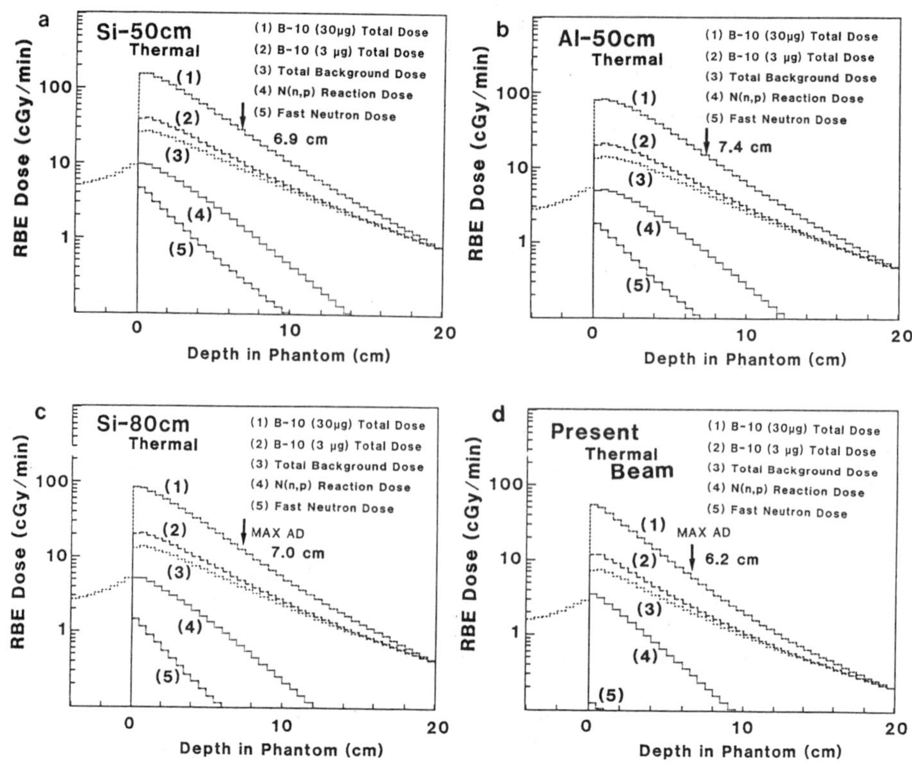

Figure 12.   RBE Dose Distributions in Phantom for Thermal Beams.

<u>Epithermal Beam from Non-Thermal Neutrons</u>: The epithermal beam is taken here to mean the five types of neutron beam with the thermal-neutron component set to zero. The five beams are the Si-50 cm beam, the Al-50 cm beam, the Si-80 cm beam, the Al_2O_3-50 cm beam, and the present beam. The thermal-neutron flux distribution in a phantom is shown in Figure 13 when the phantom is illuminated with the epithermal beams. The peak thermal-neutron flux in phantom is tabulated in Table Eight together with the incident epithermal flux. The ratio of the peak thermal-neutron flux to the incident epithermal flux is also shown in Table Eight. It can be seen that this ratio is about 1.5.

Table Eight.   Peak Thermal-Neutron Flux in a Phantom

	Si-50 cm Epithermal	Al-50 cm Epithermal	Si-80 cm Epithermal	Al_2O_3-50 cm Epithermal	Present Epithermal	
					Cal.	Exp.
(1)	6.9×10^8	6.0×10^8	3.8×10^8	9.4×10^7	1.9×10^7	1.9×10^7
(2)	1.05×10^9	9.14×10^8	5.75×10^8	1.47×10^8	3.01×10^7	2.8×10^7
(3)	1.52	1.52	1.51	1.56	1.58	1.47

(1) Incident Epithermal Total Flux (n/cm²-s)   (3) Ratio of (2)/(1)
(2) Peak Thermal-Neutron Flux in Phantom (n/cm²-s)

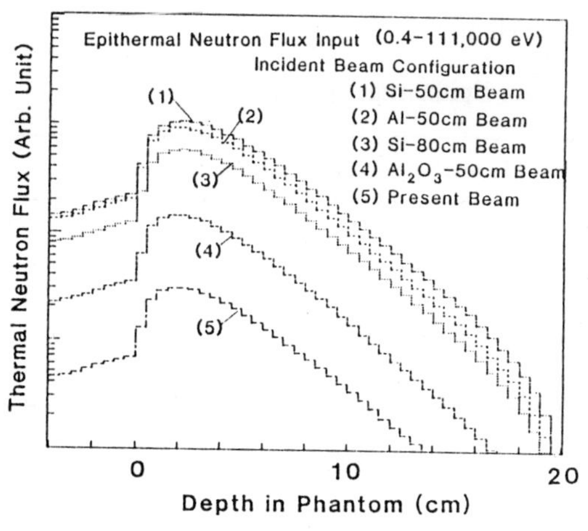

Figure 13. Thermal-Neutron Flux Distributions Across a Phantom when the Phantom is Illuminated with Various Epithermal-Neutron Beams.

The RBE dose distributions of these epithermal beams are shown in Figures 14(a), (b), (c), and (d) for a Si-50 cm beam, an Al-50 cm beam, a Si-80 cm beam, and a Al$_2$O$_3$-50 cm beam respectively. The maximum advantage depth and the advantage ratio of these epithermal beams are tabulated in Table Nine. The maximum advantage depth with RBE is about 8 cm. This value shows that a deep tumor can be treated using these epithermal beams.

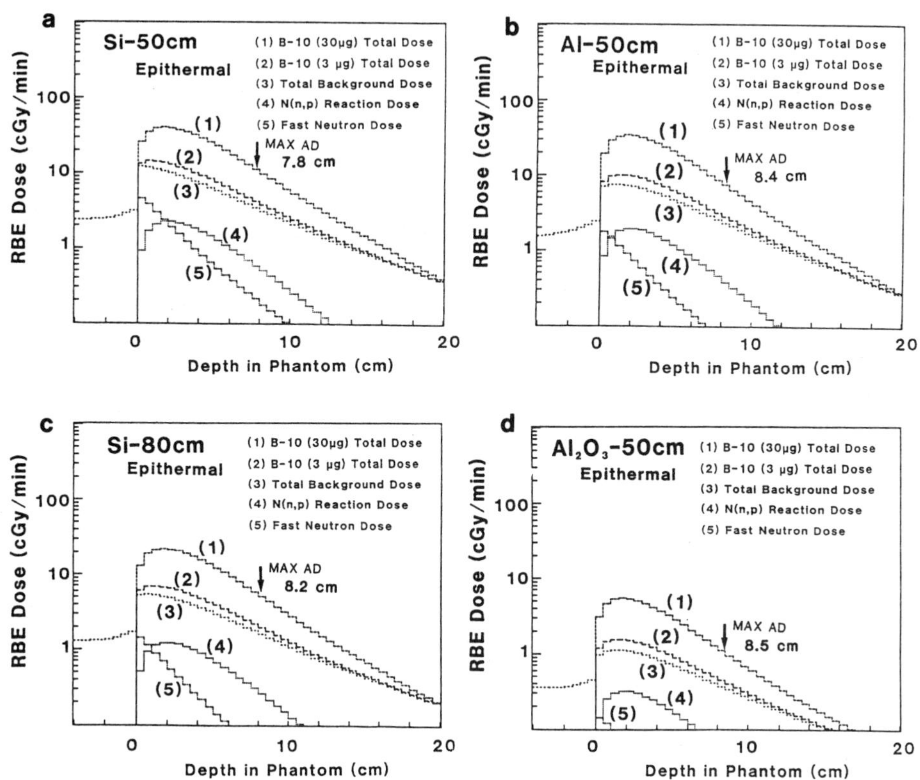

Figure 14. RBE Dose Distributions in Phantom for Epithermal Beams.

Table Nine

Dose Characteristics for Epithermal Beams

Epithermal 13 cm Diameter	Beam Configuration of Epithermal Beam			
	Si-50 cm	Al-50 cm	Si-80 cm	Al$_2$O$_3$-50 cm
Maximum Advantage Depth				
No RBE (cm)	6.8	7.4	7.1	7.3
With RBE (cm)	7.8	8.4	8.2	8.5
Advantage Dose Rate				
No RBE (cGy/min)	9.0	6.0	4.3	0.9
With RBE (cGy/min)	12.1	7.6	5.5	1.1
Advantage Ratio				
No RBE	2.4	2.7	2.6	2.7
With RBE	3.6	4.2	3.9	4.3
% Low LET Dose	13.8	13.9	14.4	15.0
% High LET Dose	13.7	9.9	11.0	8.2
% B-10 Dose (30μg/g)	72.5	76.2	74.6	76.8

DISCUSSION AND CONCLUSION

<u>Present Beam Characteristics</u>

We are currently using a 100-kW TRIGA reactor for BNCT in Japan. The maximum advantage depth is, as listed in Table Seven, 6.2 cm with RBE and 5.2 cm without RBE for a 13-cm diameter beam. The dose characteristics are shown precisely in Table Ten. The average irradiation time for the 94 treatments of brain tumors that have been performed to date was 240 minutes. As shown in Table Ten, an absorbed dose of 1150 cGy without RBE can be obtained at a depth of 6 cm in an irradiation time of 240 minutes. The maximum total background dose is 1510 cGy without RBE at the depth of 1 cm in a phantom. The acceptable doses to the skin and brain for one treatment of BNCT are also shown in Table Ten. These values constitute a suggested limit for neutron beam design. We feel that the figures given are reasonable in the case of irradiation without fractionation.

<u>Thermal Beam Characteristics</u>

The total dose characteristics of thermal beams are listed in Table Eleven for a Si-50 cm beam, an Al-50 cm beam, and a Si-80 cm beam. The maximum irradiation time in this table is the time to deliver a fast-neutron dose of 100 cGy without RBE. The value of 100 cGy is a suggested limit for the neutron beam design and is shown in Table Ten as an acceptable dose for fast neutrons. If an absorbed dose of 100 cGy for fast neutrons were really acceptable for actual BNCT treatments, the Musashi Reactor could be remodeled by using either a Si-filter or an Al-filter in order to reduce the irradiation time as shown in Table Eleven.

<u>Epithermal Beam Characteristics</u>

If the fast-neutron dose for a single irradiation must be less than 100 cGy without RBE, then the design of an epithermal beam becomes very difficult in the case of a low

Table Ten

Calculated Irradiation Doses for the Present Beam

Depth in Phantom	RBE Dose/Rad Dose		Present Beam: 100 kW, 240 min Irradiation		
	B-10 (30µg) Total Dose (cGy)	Total Background Dose (cGy)	N(n,p) Reaction Dose (cGy)	Gamma Dose (cGy)	Fast-Neutron Dose (cGy)
5 mm	13040 / 6340	1770 / 1430	850 / 530	880	29 / 18
1 cm	12000 / 5950	1810 / 1510	760 / 480	1010	25 / 15
2 cm	8360 / 4270	1490 / 1290	510 / 320	950	16 / 10
3 cm	5810 / 3060	1190 / 1060	350 / 220	830	11 / 7
4 cm	4030 / 2190	940 / 850	230 / 140	700	7 / 5
5 cm	2810 / 1580	740 / 680	160 / 100	580	5 / 3
6 cm	1970 / 1150	590 / 550	100 / 60	480	4 / 2
7 cm	1390 / 840	470 / 440	70 / 40	400	3 / 2
Acceptable Dose to Skin and Brain for One Treatment of BNCT	2500 / 1600 ~1750	1000 / 500	1000	500 / 100 ~250

Table Eleven

Calculated Irradiation Doses for Thermal Beams Using a 100-kW Reactor

Thermal Beam	Si-50 cm	Al-50 cm	Si-80 cm
Fast-Neutron Dose at Surface (cGy/min) With RBE / No RBE	4.65 / 2.91	1.78 / 1.12	1.44 / 0.90
Maximum Irradiation Time (min)	34	90	110
Advantage Depth (cm) With RBE / No RBE	6.9 / 5.8	7.4 / 6.3	7.0 / 5.9
	RBE Dose / Rad Dose (cGy)		
Total Dose at 1 cm	5230 / 2670	7450 / 3810	9080 / 4620
2 cm	4150 / 2160	6400 / 3310	7220 / 3740
3 cm	3200 / 1690	5160 / 2710	5570 / 2940
4 cm	2410 / 1300	3970 / 2120	4180 / 2250
5 cm	1790 / 990	2980 / 1630	3090 / 1700
6 cm	1310 / 740	2190 / 1230	2260 / 1280
7 cm	960 / 560	1600 / 920	1640 / 960

Table Twelve.   Calculated Irradiation Doses for an Epithermal Beam of Al_2O_3-50 cm Using a 1-MW Reactor

Al_2O_3-50 cm Epithermal Beam			RBE Dose / Rad Dose (cGy) at 1-MW, 110-min Irradiation		
Depth in Phantom	B-10 (30μg) Total Dose	Total Background Dose	N(n,p) Reaction Dose	Gamma Dose	Fast-Neutron Dose
5 mm	3350 / 1870	1070 / 880	160 / 100	560	160 / 100
1 cm	4950 / 2690	1210 / 1000	280 / 170	640	130 / 81
2 cm	5920 / 3110	1230 / 1040	350 / 220	710	83 / 52
3 cm	5470 / 2880	1140 / 980	320 / 200	710	54 / 34
4 cm	4520 / 2410	980 / 860	270 / 170	660	35 / 22
5 cm	3520 / 1910	820 / 730	200 / 130	580	24 / 15
6 cm	2650 / 1480	680 / 610	150 / 90	500	17 / 10
7 cm	1960 / 1120	550 / 510	110 / 70	430	12 / 7

power (100 kW) reactor. If the reactor power can be 1 MW, then the Al_2O_3 beam becomes the best candidate for an epithermal beam design. The beam characteristics are shown in Table Twelve.

Conclusion

A new design calculation of the irradiation field was performed using single-crystal silicon, aluminum, and Al_2O_3 instead of graphite in the Musashi Reactor. It is clear that the thermal and epithermal-neutron flux can be extracted by using either single-crystal silicon or aluminum. However, the fast-neutron flux is also enhanced. Beam optimization should be performed in accordance with an individual reactor structure.

The critical factor is the evaluation of fast-neutron doses. If the acceptable dose for fast neutrons is greater than 100 cGy without RBE for a single irradiation, then the neutron beam design work will be easy. However, a reactor power of more than 1 MW is needed to build an epithermal beam that can deliver the needed dose in a single irradiation. Such a reactor would be needed if, for medical reasons, it was not possible to administer the dose in fractions.

REFERENCES

1.   O. Aizawa, K. Kanda, T. Nozaki, and T. Matsumoto, "Remodeling and Dosimetry on the Neutron Irradiation Facility of the Musashi Institute of Technology Research Reactor for Boron Neutron Capture Therapy," Nucl. Technol., 48:150 (1980).

2.   O. Aizawa, T. Matsumoto, and H. Kadotani, "Total Neutron Cross Sections of Magnesium, Aluminum, Silicon, Zirconium, Niobium, and Molybdenum in Energy Range from 0.001 to 0.3 eV," J. Nucl. Sci. Technol., 20:717 (1983).

3.   W. A. Rhoades and F. R. Mynatt, "The DOT III Two-Dimensional Discrete Ordinates Transport Code," Oak Ridge National Laboratory, ORNL-TM-4280 (1973).

4.   "DCL-23/CASK-81: 22 Neutrons, 18 Gamma-Ray Group, P_3, Cross Sections for Shipping Cask Analysis," Radiation Shielding Information Center, Oak Ridge National Laboratory (1976).

NEUTRON BEAM STUDIES FOR A MEDICAL THERAPY REACTOR

W. A. Neuman

Idaho National Engineering Laboratory
EG&G Idaho, Inc.
Idaho Falls, ID

ABSTRACT

A conceptual design of a Medical Therapy Reactor (MTR) for neutron capture therapy (NCT) has been performed at the Idaho National Engineering Laboratory (INEL). The initial emphasis of the conceptual design was toward the treatment of glioblastoma multiforme and other presently incurable cancers. The design goal of the facility is to provide routine patient treatments both in brief time intervals (~10 minutes) and inexpensively. The conceptual study has shown this goal to be achievable by locating an MTR at a major medical facility. This paper addresses the next step in the conceptual design process: a guide to the optimization of the epithermal-neutron filter and collimator assembly for the treatment of brain tumors. The current scope includes the sensitivity of the treatment beam to variations in filter length, gamma shield length, and collimator lengths as well as exit beam aperture size. The study shows the areas which can provide the greatest latitude in improving beam intensity and quality. Suggestions are given for future areas of optimization of beam filtering and collimation.

INTRODUCTION

Interest in NCT has increased in recent years. This can be attributed to a better understanding of NCT in general, to boron compounds which are much more effectively absorbed in the tumor region, and the success of limited human trials in Japan [1-3]. The proposed near-term neutron sources are also much better suited to deliver therapeutic neutron doses with minimal background contaminants. An example of such a neutron source is the Power Burst Facility (PBF) [4]. These improvements may make NCT a viable treatment for several cancer types in the near future. Anticipating favorable results from the clinical trials expected to begin in the next five years, there will be a demand for NCT which will exceed the capabilities of current facilities. This outlook has prompted studies into future neutron sources for NCT [5,6]. This paper addresses patient treatment sensitivity to the design of an epithermal-neutron filter (energies between 1 eV and 10 keV) for an MTR.

The MTR is a pool reactor dedicated to NCT. An initial conceptual design of the facility was recently completed to address the feasibility of locating an MTR at a major medical center [7]. Because of siting requirements, the reactor must be inherently safe and designed such that no radioactivity, beyond legally permitted quantities, is released during normal or accident conditions. The current design emphasis has been directed toward treating glioblastoma multiforme. It is believed that a facility which can effectively treat this cancer type could also be utilized for other cancers, such as melanoma. A design goal of a ten-minute, single treatment period for glioblastoma multiforme has been set.

Neutron Beam Design, Development, and Performance for Neutron Capture Therapy
Edited by O. K. Harling *et al.*
Plenum Press, New York, 1990

This paper addresses the sensitivity of the treatment beam to various filter design parameters. This initial sensitivity study is meant to guide future refinements in filter design and should not be viewed as an exhaustive research program to optimize the epithermal-neutron filter. Optimization will require continued interaction between the filter design and medical treatment communities. The following sections of this paper provide an overview of the MTR facility, followed by a brief description of the modeling techniques used. The results of the study are then presented, along with an interpretation of the results for use in future design studies. The final section summarizes the current study.

OVERVIEW OF THE MTR CONCEPTUAL DESIGN

The objectives of the MTR are to routinely treat a large number of cancer patients and to serve as a facility for advanced research on NCT methods. The MTR goal is to deliver therapeutic doses of epithermal neutrons in approximately ten minutes. Additionally, the treatment beam must have minimal fast-neutron and gamma contaminants. An MTR, devoted to NCT, has been investigated and the details of the facility and reactor design have been reported elsewhere [5,7]. Hence, only a brief description of the facility is given here.

A typical arrangement for the MTR facility is shown in Figure 1. The MTR would be sited next to a major medical center, which would provide most of the patient administration, preparation, and follow-up needs. The reactor operations and medical treatment areas are physically isolated for security purposes. Interaction between the medical and reactor operations staffs will be handled by electronic communication. The conceptual design has integrated an area into the facility for the performance of advanced research on NCT. This area could handle animal studies, dosimetry, etc. It would also be isolated from the normal patient areas.

A pool reactor configuration was chosen for the MTR because of its inherent safety characteristics. The pool serves as a large heat sink to mitigate accident conditions and also eliminates accidents which might arise from pipe ruptures. The reactor is located in the narrow arm of the pool approximately 7.3 m below the pool surface. There are four radial neutron filters adjacent to the core which provide direct beams to individual treatment rooms. A vertical neutron filter is also shown, providing a neutron source for abdominal or intraoperative treatments. The reactor is cooled both by natural circulation and small enhancement pumps which direct water under Treatment Room 1 (the treatment room built into the narrow arm of the pool), through the core, and back over Treatment Room 1 through a secondary heat exchanger to the pool.

The reactor design chosen for the MTR is based on both performance criteria and the necessity for inherent safety. The fuel matrix is a 45 weight percent uranium in UZrH, 20 percent ^{235}U-enriched hydride fuel designed for current generation TRIGA reactors [8]. Note that this hydride fuel has a much higher uranium content than the UZrH fuel in most university reactors, but shares similar inherent safety characteristics with the university reactor fuel matrix. The reactor contains 1548 fuel pins in 43 assemblies, four control assemblies, and one poison and shim assembly. The core dimensions are 0.49-m on a side by 0.6-m active height. The core is surrounded by an aluminum shroud extending above the core to form a natural circulation chimney. A bismuth reflector surrounds the aluminum shroud and forms the neutron leakage surfaces for the filters. The core is symmetrically designed to provide similar neutron sources to each radial filter.

The MTR has a maximum power level of 10 MW in order to meet the definition of a research reactor as specified by the U.S. Nuclear Regulatory Commission. The reactor is designed to operate in an intermittent square-wave mode. Patients are treated for 10 minutes, with the reactor at full power. This is followed by patient monitoring, moving, and exchange during a 50-minute standby power operation, 1% full power. For single-shift operation, this yields a reactor core lifetime equal to the facility lifetime of 30 years. There is also no inherent reason that multiple-shift operation could not be utilized, if necessary, to meet treatment demands. If single-shift operation were adopted, there would be approximately 2000 treatment periods per year. If only two of the five beam ports

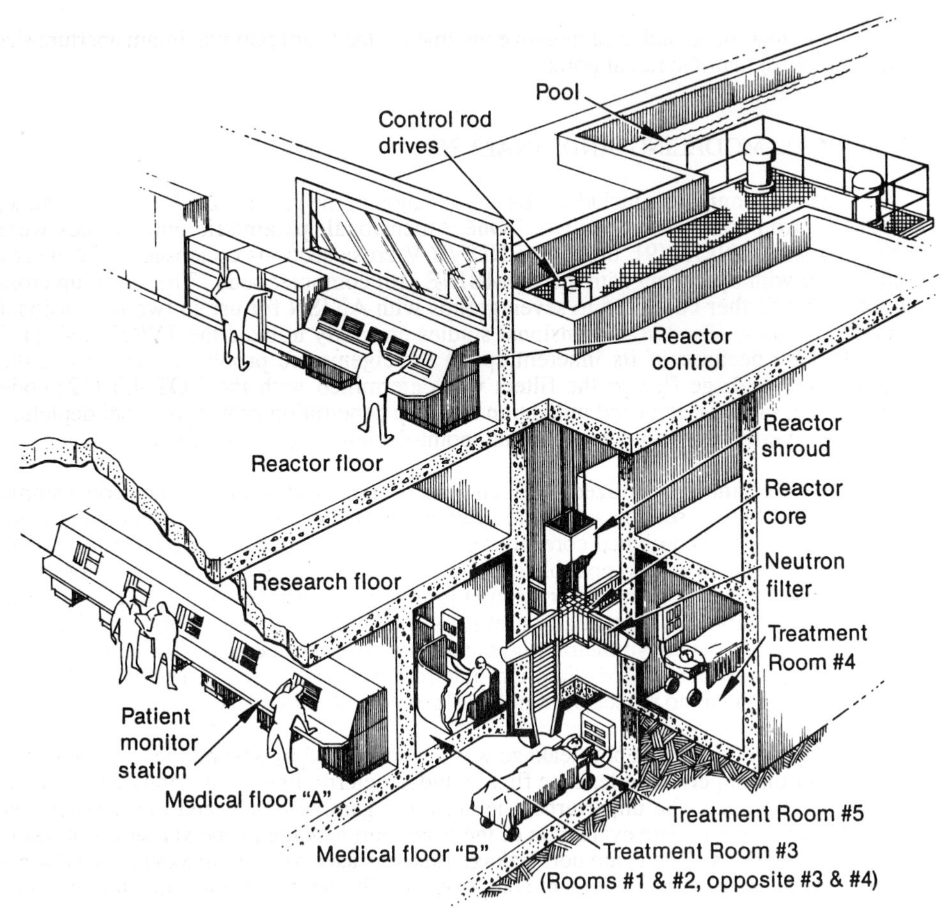

Figure 1. A Typical Arrangement for the MTR Facility.

provided were routinely utilized, more than 4000 patients could be be treated per year. The resulting reactor cost per treatment will be minimal, about $1000, making the reactor a very cost-effective neutron source.

The neutron filters moderate the core leakage flux into the epithermal-energy range and truncate the thermal-neutron component. The filter assemblies are designed to be removable in order to facilitate future advancements in filter technology and to provide a means of spectral tuning through the use of different filter assemblies. The initial filter design for the MTR is an adaptation of the proposed PBF epithermal-neutron filter [4]. The filter is cylindrical, 0.9-m long and 0.7 m in diameter, composed of 89% aluminum, 10% D_2O, and 1% lithium, by volume. A thin, 0.25-mm cadmium film is positioned past the $Al/D_2O/Li$ section to eliminate the thermal-neutron component. The filter is surrounded radially by 0.1 m of bismuth to reduce neutron loss and by a stainless-steel skin approximately 20-mm thick for structural strength. Provisions for a bismuth gamma shield between the filter and collimator have also been incorporated.

The neutrons exiting the filter assembly are collimated from the 0.7-m diameter filter to a 0.2-m beam port at the patient location in approximately 1 m. A beam-shuttering system is built into the collimator to facilitate individual control of the patient treatment duration, independent of the operation of the reactor. The apertures of the beam ports are adjustable via inserts to reduce the 0.2-m diameter openings to the desired size. The vertical filter/collimator assembly is much the same as the horizontal ones. However, its neutron source is the $Al/D_2O/Li$ section of a radial filter. Because the filter/collimator is designed for abdominal and intraoperative treatments, the beam port maximum aperture size is larger (0.5 m) than the radial ports.

NUMERICAL MODELING AND ANALYSIS

The reactor analysis includes the determination of the eigenvalue, reactivity lifetime, and the neutron source to the filters. One-dimensional, parametric core studies were performed with ANISN [9], utilizing ENDF/B-V cross sections collapsed to 39 energy groups (four with upscatter) with the COMBINE [10] computer code. The 39-group cross sections were further collapsed to seven groups with ANISN for use in two-dimensional transport analysis. The two-dimensional studies primarily utilized the TWODANT [11] transport code because of its inherent speed for eigenvalue problems. However, the resulting core leakage flux to the filters was determined with the DOT 4.3 [12] code because of its superior edits and utilization for deep penetration problems. Fuel depletion analysis was performed with the one-group, point-depletion code, ORIGEN2 [13].

The filter analysis utilized DOT and the 67-group, P_3-coupled, neutron-gamma BUGLE-80 [14] cross-section set. The filter analysis was performed independent from the core analysis. A combined core/filter calculation was not performed because of the complexity in geometry, incompatible cross-section sets, and prohibitive computer costs. The coupling of the core and filter calculations is described in [5]. DOT's multiple quadrature set option was particularly useful for modeling the filter. For the majority of the filter, a standard level-symmetric (S_8) quadrature set gave adequate angular resolution. In the collimator, a biased 166-direction quadrature set was used to finely resolve the streaming direction and to reduce the numerical problem of ray effects.

Analysis of the beam port leakage was accomplished by several means. An initial evaluation of the unperturbed leakage flux included determining the portion of the flux in the epithermal region, the unperturbed neutron and gamma doses, and the fast-neutron KERMA. A more complete evaluation of the beam port leakage included a series of head-phantom calculations. The beam port leakage flux was used as input to a cylindrical head-phantom model with brain-equivalent composition. The head-phantom calculations were performed with DOT. Various beam port aperture sizes were simulated by using only the leakage flux included in the radial extent of the desired aperture. Doses for each constituent of the phantom were determined along with the doses resulting from 3 and 30 μg/g concentrations of ^{10}B. The following sections describe these results in detail.

RESULTS

The filter/collimator parameters investigated included collimator length, filter length, gamma-shield length, and aperture size. The beam port exit quantities of interest were the leakage flux intensity, radial profile, angular divergence, spectrum, and gamma contaminant. The quantities of interest resulting from the phantom calculations were the neutron and gamma doses (for both healthy and tumor tissue) and the resulting treatment time and quality. Additionally, the advantage depth, advantage dose rate, and advantage ratio have been found, along with the dose distribution throughout the entire volume of the phantom. These quantities are discussed as they are introduced in the following sections.

Baseline Reactor/Filter Parameters

For the purposes of this study, several assumptions have been used. The reactor core and reflector have a fixed size and power level (10 MW). The reflector faces form 0.8 x 0.8-m leakage surfaces for input to the filters. The core leakage source to the filters was held fixed for all filter cases. The leakage source was determined from a reactor calculation which included a typical filter configuration to obtain the correct neutron albedo at the reflector. The specific filter configuration chosen had a 0.9-m long filter section and a 0.13-m gamma shield. The collimator was not included in the calculation. The resulting neutron source to the filter assembly is 8.45E+16 n/m^2-s. This neutron source is the

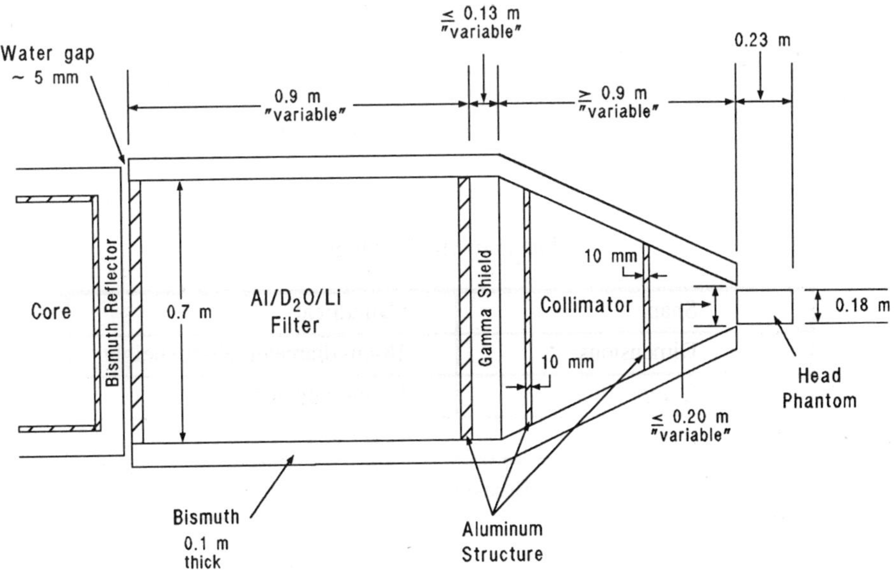

Figure 2. Filter Geometry and Head Phantom Orientation for the MTR.

minimum predicted value from either ANISN, TWODANT, or DOT analysis of the reactor/filter configuration.

The filter composition was held constant for this study at 89% aluminum, 10% D_2O, and 1% lithium, by volume. Optimization of the composition will be addressed in the future. In addition, the inside diameter of the filter was fixed at 0.7 m and was surrounded by 0.1 m of bismuth. The basic design of the filter is conservative from the standpoint that assumed water gaps and structural material, which will degrade the treatment beam, have been included in the analysis. The cadmium film at the patient side of the filter has a fixed thickness of 0.25 mm. The minimum collimator length is 0.9 m to permit ample space in the collimator section for beam shutters. The beam port aperture for the filter calculations is fixed at 0.2-m diameter. Variations in aperture size are assumed not to perturb the beam port flux inside the aperture size chosen. The filter geometry, along with the head phantom orientation, is shown in Figure 2.

The beam port leakage flux for the baseline filter is 1.37E+14 n/m²-s, of which 90.9% is in the epithermal-energy range. The resulting energy spectrum is shown in Figure 3. The gamma flux is 4.14E+11 n/m²-s, or 0.3% of the neutron flux. The flux is very uniform over the central 0.1-m diameter portion of the beam (1% variation) and falls off less than 10% over the entire 0.2-m beam diameter. The exiting beam has reasonable angular collimation with a flux-to-current ratio of $\phi/J < 1.4$. An isotropic distribution has $\phi/J = 2.0$ and a monodirectional beam would yield $\phi/J = 1.0$.

The leakage flux from the baseline filter was input to a 23-cm long, 18-cm diameter, cylindrical head phantom of brain-equivalent composition (see Table One - note that the diameter of the phantom is 2 cm less than the maximum beam port aperture). Neutron and gamma doses were determined for each component of the composition and ^{10}B. These doses are summarized in Figure 4 along the centerline of the phantom. The doses with relative biological effectiveness (RBE) factors of 2.3 for the $^{10}B(n,\alpha)$ reaction, 1.6 for fast neutrons and $^{14}N(n,p)$ protons, and 1.0 for gamma rays are shown in Figure 5. Several figures of merit for the phantom calculation were determined. The advantage depth is defined as the depth in the phantom where the total therapeutic dose rate equals the maximum total background dose rate, with and without 3 μg/g ^{10}B in normal tissue. The

Table One

Phantom Specifications

Shape:	Cylindrical
Dimensions:	18-cm diameter, 23-cm height
Composition:	Brain-equivalent
Density:	1.047 g/cc
Elemental Breakdown (weight percent)	

H - 10.57	Na - 0.14
C - 13.97	P - 0.39
N - 1.84	Cl - 0.14
O - 72.59	K - 0.39

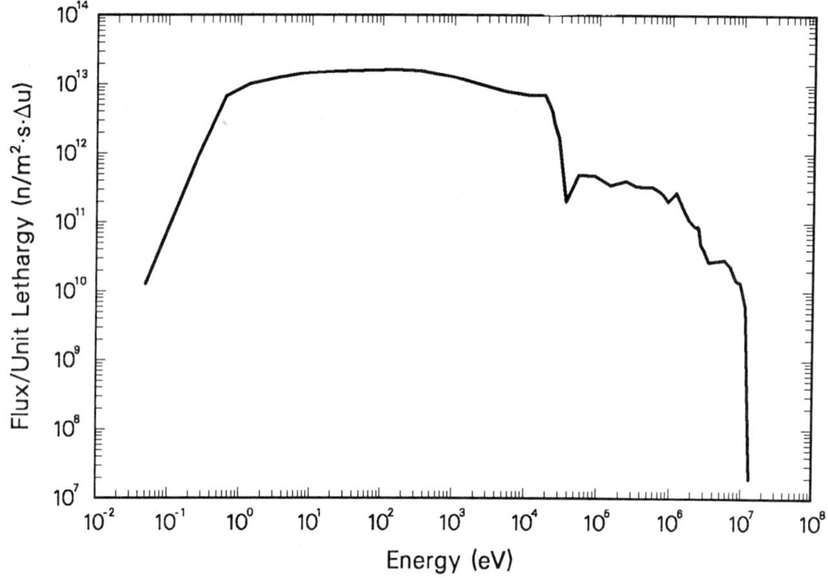

Figure 3.   Resulting Beam Port Energy Spectrum for Baseline Filter Configuration.

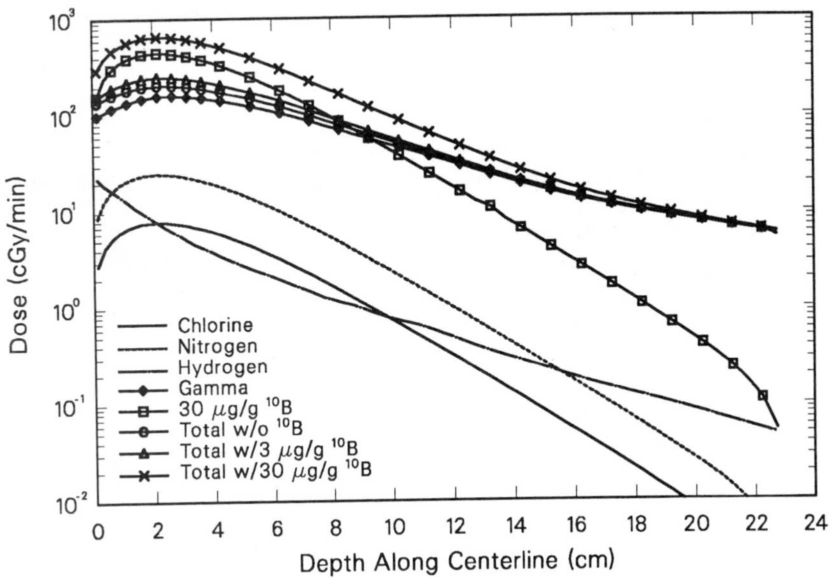

Figure 4. Resulting Axial Dose Rates for Baseline Filter Configuration.

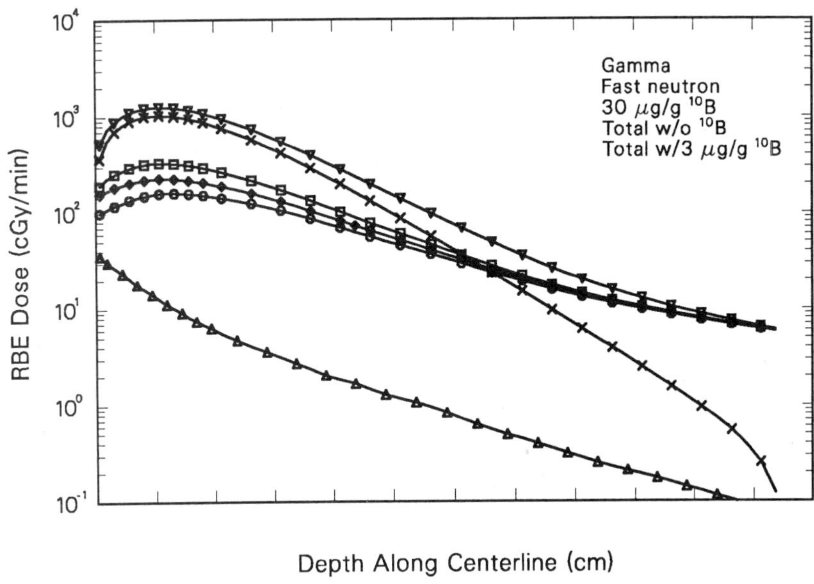

Figure 5. Resulting Axial RBE-Dose Rates for Baseline Filter Configuration.

Table Two

Baseline Filter Configuration and Performance

Filter length (m)	0.9
Collimator length (m)	0.9
Gamma shield length (m)	0.13
Aperture diameter (m)	0.2
Peak thermal flux in phantom (n/m^2-s)	2.98E+14
Peak thermal flux in location (cm)	2.13
Peak ^{10}B dose at 30 μg/g (cGy/min)	450
^{10}B dose at 6 cm, 30 μg/g (cGy/min)	200
Treatment duration (min)[a]	5
Advantage depth (max/min)[b]	
No RBE (cm)	7.5/6.9
With RBE (cm)	8.9/7.8
Advantage dose rate [b]	
No RBE (cGy/min)	207/252
With RBE (cGy/min)	226/329
Advantage ratio [b]	
No RBE	2.9/2.5
With RBE	5.0/3.6
% low-LET dose [c]	26.9/26.6
% high-LET dose	7.3/7.4
% ^{10}B dose (30 μg/g)	65.8/66.0

[a] 1000 cGy assuming 30 μg/g ^{10}B dose at 6 cm.

[b] First number is with no ^{10}B in healthy tissue;
second is with 3 μg/g ^{10}B in healthy tissue.

[c] Non-RBE values.

advantage dose rate is the peak background dose, or the therapeutic dose at the advantage depth. The advantage ratio is defined as the integral of the total therapeutic dose divided by the integral of the total background dose, from the beam entrance to the advantage depth.

These results, along with the peak thermal-neutron flux, the maximum ^{10}B dose, and the ^{10}B dose at a depth of 6 cm in the phantom, are summarized in Table Two.

The results of the baseline filter given in Table Two show the utility of the beam for NCT. The beam has good penetration, low contaminant dose levels, and substantial treatment intensity. The incident gamma-dose rate at the surface is 2.3 cGy/min and the incident fast-neutron dose rate is 22.3 cGy/min. If one assumes that a ^{10}B dose of 1000 cGy (no RBE factors) is necessary to treat the tumor and wishes to treat to 6 cm, a single treatment of only 5-min duration is necessary. For this treatment period, a peak fast-neutron dose of 1.2 Gy (<1.8 with RBE factors) and incident gamma dose of 0.12 Gy are delivered. The short duration of the treatment period suggests that the design incorporates a great deal of flexibility. The radial flatness of the treatment beam can be seen from the thermal-flux contours shown in Figure 6. At the peak flux depth, ~2 cm, the flux only drops by ~15% in the inner 10-cm diameter.

<u>Parametric Studies</u>

The initial parameter varied was the beam port aperture size. This parameter is applicable to any of the various filter configurations utilized. The figure of merit quantities

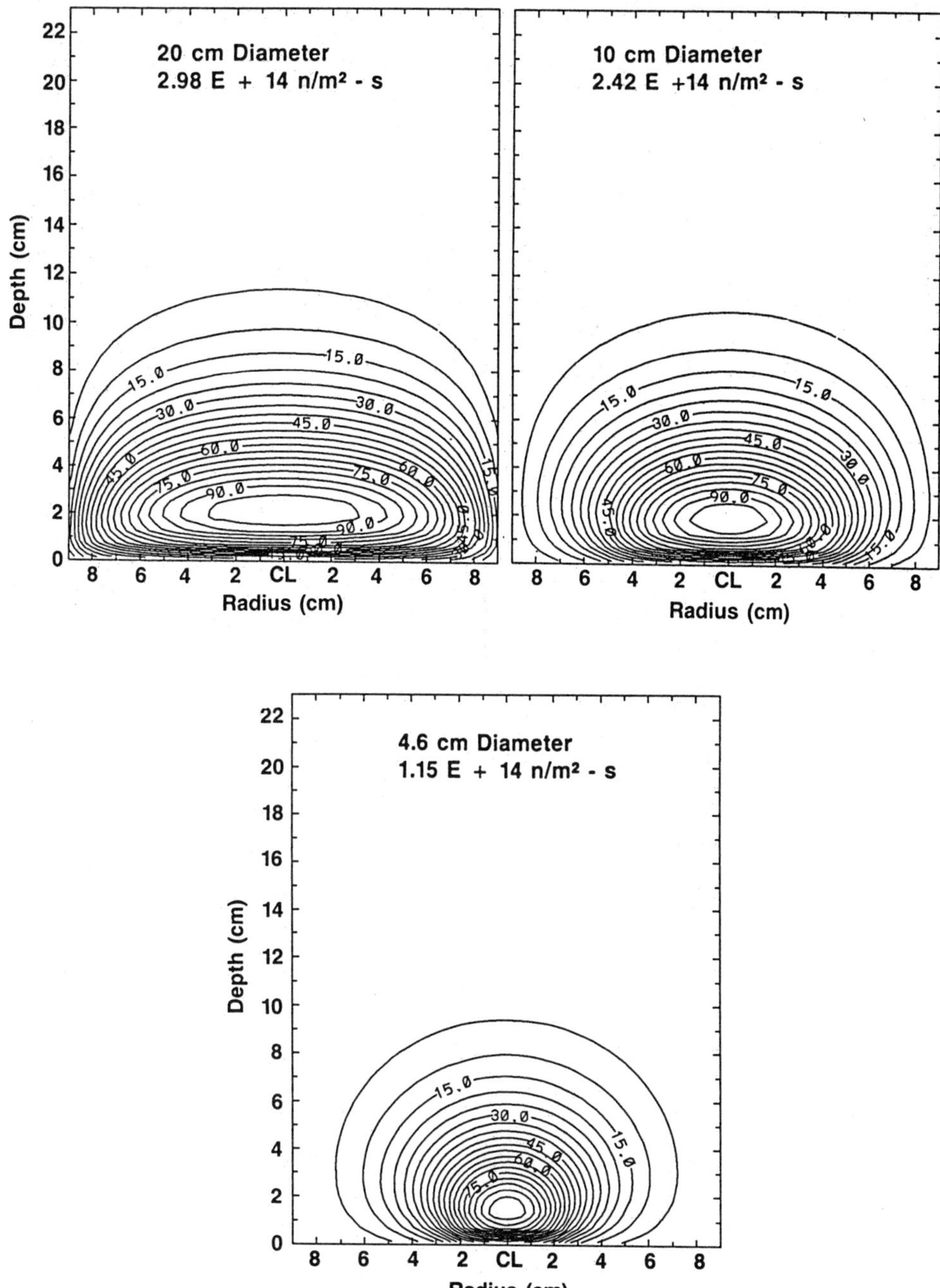

Figure 6. Thermal Flux Sensitivity to Variations in Aperture Size for the 20, 10, and 4.6-cm Diameter Beams, Respectively.

Table Three

Effect of Aperture Size Variation

	20	10	4.6
Aperture diameter (cm)	20	10	4.6
Peak thermal flux in phantom (n/m^2-s)	2.98E+14	2.42E+14	1.15E+14
Peak thermal flux in location (cm)	2.13	1.88	1.38
Peak ^{10}B dose at 30 μg/g (cGy/min)	450	360	170
^{10}B dose at 6 cm 30 μg/g (cGy/min)	200	133	43
Treatment duration (min)	5	8	23
Advantage depth (max/min)			
No RBE (cm)	7.5/6.9	7.1/6.5	6.2/5.5
With RBE (cm)	8.9/7.8	4.8/7.3	7.2/6.2
Advantage dose rate			
No RBE (cGy/min)	207/252	145/181	63/79
With RBE (cGy/min)	226/329	162/245	75/113
Advantage ratio			
No RBE	2.9/2.5	3.2/2.6	3.4/2.8
With RBE	5.0/3.6	5.5/3.8	5.8/3.9
% low-LET dose	26.9/26.6	23.1/22.8	18.5/18.2
% high-LET dose	7.3/7.4	8.1/8.1	10.6/10.7
% ^{10}B dose (30 μg/g)	65.8/66.0	68.8/69.1	70.9/71.1

for the three different aperture sizes are summarized in Table Three. Aside from the obvious drop in intensity, several other observations can be made: (1) the smaller aperture beams diverge more rapidly, (2) both the peak thermal flux location and the advantage depth decrease with decreased aperture size, and (3) the length of the treatment period for a 1000-cGy ^{10}B dose at 6 cm increases from five minutes with the 20-cm aperture to eight and 24 minutes with the 10 and 4.6-cm apertures, respectively. Notice that the treatment period increases considerably faster than the ratio of diameters for the 10 and 4.6-cm cases. The 20-cm aperture is less comparable in this way because the phantom diameter is 18 cm and does not intersect the entire flux from the largest aperture. The thermal-flux contours for the three different aperture sizes are shown in Figure 6. The contours show the increased fall-off of the flux, both axially and radially. The thermal-flux intensity inside a 10-cm diameter disk at 2 cm has decreased from 85% of the peak value for the 20-cm beam to 50% for the 10-cm beam and 18% for the 4.6-cm beam.

The length of the collimator section of the filter configuration affects beam intensity, angular divergence, and radial flatness. The collimator length was varied from the baseline value of 0.9 m to 2.0 m. Results are summarized in Table Four. The two-meter collimator decreased the flux-to-current ratio from 1.39 to 1.27 and the radial variation of the exiting flux was 6% vs. 10% when the 0.9-m collimator was used. The 2-m collimator case does, however, require a 25-minute treatment time compared to the 5-minute treatment of the baseline case. The longer collimator did not improve the advantage depth or relative background doses, nor did it appreciably change the overall flatness of the thermal-flux contours in the phantom. It is, therefore, not advantageous to increase the collimator length beyond 0.9 m to decrease the beam's angular divergence for the MTR configuration. This result effectively shows that the baseline configuration already yields a well-collimated beam and the benefits that can be gained from further increasing the collimation are minimal.

The purpose of the bismuth on the patient side of the filtering section is to reduce the gamma component of the treatment beam. Because of the minimal incident gamma dose obtained with the 13-cm thick shield of the baseline configuration, reduction of the thickness was investigated. The results are summarized in Table Five. With the gamma

Table Four

Effect of Collimator Length Variation

Collimator length (m)	0.9	1.5	2.0
Peak thermal flux in phantom (n/m^2-s)	2.98E+14	1.16E+14	6.16E+13
Peak thermal flux in location (cm)	2.13	2.13	2.13
Treatment duration (min)	450	175	93
Peak ^{10}B dose at 30 µg/g (cGy/min)	200	76	40
^{10}B dose at 6 cm 30 µg/g (cGy/min)	5	13	25
Advantage depth (max/min)			
No RBE (cm)	7.5/6.9	7.5/6.9	7.5/6.9
With RBE (cm)	8.9/7.8	8.8/7.8	8.8/7.7
Advantage dose rate			
No RBE (cGy/min)	207/252	80/98	43/52
With RBE (cGy/min)	226/329	87/128	47/68
Advantage ratio			
No RBE	2.9/2.5	2.9/2.5	2.9/2.5
With RBE	5.0/3.6	5.0/3.6	5.0/3.6
% low-LET dose	26.9/26.6	26.9/26.6	26.9/26.7
% high-LET dose	7.3/7.4	7.3/7.4	7.4/7.4
% ^{10}B dose (30 µg/g)	65.8/66.0	65.8/66.0	65.7/65.9

Table Five

Effect of Gamma Shield Length Variation

Gamma shield length (cm)	13	7	0
Peak thermal flux in phantom (n/m^2-s)	2.98E+14	4.24E+14	7.14E+14
Peak thermal flux in location (cm)	2.13	2.13	2.13
Peak ^{10}B dose at 30 µg/g (cGy/min)	450	640	1080
^{10}B dose at 6 cm 30 µg/g (cGy/min)	200	285	495
Treatment duration (min)	5	3.5	2
Incident gamma dose (cGy)	0.12	0.26	1.2
Advantage depth (max/min)			
No RBE (cm)	7.5/6.9	7.5/6.9	7.5/6.9
With RBE (cm)	8.9/7.8	8.9/7.8	8.9/7.8
Advantage dose rate			
No RBE (cGy/min)	207/252	300/363	555/663
With RBE (cGy/min)	226/329	327/474	602/850
Advantage ratio			
No RBE	2.9/2.5	2.9/2.4	2.7/2.3
With RBE	5.0/3.6	4.9/3.6	4.5/3.4
% low-LET dose	26.9/26.6	27.2/27.0	29.9/29.6
% high-LET dose	7.3/7.4	7.4/7.4	7.2/7.2
% ^{10}B dose (30 µg/g)	65.8/66.0	65.4/65.6	62.9/63.2

Table Six

Effect of Filter Length Variation

Filter length (m)	0.7	0.9	1.1
Peak thermal flux in phantom ($n/m^2 \cdot s$)	6.97E+14	2.98E+14	1.23E+14
location (cm)	2.13	2.13	2.13
Peak ^{10}B dose at 30 $\mu g/g$ (cGy/min)	1050	450	185
^{10}B dose at 60 mm 30 $\mu g/g$ (cGy/min)	485	200	78
Treatment duration (min)	2	5	13
Scalp fast neutron dose (cGy)	220	110	65
Total scalp dose w/o ^{10}B (cGy)	780	690	670
Advantage depth (max/min)			
No RBE (cm)	7.5/6.9	7.5/6.9	7.5/6.8
With RBE (cm)	8.9/7.8	8.9/7.8	8.8/7.7
Advantage dose rate			
No RBE (cGy/min)	514/619	207/252	83/102
With RBE (cGy/min)	576/818	226/329	90/133
Advantage ratio			
No RBE	2.8/2.4	2.9/2.5	3.0/2.5
With RBE	4.6/3.4	5.0/3.6	5.1/3.7
% low-LET dose	26.4/26.1	26.9/26.4	27.1/26.8
% high-LET dose	9.3/9.4	7.3/7.4	6.5/6.6
% ^{10}B dose (30 $\mu g/g$)	64.3/64.5	65.8/66.0	66.4/66.6

shield completely removed, the surface gamma-dose rate increased to 65 cGy/min. A substantial fraction of the increase is due to an overall increase in beam intensity from 1.37E+14 n/m^2-s for the baseline design to 3.35E+14 n/m^2-s without the shield. This increased intensity would necessitate a mere two-minute treatment time for a 1000-cGy ^{10}B dose at 6 cm. The resulting incident gamma dose at the surface is 1.3 cGy. The advantage depth was not affected by variations in shield thickness. However, the advantage ratio did decrease by ~7% with the shield removed and the percentage of the integral dose (advantage dose ratio) due to gammas increased from 27 to 30% of the total.

The final parameter varied in the study was the length of the Al/D$_2$O/Li filter section. Reducing the high-LET fast-neutron component of the treatment beam requires adequate filtering of the neutrons into the epithermal-energy range. If too much filtering is done, there will be a penalty in beam intensity. The length of the filter section was varied from 0.7 to 1.1 m. The results are summarized in Table Six. The change in the beam port flux spectrum is shown in Figure 7. The benefit derived from increasing the filtering length diminishes above 0.9 m if the treatment time is accounted for. Figure 8 shows that the percentage of the beam exit flux in the epithermal range begins to approach a plateau near 0.9 m. However, the treatment duration increases more rapidly above 0.9 m. Therefore, reducing the high-LET healthy tissue dose becomes more and more difficult as the beam is purified. The impact on filter design is significant because the obtainable tumor treatment dose may be limited by the high-LET healthy tissue dose accrued by the patient. Because of the sensitivity of the neutron beam to variations in the filter length, careful attention to the optimization of this parameter will be necessary once healthy tissue tolerances are better defined.

SUMMARY AND CONCLUSIONS

This work represents an initial study into the sensitivity and optimization of an epithermal-neutron beam filter for a Medical Therapy Reactor (MTR). The baseline configuration was based on the proposed Power Burst Facility (PBF) filter design. The

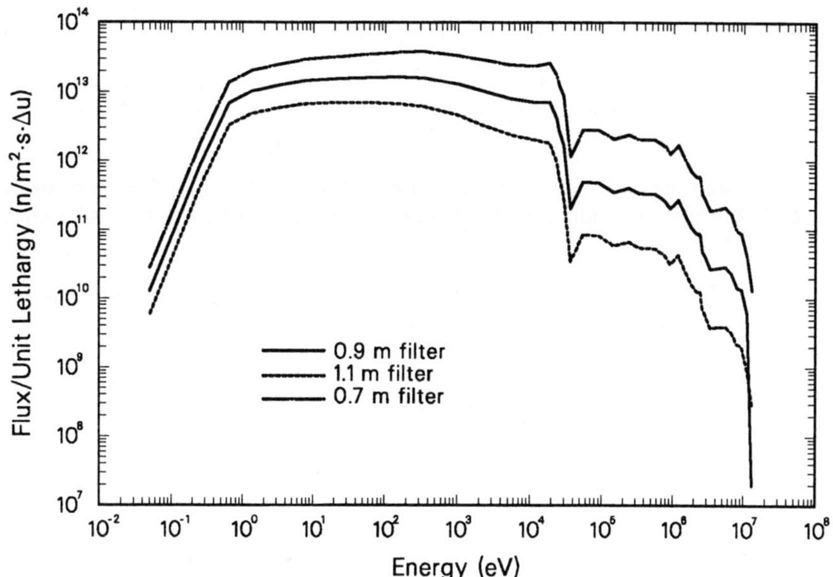

Figure 7. Sensitivity of the Beam Port Spectrum to Changes in Filter Length.

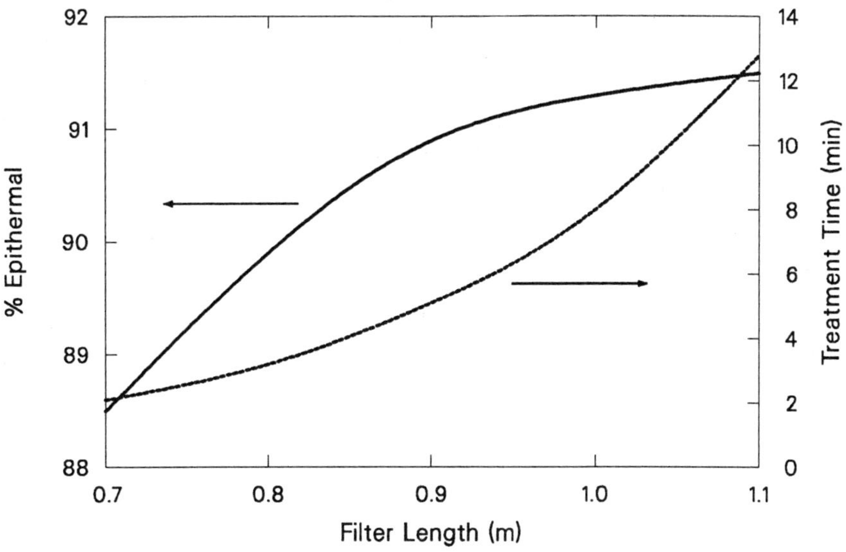

Figure 8. Beam Purity and Corresponding Treatment Time Variations with Filter Length.

parametric study investigated variations in aperture size, collimator length, gamma shield length, and filter length. The evaluation of the resulting beam port leakage flux included analysis of the unperturbed leakage flux and cylindrical head phantom calculations utilizing brain-equivalent composition.

The parametrics can be summarized as follows:

1.   Decreasing the aperture size increases the required treatment time more rapidly than the decrease in diameter and delivers a much less uniform dose both radially and axially in the phantom;

2.   Increasing the collimator length beyond 0.9 m yields a less divergent beam port flux, but does not appreciably change the treatment characteristics based on the phantom results;

3.   Complete elimination of the gamma shield increases the incident gamma dose at the phantom surface from 0.12 to 1.3 Gy for a treatment period; and

4.   Increasing the filter length increases the fraction of the beam port flux in the epithermal range less efficiently as the total filter length increases.

These studies have begun to determine the areas which can lead to substantial improvements in beam intensity and treatment quality. To ensure uniform tumor irradiation, the beam port aperture should be as large as possible. Reduction in the gamma shield is potentially beneficial because the baseline design has a minimal incident gamma dose. The potential gain in beam intensity is 250% and should be pursued further. The changes in filter length show that the benefits obtainable by reducing high-LET contaminants level off around the baseline length of 0.9 m, but the drop in beam intensity does not. The susceptibility of healthy tissue damage to the high-LET reactions suggests that the degree of filtering is also an area which deserves future investigation. The intensity variations found by varying the filter thickness from 0.7 to 1.1 m were 600%.

The results of the study indicate that a doubling of the beam intensity can be achieved without significantly impacting the purity and utility of the epithermal beam. This will positively impact the design requirements of the reactor, enabling a greater margin of safety and public acceptance to be achieved for the MTR. Utilization of the sensitivity analysis presented will necessarily require guidance from the medical community on required tumor doses and healthy tissue tolerance levels. Future areas of study should include filter composition, radial size (partially governed by mechanical constraints), and optimization of the core reflector and filter interface. Additionally, other treatment modalities will need to be investigated to ensure the flexibility of the MTR facility.

ACKNOWLEDGMENT

This work was performed under the auspices of the U.S. Department of Energy, DOE Contract No. DE-AC07-76ID01570.

COPYRIGHT

REFERENCES

1.   Proc. Second Int. Symp. on Neutron Capture Therapy, Tokyo, 1985, H. Hatanaka, ed., Nishimura Co., Ltd., Niigata, Japan (1986).

2.   Proc. Third Int. Symp. on Neutron Capture Therapy, <u>Strahlenther. Onkol.</u>, D. Gabel, ed., 165(2/3):5-257 (1989).

3.   <u>Proc. 1988 Workshop on Clinical Aspects of Neutron Capture Therapy</u>, R. G. Fairchild, V. P. Bond, and A. D. Woodhead, eds., Basic Life Sciences Series, Vol. 50, Plenum Press, New York (1989).

4.   D. K. Parsons, F. J. Wheeler, B. L. Rushton, and D. W. Nigg, "Neutronics Design of the INEL Facility for Boron Neutron Capture Therapy Clinical Trials," in <u>Proc. 1988 ANS Int. Reactor Physics Conf.</u>, Jackson Hole, WY, Vol. II, p. 2-433 (Sept. 1988).

5.   W. A. Neuman, D. K. Parsons, and J. A. Lake, "Neutronics Design of a Medical Therapy Reactor," in <u>Proc. 1988 ANS Int. Reactor Physics Conf.</u>, Jackson Hole, WY, Vol. II, p. 2-443 (Sept. 1988).

6.   C. K. Wang, T. E. Blue, and R. Gahbauer, "A Neutronic Study of an Accelerator-Based Neutron Irradiation Facility for Boron Neutron Capture Therapy," <u>Nucl. Technol.</u>, 84:93 (1989).

7.   W. G. Lussie, W. A. Neuman, J. L. Jones, R. L. Drexler, J. A. Lake, M. L. Griebenow, R. R. Hobbins, D. R. DeDoisblance, and C. F. Leyse, "Medical Therapy Reactor Preconceptual Design Studies," EG&G Idaho, Inc., Idaho National Engineering Laboratory, <u>EG&G Informal Report #EGGNERD-8337</u> (Dec. 1988). See also W. A. Neuman and J. L. Jones, "Conceptual Design of a Medical Reactor for Neutron Capture Therapy," <u>Nucl. Technol.</u> (1990) to be Published.

8.   M. T. Simnad, "The UZrH: Its Properties and Use in TRIGA Fuel," <u>General Atomics Report #4314</u> (Feb. 1980).

9.   D. K. Parsons, "ANISN/PC Manual," Idaho National Engineering Laboratory, <u>EGG-2500</u> (1987).

10.   COMBINE is an updated and combined version of the PHROG and INCITE cross-section generation codes using ENDF/B-V data. R. L. Curtis, F. J. Wheeler, G. L. Singer, and R. A. Grimesey, Idaho National Engineering Laboratory, <u>IN-1435</u> (1971). See also R. L. Curtis and R. A. Grimesey, Idaho National Engineering Laboratory, <u>IN-1062</u> (1967).

11.   R. E. Alcouffe, F. W. Brinkley, D. Marr, and R. O'Dell, "User's Manual for TWODANT: A Code Package for Two-Dimensional, Diffusion-Accelerated, Neutral-Particle Transport," Los Alamos National Laboratory, <u>LA-10049-M</u>, Rev. 1 (1984).

12.   W. A. Rhoades and R. L. Childs, "Updated Version of DOT 4 One- and Two-Dimensional Neutron/Photon Transport Code," Oak Ridge National Laboratory, <u>ORNL-5851</u> (1982).

13.   A. G. Croft, "ORIGEN2: A Revised and Updated Version of the Oak Ridge Isotope Generation and Depletion Code," Oak Ridge National Laboratory, <u>ORNL-5621</u> (July 1980).

14.   R. W. Roussin, "BUGLE-80: Coupled 47-Neutron, 20-Gamma-Ray, P_3, Cross-Section Library for LWR Shielding Calculations," Radiation Shielding Information Center, <u>DLC-75</u> (1980).

INVESTIGATION OF NEUTRON BEAMS FOR THE REALIZATION OF
BORON NEUTRON CAPTURE THERAPY

Gy. Csom, É. M. Zsolnay, and E. J. Szondi

Institute of Nuclear Technics
Budapest Technical University
Budapest, Hungary

ABSTRACT

Results of the initial theoretical and experimental work performed at the Institute of Nuclear Technics of the Budapest Technical University (BTU NTI) regarding the design of a pure thermal neutron field for neutron capture therapy experiments are reported. Calculations show that a thermal neutron flux of approximately 10^9 neutrons cm^{-2}s^{-1} will be generated using a filter made of graphite and bismuth.

INTRODUCTION

Research on boron neutron capture therapy (BNCT) in Hungary is being performed cooperatively at several different institutions including the National Institute of Oncology (Prof. Dr. S. Eckhardt, Prof. Dr. J. Sugár, Dr. O. Csuka), the Budapest Technical University, Institute of Nuclear Technics (Prof. Dr. Gy. Csom, Dr. É. M. Zsolnay, Dr. R. J. Szondi, Dr. E. Virágh, Dr. S. Fehér), the Budapest Technical University, Department of Organic Chemical Technology (Prof. Dr. L. Töke, Dr. I. Bitter), Eötvös University, Institute of Organic Chemistry (Dr. M. Szekerke, Dr. Ferenc Hudecz, Dr. G. Mezö, Dr. J. Kajtár), and Eötvös University, Department of Immunology (Prof. Dr. J. Gergely, Dr. F. Uher, Dr. G. Sármai).

The BTU NTI has a nuclear reactor with a facility suitable for BNCT irradiations. This paper summarizes the efforts of NTI to design and develop a radiation field of the appropriate characteristics.

NUCLEAR REACTOR OF BTU NTI

The Institute's nuclear reactor is a swimming pool type facility with a maximum power of 100 kW [1]. The maximum thermal neutron flux in the core is 2.7×10^{12} neutrons cm^{-2}s^{-1}. The reactor has five horizontal channels for experimental purposes. Four of these terminate at the core edge in a radial configuration while the fifth is arranged tangentially. Figure 1 is a horizontal cross section of the reactor.

The thermal neutron flux is highest and the intensity of the gamma radiation lowest at the output of the tangential port. This port was therefore selected for development of an irradiation field for BNCT. Also, it is the port in which experiments have been and are now being carried out [2].

The facility which might be suitable for BNCT experiments is the large irradiation tunnel which is shown in Figures 1 and 2. This tunnel begins at the outer boundary of the

1 - Fuel Assemblies
2 - Graphite Reflector
3 - Irradiation Tunnel
4 - Railway
5 - Heavy Concrete Shield
6 - Normal Concrete Shield
7 - Horizontal Radial Channels I, II, III, and IV
8 - Horizontal Tangential Channel V

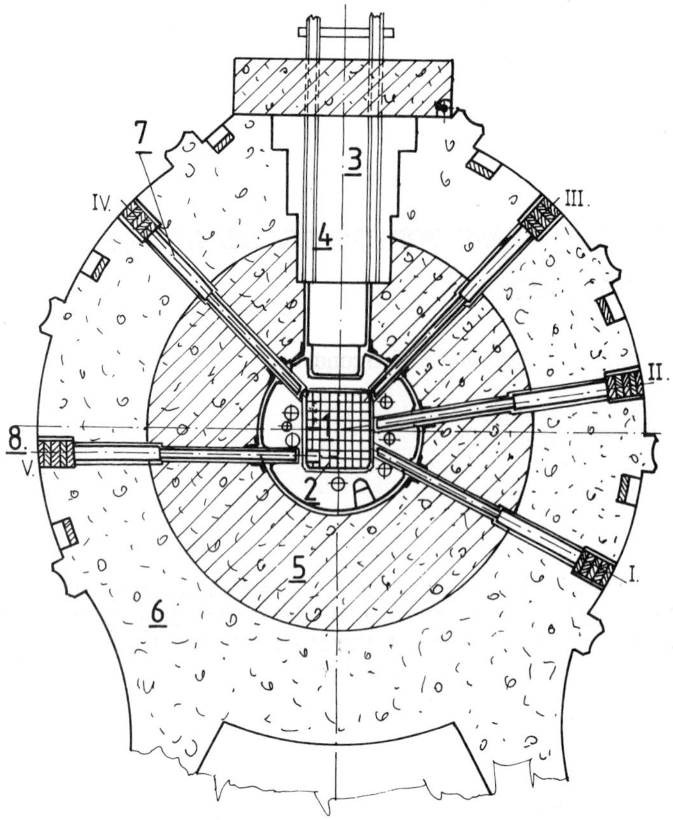

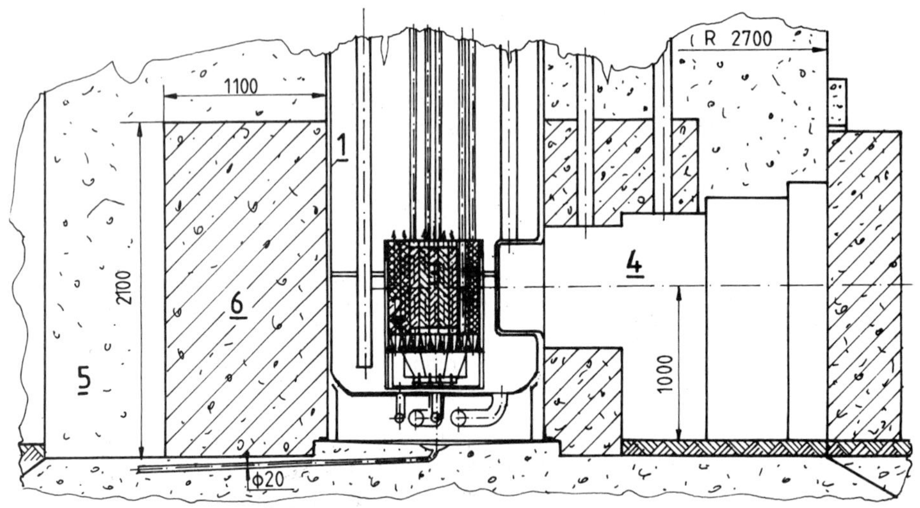

Figure 1. Horizontal Section of the Reactor Facility.

Figure 2. Vertical Section of the Reactor Facility.
1 - Reactor Tank; 2 - Graphite Reflector; 3 - Fuel; 4 - Irradiation Tunnel; 5 - Heavy Concrete Shield; 6 - Normal Concrete Shield.

biological shield with an aperture of 130x170 cm^2 and, after stepwise decrease, approaches the active core to within 23 cm of a surface that is 50x50 cm^2. It is equipped with a carriage running on a railway and is capable of transporting different experimental arrangements or large samples (e.g. animals) into the irradiation zone. The 23 cm layer between the core and the internal surface of the tunnel is composed (starting from the core) of 7.2 cm water, 7.2 cm graphite, and 9 cm water. The 9 cm water gap was replaced with air, thereby increasing the neutron flux by a factor of 3-4, depending on the neutron energy [3]. This of course improves irradiation conditions within the tunnel.

PROGRAM FOR DEVELOPMENT OF THE IRRADIATION FIELD

The basic requirements for obtaining a radiation source suitable for selective (boron) neutron capture therapy are [4, 5]:

- Availability of an intense and collimated neutron beam at the site of irradiation (a thermal neutron flux greater than 10^9 neutrons cm^{-2}s^{-1} is needed),

- Reduction of fast and, for a pure thermal neutron beam, also epithermal neutron components, and

- Elimination of gamma radiation from neutrons.

A pure thermal neutron beam is applicable only to tumors situated near the skin surface. For the treatment of deep tumors, epithermal neutrons are needed.

Considering the BTU NTI reactor's maximum power and neutron characteristics, the development of a pure thermal neutron beam is realistic and extensive research work, including theoretical calculations and experimental investigations, has been started [6]. The objective is to determine the optimum neutron and gamma filter and/or neutron converter arrangement resulting in favorable irradiation conditions for BNCT.

Parallel with this work, an irradiation field for preliminary biological experimentation was also developed [7].

IRRADIATION FIELD FOR PRELIMINARY BNCT EXPERIMENTS

A neutron beam source with improved parameters as compared to the port's original leakage spectrum was created at the 5th horizontal port for preliminary BNCT experiments. Port No. 5 was selected because of its tangential alignment. It offers the opportunity to exclude some of the more direct unwanted fast neutrons and gamma rays. On the other hand, this port has the largest beam diameter and a relatively high (thermal) neutron flux (see Figure 3).

Several combinations of graphite and bismuth plugs were investigated to obtain a fast neutron plus gamma-ray filter with parameters optimized for the intended irradiations [7]. The parameters of the irradiation field for three different filter arrangements (Figure 4) are given in Table One, together with the characteristics of the port's original leakage spectrum. These measurements were made by means of activation detectors and TLDs.

NEUTRON FLUX MEASUREMENTS IN THE IRRADIATION TUNNEL

Experiments were performed in order to determine the leakage neutron spectrum in the irradiation tunnel. Also, the neutron characteristics of different graphite configurations were investigated [3, 8].

The neutron flux values were measured using activation detectors (Dy-164, Au-197, In-115, S-32), semiconductor detectors (Si (Li)), and proton recoil detectors [3]. Measurements were made both with and without the water displacer mentioned before. The results given below pertain to a reactor power of 100 kW. They were obtained with the water displacer being present between the core and the tunnel surface.

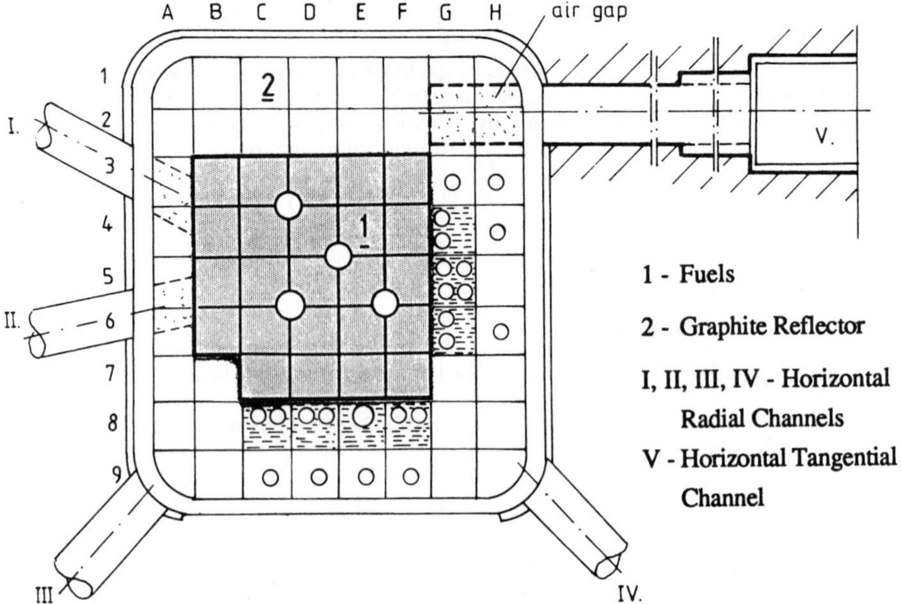

Figure 3.   Relation of the Horizontal Tangential Channel to the Core.

a) Filter I   b) Filter II

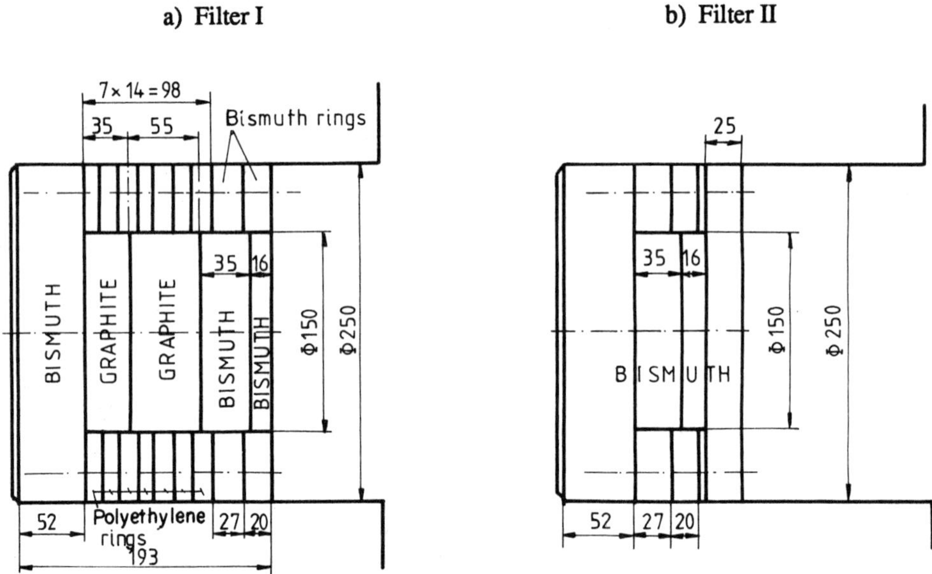

c) Filter III is Similar in Arrangement to Filter II, but the Last 25 mm of Bismuth is Missing, i.e., Thickness of Bi in this Case is 103 mm.

Figure 4.   Filter Arrangements in the 5th Horizontal Port.

Table One

Characteristics of the Irradiation Field at the 5th Horizontal Port at a
Reactor Power of 100 kW.

Position	Φ_{th} $(cm^{-2}s^{-1})$	$\Phi_{epi}^{(1/E)}$ [0.5 eV - 500 keV] $(cm^{-2}s^{-1})$	Φ_f [E>1 MeV] $(cm^{-2}s^{-1})$	$\dot{D}_\gamma$ (cGy/min)
At the outlet of the port (original leakage spectrum)	4.36 E+07	4.17 E+07	0.93 E+07	87
Behind filter I (193 mm; graphite-bismuth combination)	4.05 E+06	2.11 E+06	0.59 E+06	0.15
Behind filter II (128 mm; bismuth)	1.08 E+07	7.76 E+06	1.48 E+06	0.22
Behind filter III (103 mm; bismuth)	1.50 E+07	1.29 E+07	2.24 E+06	0.52
The uncertainty of the data is less than 10%.				

Consider the internal 50x50 cm^2 surface of the irradiation tunnel to be the x-y plane (x being the horizontal axis while y is the vertical one). The axis normal to the x-y plane is denoted by z. The origin is the center of the plane. The neutron flux values measured in this plane in the direction of x (for y = 0) and in the direction of y (for x = 0), are given in Tables Two and Three, respectively. It can be seen that the flux distribution along the internal surface of the tunnel is sufficiently uniform to result in an irradiation field of the required size for BNCT experiments.

The neutron flux values in the tunnel as determined along the z-axis (for x = y = 0) are shown in Table Four.

Graphite blocks of different thicknesses and configurations have been placed in the tunnel in order to investigate their effect on the neutron spectrum. Figure 5 shows the configurations of interest together with the position of the activation detectors used in the experiment. The results of the measurements are given in Table Five.

CALCULATIONS TO OPTIMIZE THE IRRADIATION FIELD WITHIN THE TUNNEL

Taking into consideration the software and hardware available to the program, the following computer codes were selected for the calculations:

a) One dimensional neutron and gamma shielding code SABINE-3 [9].

b) One dimensional neutron and gamma transport code ANISN [10].

c) Three dimensional gamma shielding code MERCURE-3 [11].

Table Two

Neutron Flux Values in Horizontal Direction (y = 0) Along the Internal Surface of the Irradiation Tunnel at Medium Height of the Core at a Reactor Power of 100 kW (in units of 10^{10} cm^{-2}s^{-1}).

x, (cm) Energy Range	+15	+5	0	-5	-15
Thermal	4.42	3.91	4.26	4.05	4.37
0.5 eV - 100 keV	1.59	–	1.52	–	1.73
E > 1 MeV	1.64	–	1.78	–	1.37
E > 3 MeV	0.23	–	0.24	–	0.20

Table Three

Neutron Flux Values in Vertical Direction (x = 0) Along the Internal Surface of the Irradiation Tunnel at a Reactor Power of 100 kW (in units of 10^{10} cm^{-2}s^{-1}).

y, (cm) Energy Range	+25	+15	+5	0	-5	-15	-25
Thermal	4.16	4.22	3.89	4.26	4.28	4.31	3.80
0.5 eV - 100 keV	1.03	1.22	–	1.52	–	–	1.42
E > 1 MeV	1.42	1.53	–	1.78	–	1.67	–
E > 3 MeV	–	0.22	–	0.24	–	0.24	–

Table Four

Neutron Flux Values in the Irradiation Tunnel in Direction z, Perpendicular to the Inner Surface, for x = y = 0. Reactor Power of 100 kW (in units of 10^{10} cm^{-2}s^{-1}).

z, (cm) Energy Range	23.4	37.6	51.9	65.9	79.9	93.9	106.4
Thermal	4.16	3.48	2.71	1.89	1.49	1.43	1.80
0.5 eV - 100 keV	1.53	1.21	1.07	0.67	0.53	0.46	0.47
E > 1 MeV	1.78	1.09	0.82	0.57	0.41	0.37	0.40
E > 3 MeV	0.24	0.15	0.11	0.082	0.057	0.046	0.039

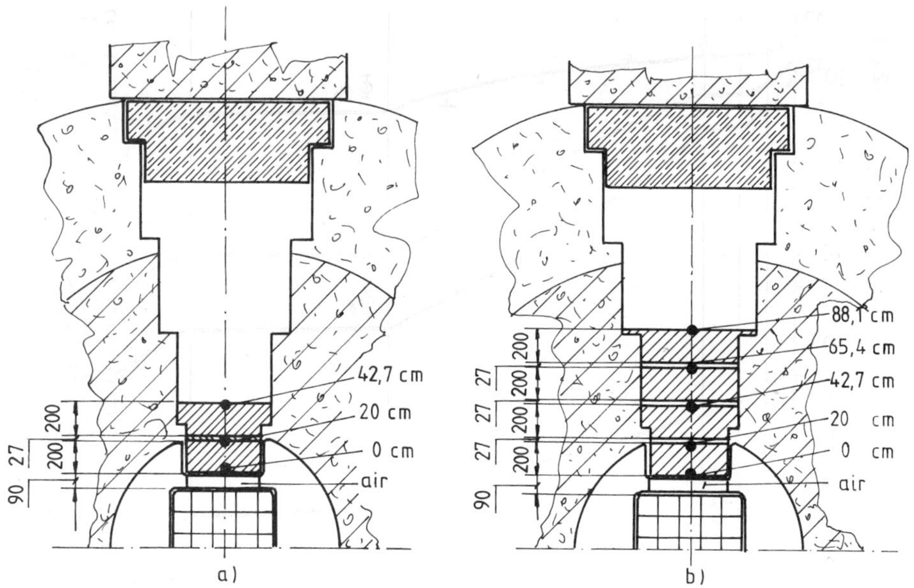

Figure 5.   Cross Section of the Tunnel with the Various Graphite Filters.

Table Five

Neutron Flux Values for the Configurations Shown in Figures 5a and 5b.
Reactor Power of 100 kW (in units of 10^{10} cm^{-2}s^{-1}).

Configuration	Energy Range	23.4	43.4	66.1	88.8	111.5
see Fig. 5a	Thermal	23.4	8.98	0.62	–	–
	E > 1 MeV	3.31	–	0.042	–	–
	E > 3 MeV	0.73	0.097	0.010	–	–
see Fig. 5b	Thermal	22.9	8.97	2.56	0.69	0.038
	E > 1 MeV	4.08	0.45	0.056	0.0069	–
	E > 3 MeV	0.68	0.091	0.011	0.0016	0.0002

The code SABINE-3 was considered sufficient for performing the initial calculations because, by proper description of the geometrical conditions (buckling), an acceptable agreement between the calculated and measured results could be obtained.

SABINE-3, with its combined removal-diffusion method, uses 26 diffusion groups and 19 removal groups.   Three-dimensional neutron diffusion can be taken into consideration by this one-dimensional code, with the aid of the buckling parameter.   The diffusion equation system used was:

$$D_i \left[\Phi_i''(r) + \frac{p}{r} \Phi_i'(r) \right] - \Sigma_i \, \Phi_i(r) + S_i(r) = 0, \qquad i = 1,2,3,\dots \qquad (1)$$

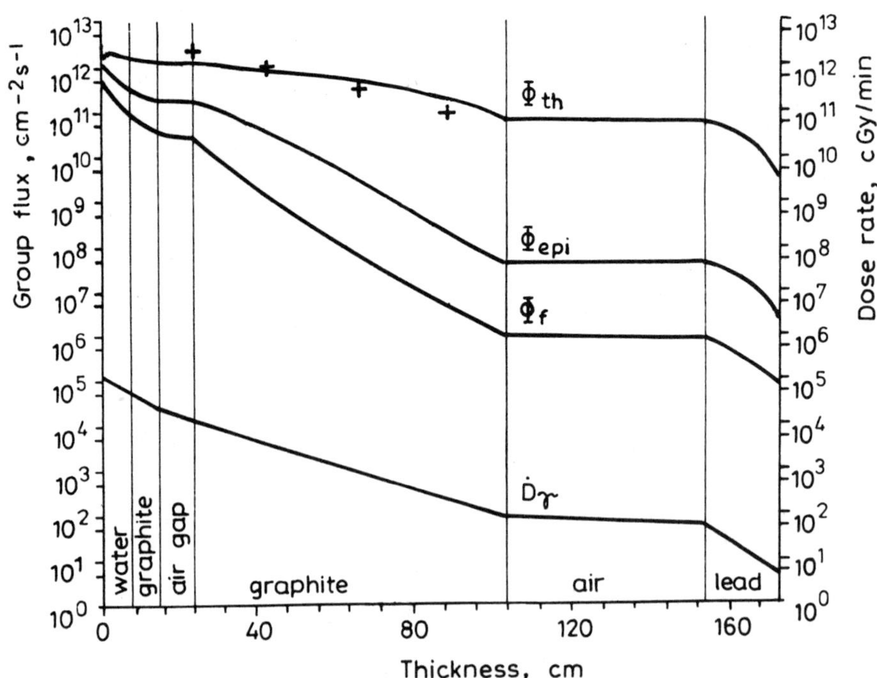

Figure 6. Neutron Flux and Gamma Dose Rate Along the Geometrical Axis of the Tunnel (d_{gr} = 80 cm, d_{lead} = 20 cm).

$$D_i = 1/(3\Sigma_{tr,i})$$
(2)

$$\Sigma_i = \Sigma_{out,i} + D_i B^2$$
(3)

where

D	is the diffusion coefficient,
Φ	is the neutron spectrum in the form of group flux values,
r	is the space coordinate,
p	is the geometry index (plane: p = 0, cylindrical: p = 1, spherical: p = 2),
S	is the source term,
Σ_{tr}	is the transport cross section,
Σ_{out}	is the cross section that accounts for all collisions that remove neutrons from the ith group either by slowing down or by absorption,
B	is the geometrical buckling parameter, and
i	is the energy group index.

Fitting of calculated results to measurements can be achieved by the proper selection of the geometrical buckling parameter, B.

The cross-section library (SABINE-3, Version B) available for the calculations did not contain nuclear data for bismuth. Therefore, that for lead was used instead of bismuth.

Some results from the SABINE-3 calculations are presented in Figure 6. Space

Table Six

Characteristics of the Irradiation Field Behind a 20-cm Lead Shield (SABINE-3).

Thickness of the Graphite (cm)	30	40	50	60
Thermal Neutron Flux, Φ_{th} [E < 0.5 eV] $(cm^{-2}s^{-1})$	8.27 E+09	6.28 E+09	5.57 E+09	3.01 E+09
Epithermal Neutron Flux, Φ_{epi} [0.5 eV < E < 500 keV] $(cm^{-2}s^{-1})$	9.90 E+08	3.45 E+08	1.11 E+08	2.93 E+06
Fast Neutron Flux, Φ_f [E>1 MeV] $(cm^{-2}s^{-1})$	2.02 E+07	6.07 E+06	1.87 E+06	7.84 E+04
Gamma Dose Rate, $\dot{D}_\gamma$ (cGy/min)	11.6	8.53	7.02	3.87
Φ_{th}/Φ_{epi}	8.35	19.8	50.2	1030
Φ_{th}/Φ_f	4.09 E+02	1.12 E+03	2.98 E+03	3.84 E+04
$\Phi_{th}/\dot{D}_\gamma$ (min cGy^{-1} cm^{-2}s^{-1})	7.13 E+08	8.00 E+08	7.93 E+08	7.78 E+08

dependent neutron and gamma characteristics of the irradiation field obtained for 80 cm of graphite and 20 cm of lead are shown. The experimental data corresponding to Table Five are also given in this figure. It can be seen that the calculations predict a more moderate neutron flux attenuation both in the graphite and in the air gap than was measured in the experiments. Therefore, the neutron flux values behind the lead shield will be smaller in practice than would be expected based on the calculated results given in Table Six. The deviation in the case of the thermal neutron flux approaches an order of magnitude.

Table Six summarizes the calculated fluxes, being the most important ones from our point of view, obtained behind the lead shield for different graphite thicknesses. From these data, it is clear that the ratio of the thermal neutron flux to the other quantities of interest is increasing as the thickness of the graphite is increased. In contrast, the absolute value of the thermal neutron flux is decreasing. As a result, for the conditions present in the irradiation tunnel (geometry, reactor power, and spectral distribution of the neutron beam) the optimum thickness of the graphite filter seems to be between 50 and 80 cm. This result is especially remarkable if one considers the fact that the expected values of the thermal neutron flux are about one order of magnitude smaller than the ones shown by Table Six.

The neutron and gamma characteristics of the irradiation field were also investigated as a function of the thickness of the lead. In these calculations, the thickness of the graphite filter was fixed at 50 cm. The results shown in Table Seven suggest that a lead thickness of 15 cm be used.

Irradiation conditions could be further improved by replacing the lead shield with one of bismuth. Measurements were performed at one of the horizontal channels of the

Table Seven

Characteristics of the Irradiation Field Behind the Lead Shield (SABINE-3).[a]

Thickness of the Lead Shield (cm)	10	15	20	25
Thermal Neutron Flux, Φ_{th} [E < 0.5 eV] $(cm^{-2}s^{-1})$	1.07 E+10	7.82 E+09	5.57 E+09	3.91 E+09
Epithermal Neutron Flux, Φ_{epi} [0.5 eV < E < 0.5 MeV] $(cm^{-2}s^{-1})$	1.70 E+08	1.36 E+08	1.11 E+08	9.18 E+07
Fast Neutron Flux, Φ_f [E>1 MeV] $(cm^{-2}s^{-1})$	5.17 E+06	3.11 E+06	1.87 E+06	1.13 E+06
Gamma Dose Rate, $\dot{D}_\gamma$ (cGy/min)	13.80	9.90	7.02	5.27
Φ_{th}/Φ_{epi}	62.9	57.5	50.2	42.6
Φ_{th}/Φ_f	2.07 E+03	2.51 E+03	2.98 E+03	3.46E+03
$\Phi_{th}/\dot{D}_\gamma$ (min cGy^{-1} cm^{-2}s^{-1})	7.75 E+08	7.90 E+08	7.93 E+08	7.42 E+08

[a] Thickness of the graphite layer is 50 cm.

reactor in order to compare the neutron and gamma attenuation properties of lead and bismuth. The experimentally determined neutron and gamma attenuation coefficients together with calculated results based on the available (incomplete) cross-section data indicate that in the case of equal thicknesses, the replacement of the lead by bismuth will result in an increase of the neutron flux values in all the energy regions of interest. The largest improvement can be expected in the value of the thermal neutron flux, which is a favorable event in our case.

CONCLUSION

A pure thermal neutron beam in the irradiation tunnel of the BTU NTI reactor can be developed by utilizing a filter composed of graphite and lead (bismuth). Initial calculations have shown that:

1) The optimum thickness of the graphite plug giving favorable neutron beam characteristics lies between 50 and 80 cm, while that for the lead is ~15 cm.

2) The thermal neutron flux behind the filter is expected to be in the vicinity of 10^9 neutrons cm^{-2}s^{-1} at a reactor power of 100 kW, while the gamma and fast neutron backgrounds remain relatively low (refer to the corresponding data of Tables Six and Seven).

3) Further improvement in the neutron beam's characteristics can be achieved by replacing the lead with bismuth.

Calculations with more sophisticated computer codes (ANISN and MERCURE-3) are needed to refine the results and to arrive at a decision on the final filter configuration. These calculations are now in progress.

REFERENCES

1. Gy. Csom, F. Lévai, and G. Keömley, "Educational and Research Activities of the Nuclear Training Reactor of the BTU," IAEA Seminar on Appl. Res. and Service Activity for Research Reactor Operations, Copenhagen, Denmark, 9-13 Sept. 1988.

2. Gy. Csom, É. M. Zsolnay, S. Fehér, and E. Virágh, "Radiation Field for BNCT Experiments," Strahlenther. Onkol., 165(2/3):78 (1989).

3. É. M. Zsolnay, "Experimental Investigation of Radiation Shielding Arrangements," Budapest, Hungary, BME-TR-121/81 (1981) (in Hungarian).

4. K. Kanda and T. Kobayashi, "Progress in Physics for Thermal Neutron Capture Therapy," in Proc. Second Int. Symp. on Neutron Capture Therapy, Tokyo, 1985, H. Hatanaka, ed., Nishimura Co., Ltd., Niigata, Japan, p. 70 (1986).

5. Y. Oka, K. Kanasugi, and N. Aoyagi, "Nuclear Design for a Pure and High Thermal Neutron Field at JRR-4," in Proc. Second Int. Symp. on Neutron Capture Therapy, Tokyo, 1985, H. Hatanaka, ed., Nishimura Co., Ltd., Niigata, Japan, p. 97 (1986).

6. Gy. Csom, É. M. Zsolnay, S. Fehér, and E. Virágh, "Preliminary Experiments for Application of Boron Neutron Capture Therapy at the NTI of the Budapest Technical University," Budapest, Hungary, BME-TR-RES-16/88 (1988) (in Hungarian).

7. Gy. Csom, É. M. Zsolnay, S. Fehér, and E. Virágh, "Preliminary BNCT Experiments at the Nuclear Reactor of the NTI of BTU," Budapest, Hungary, BME-NTI-168/87 (1987) (in Hungarian).

8. Gy. Csom, É. M. Zsolnay, E. J. Szondi et al., "Results of Radiation Shielding Experiments Corresponding to the International (CMEA) Test Sample No. 2," Budapest, Hungary, BME-TR-61/77 (1977) (available in Russian and Hungarian).

9. C. Ponti, H. Preusch, and H. Schubert, "SABINE – A One Dimensional Bulk Shielding Program," EURATOM, ISPRA, EUR 3636e (1967) and C. Ponti and R. Hensen, "SABINE-3 – An Improved Version of the Shielding Code SABINE," ESIS, ISPRA (1974).

10. W. W. Engle, Jr., "A User's Manual for ANISN; One Dimensional Discrete Ordinates Transport Code with Anisotropic Scattering," Oak Ridge National Laboratory, K-1693 (1967).

11. C. Devilleres, "Programme MERCURE-3," Fontenay-aux-Roses, CEA-R3262 (1967).

INTERMEDIATE ENERGY NEUTRON BEAMS FROM THE MURR

R. M. Brugger and W. H. Herleth

Nuclear Engineering Department and Research Reactor
University of Missouri
Columbia, MO

ABSTRACT

Several reactors in the United States are potential candidates to deliver beams of intermediate energy neutrons for NCT. At this time, moderators, as compared to filters, appear to be the more effective means of tailoring the flux of these reactors. The objective is to sufficiently reduce the flux of fast neutrons while producing enough intermediate energy neutrons for treatments. At the University of Missouri Research Reactor (MURR), the code MCNP has recently been used to calculate doses in a phantom. First, "ideal" beams of 1, 35, and 1000 eV neutrons were analyzed to determine doses and advantage depths in the phantom. Second, a high quality beam that had been designed to fit in the thermal column of the MURR, was reanalyzed. MCNP calculations of the dose in phantom in this beam confirmed previous calculations and showed that this beam would be a nearly ideal one with neutrons of the desired energy and also a high neutron current. However, installation of this beam will require a significant modification of the thermal column of the MURR. Therefore, a second beam that is less difficult to build and install, but of lower neutron current, has been designed to fit in MURR port F. This beam is designed using inexpensive Al, S, and Pb. The doses calculated in the phantom placed in this beam show that it will be satisfactory for sample tests, animal tests, and possible initial patient trials. Producing this beam will require only modest modifications of the existing tube.

INTRODUCTION

Neutron Capture Therapy (NCT) requires beams of intermediate energy neutrons to treat cancers that are several centimeters deep within the body. These neutrons will penetrate to the cancer before moderating into thermal neutrons that will then ignite the reactions that will destroy the cancer cells. The label "intermediate neutrons" as used here means neutrons that have energies from ~1 eV to ~10 keV. Either reactors or accelerators could be sources of such neutrons, but in this paper only neutrons from a reactor will be considered [1-4].

The primary neutrons from the fission process in a reactor have an average energy of 1.5 MeV. These neutrons are usually moderated to thermal energies to sustain the fission reaction and in some reactors to make them more useful for experiments. Many reactors have beams of thermal neutrons. In contrast, beams of intermediate neutrons have seen limited development and use. In the United States, there remain only a few research reactors that have both high enough neutron fluxes and the necessary flexibility to be candidates for producing beams of intermediate neutrons with sufficient intensity so that NCT can be expedient. Table One lists those reactors that might be capable of producing a beam of intermediate neutrons intense enough to be used for NCT [5].

Neutron Beam Design, Development, and Performance for Neutron Capture Therapy
Edited by O. K. Harling *et al.*
Plenum Press, New York, 1990

Table One.   Reactors in United States Capable of Being Used for NCT.

NAME†	DESIGNATION	LOCATION	TYPE	POWER (kW)	STARTUP
High Flux Isotope Reactor (DOE)	HFIR	Oak Ridge, TN	Tank flux trap	85000	1965
Brookhaven High Flux Beam Research Reactor (DOE)	HFBR	Upton, NY	Heavy water	60000	1965
National Bureau of Standards Reactor	NBSR	Gaithersburg, MD	Heavy water	20000	1967
Power-Burst Facility (DOE)	PBF	INEL, ID	Open tank	*20000	1973
Missouri, University of	MURR	Columbia, MO	Tank	10000	1966
Omega West Reactor (DOE)	OWR	Los Alamos, NM	Tank	8000	1956
Brookhaven Medical Research Reactor (DOE)	BMRR	Upton, NY	Tank	5000	1959
Georgia Institute of Technology	GTRR	Atlanta, GA	Heavy water	5000	1964
Massachusetts Institute of Technology	MITR	Cambridge, MA	Heavy-water reflected	5000	1958
Union Carbide Corporation Reactor	UCNR	Sterling Forest, NY	Pool	5000	1961
Bulk Shielding Reactor	BSR	Oak Ridge, TN	Pool	2000	1950
Michigan, University of (Ford Nuclear Reactor)	FNR	Ann Arbor, MI	Pool	2000	1957
Rhode Island Nuclear Science Center	RINSC	Fort Kearney, RI	Pool	2000	1964
State University of New York (Western NY Nuclear Res. Ctr., Inc.)	SUNY	Buffalo, NY	Pool	2000	1961
Virginia, University of	UVAR	Charlottesville, VA	Pool	2000	1960
GA Technologies, Inc., Adv. TRIGA-Mk F Prototype Reactor	TRIGA-Mk F	La Jolla, CA	U-Zr hydride	1500	1960
Illinois, University of	TRIGA-Mk II	Urbana-Champaign, IL	U-Zr hydride	1500	1960
Lowell, University of		Lowell, MA	Pool	1000	1974
Neutron Radiography Facility	TRIGA-Mk I	FMEF-Richland, WA	U-Zr hydride	1000	1984
North Carolina State University	PULSTAR	Raleigh, NC	Pool	1000	1972
Northrop Corporate Laboratories (Space Radiation Lab.)	TRIGA-Mk F	Hawthorne, CA	U-Zr hydride	1000	1963
Oregon State University	TRIGA-Mk II	Corvallis, OR	U-Zr hydride	1000	1967
Penn State TRIGA Reactor (Pennsylvania State University)	PSTR	University Park, PA	Pool-TRIGA Core	1000	1965
Texas A&M University	TRIGA	College Station, TX	U-Zr hydride	1000	1961
US Geological Survey Laboratory (Dept. of the Interior)	TRIGA-Mk I	Denver, CO	U-Zr hydride	1000	1969
Washington State University	WSTR	Pullman, WA	Pool-TRIGA Core	1000	1967
Wisconsin, University of	TRIGA	Madison, WI	Pool-TRIGA Core	1000	1967
Cornell University	TRIGA-Mk II	Ithaca, NY	U-Zr hydride	500	1962
Arizona, University of	TRIGA-Mk I	Tucson, AZ	U-Zr hydride	250	1958
California, Irvine, University of	TRIGA-Mk I	Irvine, CA	U-Zr hydride	250	1969
GA Technologies, Inc., TRIGA-Mk I Prototype Reactor	TRIGA-Mk I	La Jolla, CA	U-Zr hydride	250	1958
Kansas State University	TRIGA-Mk II	Manhattan, KS	U-Zr hydride	250	1962
Neutron Radiograph Facility (DOE)	NRAD	INEL, ID	Pool-TRIGA Core	250	1977
Neutron Radiograph Facility (DOE)	TRIGA-Mk I	Richland, WA	U-Zr hydride	250	1977
Texas at Austin, University of	TRIGA-Mk I	Austin, TX	U-Zr hydride	250	1963
Utah, University of	TRIGA-Mk I	Salt Lake City, UT	U-Zr hydride	250	1975
Missouri at Rolla, University of	UMRR	Rolla, MO	Pool	200	1961
Florida, University of	UFTR	Gainesville, FL	Graphite/water	100	1959
Washington, University of	Educator	Seattle, WA	Graphite/water	100	1961
Ohio State University		Columbus, OH	Pool	**10	1961

* proposed modification
** proposed upgrade to 500 kW
† Nuclear Reactors Built, Being Built, or Planned in the US as of 3/85

Producing a beam of intermediate neutrons from the fission and moderated neutrons in a reactor requires either filters or additional moderators. Filters stop the fast and thermal neutrons and the gamma rays while allowing some intermediate neutrons to pass through. In contrast, moderators slow down the fast neutrons to the intermediate range faster than they remove the intermediate neutrons to the thermal range. Thus, moderators yield a flux of intermediate neutrons as the source for an epithermal beam. Both filters and moderators have been tried with some success to produce beams for NCT.

At the University of Missouri Research Reactor (MURR), we have been designing and testing different filters and moderators to produce beams of intermediate neutrons [6]. Several encouraging designs have been realized with filters and our attention has now been directed toward moderators. The challenge is to reduce the number of fast neutrons and gamma rays as compared to the number of intermediate neutrons so that the dose to a patient from the fast neutrons and the gamma rays will be small compared to the dose from the capture of neutrons in boron-10 in the cancers. In addition, enough intermediate neutrons must remain in the beam so that a sufficient number of B-10 reactions will occur to give a lethal dose to the cancer in a reasonable time. In this paper, we describe recent evaluations of several beams that can be built at the MURR. The code MCNP has been used in the design process and the analyzed beams have been compared to "ideal" beams of 1, 35, and 1000 eV energy neutrons that would give the optimum advantage depths and advantage ratios.

MCNP

The Monte Carlo code MCNP has been used to guide the design of the beams considered in this paper. The beam for the thermal column was analyzed using coupled neutron-photon calculations to follow neutrons from the core to the phantom and to study the flux profiles and radiation doses produced in the area of interest. 'Photon-only' calculations were also used to trace core gamma rays. In the case of the design of the beam for MURR port F, the neutron energy spectrum (41 groups) was determined at the edge of the beryllium reflector, and this was used as the source for the beam design. Because of geometrical considerations, this beam is monodirectional.

MCNP was obtained from the Radiation Shielding Information Center (RSIC) at ORNL and was run on a Micro-VAX II at the MURR. Every effort was made to use the most up-to-date cross-section data available, including using $S(\alpha, \beta)$ treatment on material when possible. A typical computer run would last two days and follow 50,000 to 100,000 particles in order to achieve a stable figure of merit and low relative errors. The variance reduction techniques employed were very conservative in order to minimize the possibility of introducing errors in the calculations.

PHANTOM

A phantom was used for each beam calculation to determine the effectiveness of each beam. For the beam designed for MURR port F and for the "ideal" beams, the phantom was a right cylinder, 11.3 cm in radius and 10 cm long. This phantom was divided into several 1 cm thick disks followed by several 2 cm disks. The fluxes of neutrons, gamma rays, and doses were calculated in each disk. The phantom was coaxial with the beam. The phantom was composed of water with 0 μg, 3 μg, or 30 μg of B-10 per gram of water. For the beam designed for the thermal column, the phantom was a right parallelepiped with a face 20 cm x 20 cm toward the beam and 10 cm long along the beam. Figure 1 shows these two phantoms.

IDEAL BEAMS

The MCNP code was run with the phantom only for beams of 0.025 eV, 1 eV, 35 eV, 1 keV, and 35 keV to obtain a base line of the best possible beam conditions. The

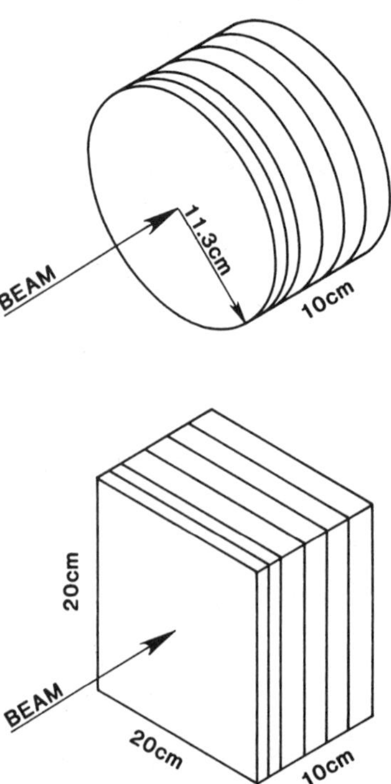

Figure 1.   Diagrams of the two phantoms that were used in the MCNP calculations. The upper phantom was used with the "ideal beams" and the MURR port F beam. The lower phantom was used with the thermal column beam.

beams were centered on the face of the phantom with all the neutrons directed along the axis. The beams contained only neutrons with no gamma rays. The doses in the cells of the phantom were calculated for water only, 3 µg B-10/g tissue, and 30 µg B-10/g tissue.

Figure 2 shows the results of some of these calculations for the phantom with water only; that is, no boron. For the 1 eV, 35 eV, and 1 keV beams of neutrons, the major part of the dose comes from induced gamma rays generated by capture of neutrons in the phantom. Because it is necessary to have neutrons to initiate the B-10 reaction, this induced gamma dose will be present even under the best conditions. These induced gamma rays therefore set the lower limit of background dose that can be achieved. The neutron dose of Figure 2 comes from the exchange of kinetic energy from the neutrons to particles in the phantom and increases as the energy of the neutrons increases. For the 35 keV beam, a significant dose is added to the induced gamma dose from proton recoils, particularly at the front surface of the phantom.

The composition of the phantom used for the calculations was water only (11.11 wt% H and 88.88 wt% O, density = 1.000 g/cm^3) which was assumed to be a good approximation of tissue and skull. Harling (MIT) has suggested that a more realistic material might be 10.57 wt% H, 13.97 wt% C, 1.84 wt% N, 72.59 wt% O, 0.14 wt% Na, 0.39 wt% P, 0.14 wt% Cl, 0.39 wt% K, density = 1.047 g/cm^3. To compare the doses from these two compositions, the beams of 1 and 35 eV were rerun with this brain composition instead of water. A comparison of the induced doses produced by the water phantom to the brain phantom shows that the induced dose with this brain material is no more than 10% higher than that from the water. Thus, the water phantom is a good approximation for these designs.

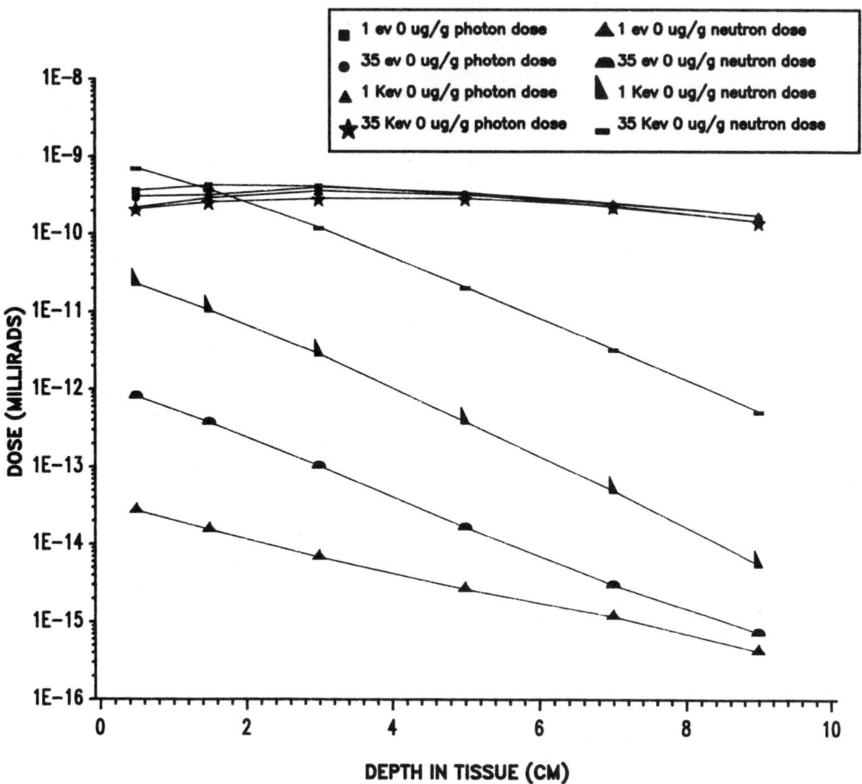

Figure 2.   The induced gamma doses and the neutron doses from the ideal beams. Each curve is for a beam of one neutron.

Continuing the discussion of the beam calculations with the water phantom, Figures 3-6 show the total doses calculated in the phantom for the three different B-10 concentrations for the four beams. Several trends are apparent. First, the penetration depth increases somewhat as the energy of the neutrons increases. Second, the dose from the B-10 captures decreases slightly as the energy increases. This indicates that the neutrons penetrate further as the energy increases but that there are fewer of them at any location. Third, the 1 eV, 35 eV, and the 1 keV beams have a relatively low contribution from the proton recoils but, for the 35 keV neutrons, the surface dose from the neutrons is significant. The 35 keV neutrons are too energetic to be "ideal" for therapy. The upper limit of energy of an effective intermediate energy neutron beam is below 35 keV. This suggests that filters or moderators which use Al are a compromise to some extent because of the window in the cross section of Al at 30 keV. One additional beam was run with the above described geometry; a beam of 0.025 eV neutrons. Figure 7 shows results of this calculation. The maximum dose is at the front surface; there is no peaking of the dose as the beam travels through the phantom. These neutrons do not penetrate to treat deep cancers.

Table Two lists the advantage parameters for these beams. The maximum advantage depth that can be achieved before a limiting dose is induced in the phantom material is about 8 cm. These depths are reached with the 35 eV and the 1 keV beams. The 1 eV beam does not penetrate quite as far and the 35 keV beam is limited by the neutron dose. The best ratio of dose with 30 μg of boron to the dose with water only in the phantom is about 4 achieved with the 1 eV, 35 eV, and the 1 keV beams while the ratio for the 0.025 eV and the 35 keV beams is much smaller, 1.5. The 1 eV, 35 eV, and 1 keV beams are "ideal" and establish a design target. (Note: Advantage dose is in units of 10^{-10} mrad per incident neutron.)

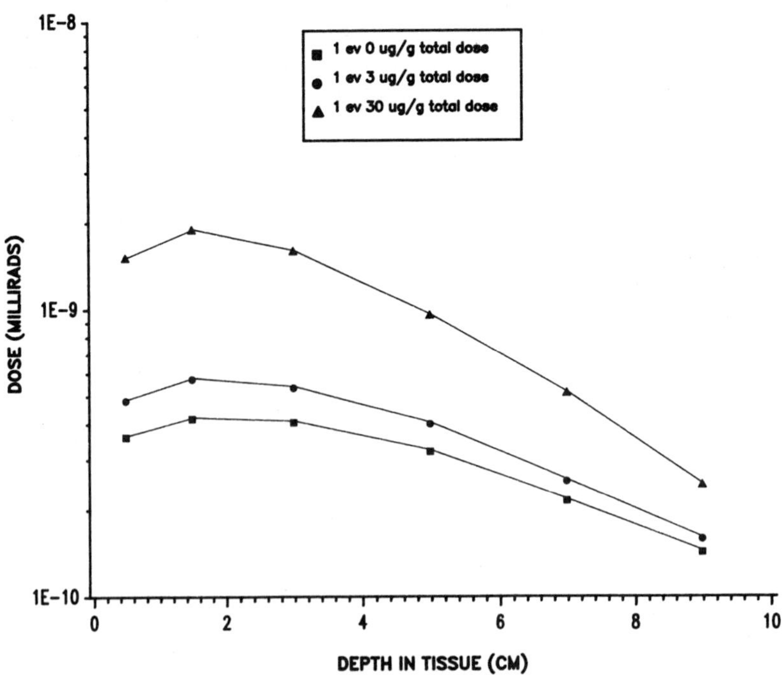

Figure 3. The total doses in the phantom for the three loadings of B-10 for the 1 eV-beam.

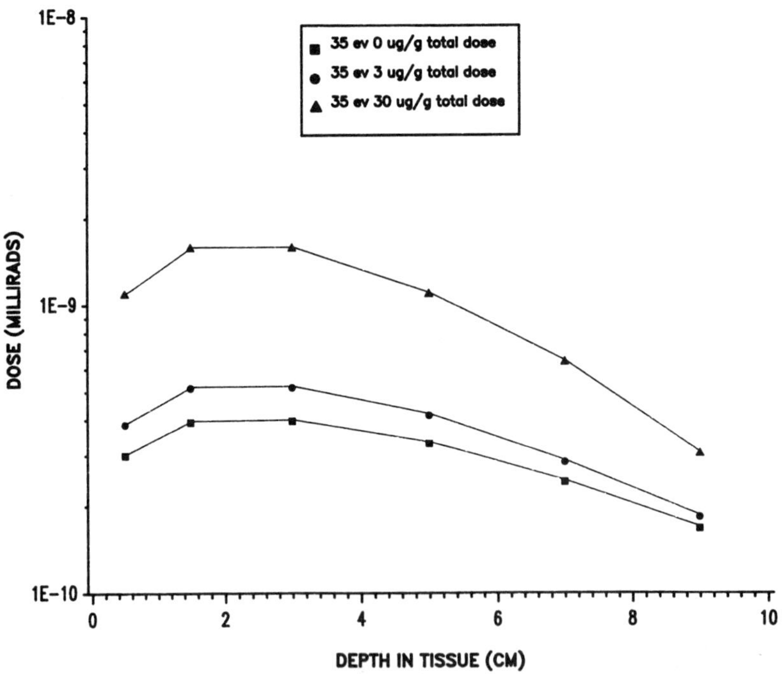

Figure 4. The total doses in the phantom for the three loadings of B-10 for the 35 eV-beam.

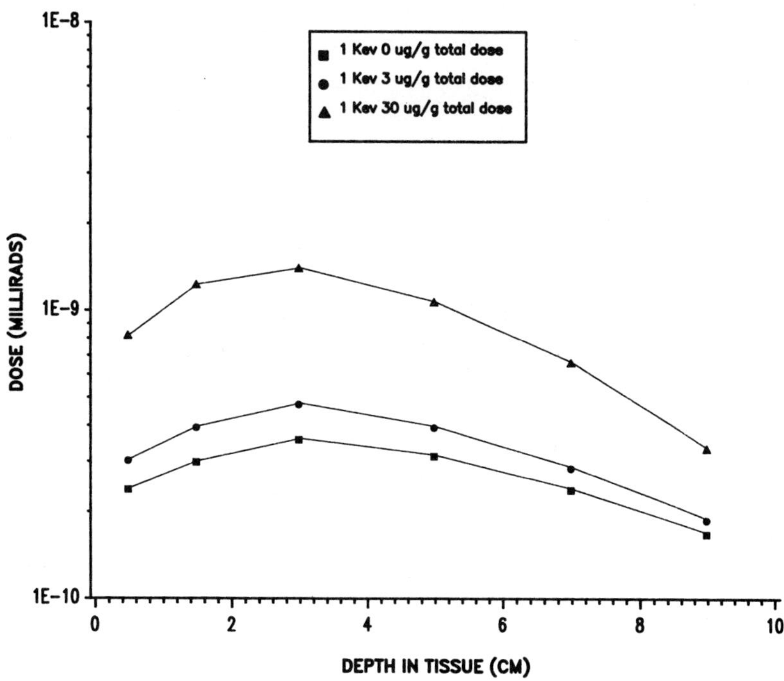

Figure 5.   The total doses in the phantom for the three loadings of B-10 for the 1 keV-beam.

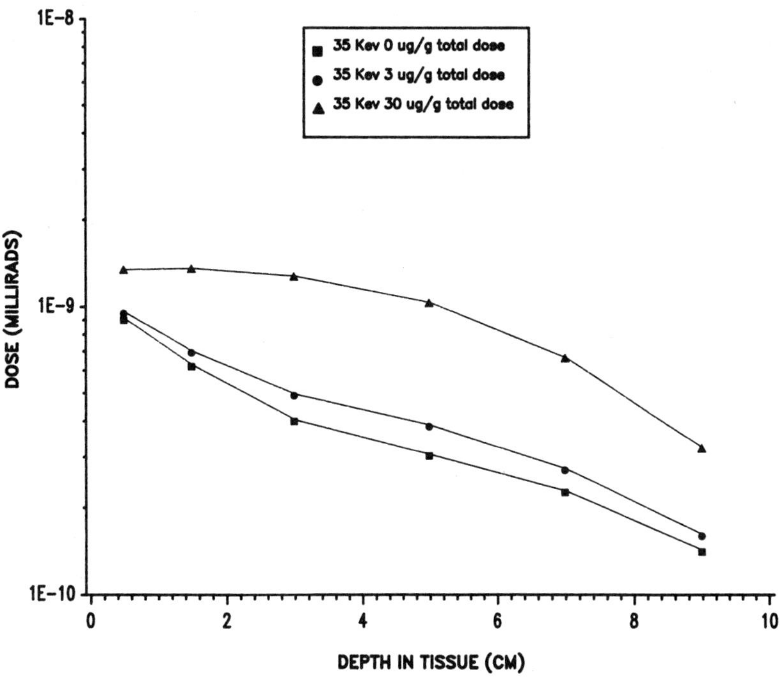

Figure 6.   The total doses in the phantom for the three loadings of B-10 for the 35 keV-beam.

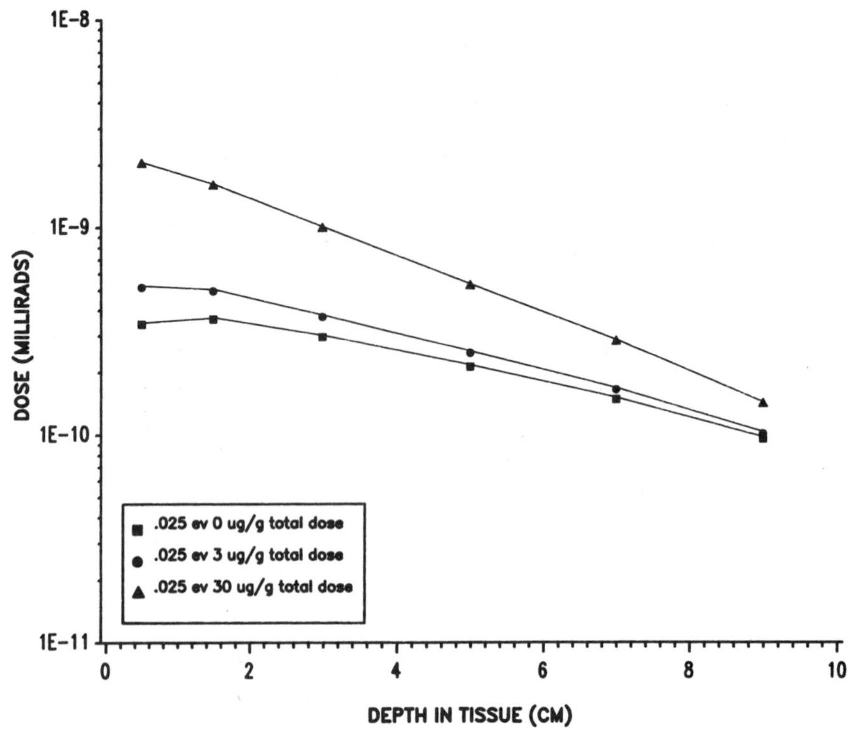

Figure 7.   The total doses in the phantom for the three loadings of
B-10 for the 0.025-eV beam.

THERMAL COLUMN BEAM

Even with a high power research reactor such as the MURR, a large source area is needed in order to have a beam that is of high intensity. The only beam position in the MURR that has a large source area is in the thermal column. Several years ago, an NCT beam was designed for the thermal column. In this design, the flux of neutrons leaking from the core of the MURR is tailored with a moderator of Al_2O_3 and a gamma ray shield of bismuth. The design was based on calculations using the diffusion code DISNEL for neutrons and the code QAD for gamma rays. The results were promising [7].

Table Two

Advantage Parameters of "Ideal" and Other Beams.

BEAM	0.025 eV	1 eV	35 eV	1 keV	35 keV
Advantage Depth (Max/Min) No RBE (cm)	6.4/5.2	7.6/6.6	8.4/7.7	8.7/7.9	5.6/5.4
Advantage Dose Rate No RBE (relative)	4.0	4.2	4.2	3.2	9

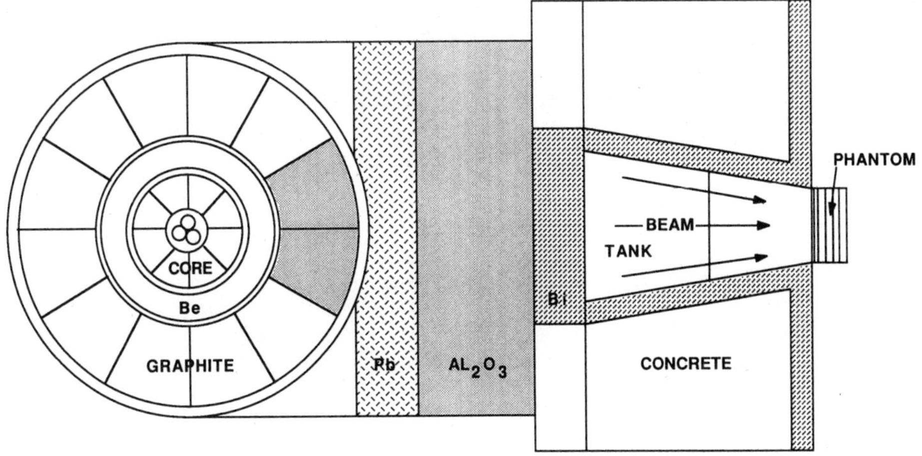

Figure 8. Diagram of the NCT beam that could be assembled in the thermal column of the MURR. The moderator consists of 22.3 cm of Al_2O_3, 13 cm of Pb, 32 cm of Al_2O_3, and 15 cm of Bi. (<u>Note</u>: Distance from center of core to face of phantom is 174 cm.)

Recently, this design was reanalyzed using the MCNP code. Figure 8 shows the cross section of this beam. Two of the graphite wedges near the core of the reactor are replaced with wedges of Al_2O_3. The permanent lead that was placed in the pool to shield the thermal column from the core is left in place. More Al_2O_3 moderator follows this lead. The beam cavity is lined with bismuth to reduce the flux of induced gamma rays that could reach the phantom. The bismuth, and the Al_2O_3 that were located away from the core had 0.1% Li-6 added to reduce the thermal neutron flux and the induced gamma rays. A tank which can be filled with water to act as a shutter is shown but is empty for the calculations. The phantom is placed at the exit of the beam as shown.

Figures 9 and 10 show the results of the calculations. The calculations, which start with one neutron, are normalized to a MURR operating power of 10 MW by matching the magnitudes of the neutron fluxes at the outer surface of the beryllium reflector. The flux for neutrons with energies from 1 eV to 30 keV is known from both calculations and measurements to be 8×10^{13} at the midplane of the core. The core gamma ray flux was normalized in the same way.

The effect of gamma rays generated in the core of the reactor was evaluated to determine how much these contribute to the dose in the phantom. Because the lead shield is left in place, the number of core gamma rays getting past this shield is insignificant compared to the induced gamma rays produced by neutrons near the phantom. The core gamma rays are not a significant contribution in this design.

Table Three lists the advantage parameters for the NCT beam in the thermal column. The advantage depth is 6.4 cm and the ratio of the dose with 30 μg B-10 vs. water only is 3. These are approaching the 8 cm advantage depth and the ratio of 4 of the ideal beams. The peak dose rate with 30 μg B-10 and the advantage dose rate are 666 and 210 rad/min respectively. Thus, a therapeutic dose could be delivered to the phantom or a patient in 10 to 30 minutes.

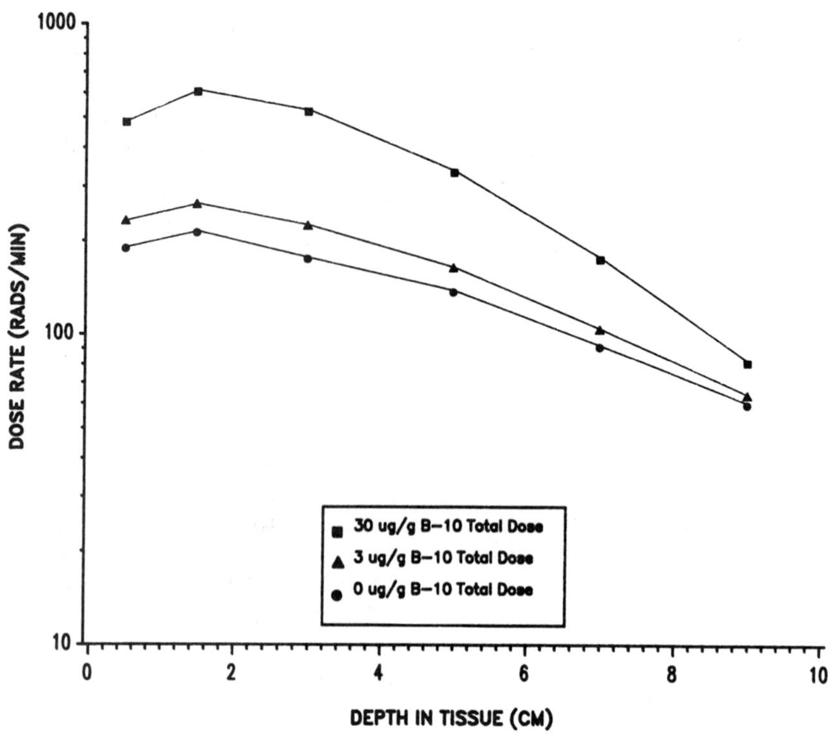

Figure 9.   The total doses from the thermal column beam.

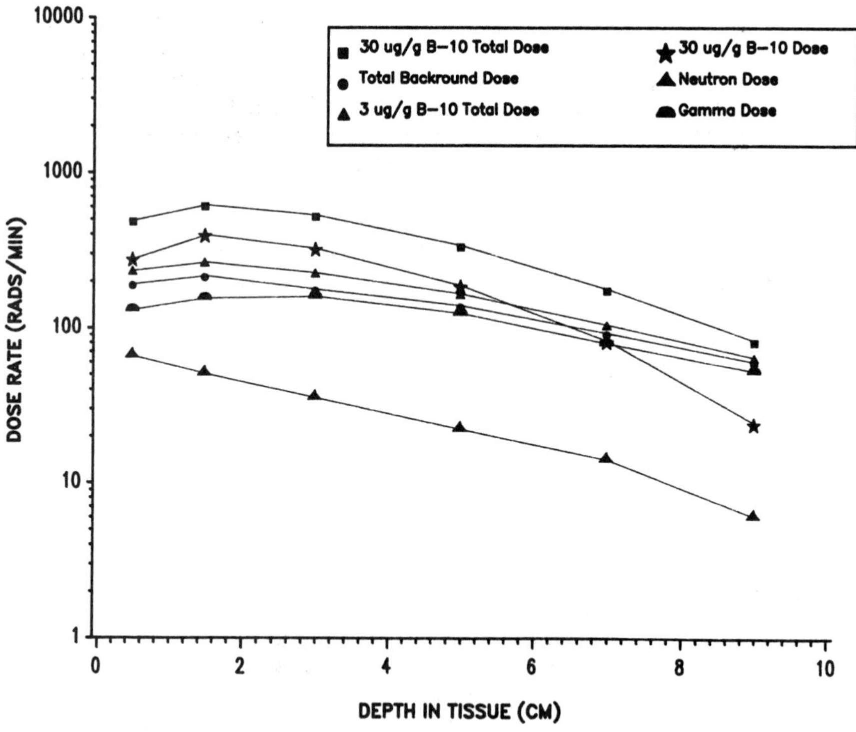

Figure 10.   The contributing parts of the total dose for the thermal column beam.

Table Three

Advantage Parameters of Beams.

BEAM	THERMAL COLUMN	PORT F
Advantage Depth (Max/Min) No RBE (cm)	6.4/5.8	6.0/5.6
Advantage Dose Rate No RBE (rad/min)	210	0.46/0.44

MURR PORT F BEAM

Because installation of an NCT beam in the thermal column will require significant changes in the structure which will in turn require money, manpower, and interruption of other research, the other beam ports of the MURR were studied to determine if a filter or moderator could be placed in one of these and produce a usable beam for continuing studies and tests. MURR port F was selected for this design because it could be made available without interrupting other research. However, this port is not ideal. Although it does have a strong source flux, the source area is small, only 10 cm in diameter, and the tube is long, 3.24 m from the source to the outer edge of the biological shielding.

Several designs were considered in which beam intensity was traded against beam purity, cost, and feasibility. The design that was selected is described here. This design is based on the moderator concept. However, in this case, the moderator is not close to the source of neutrons, but is placed close to the phantom or patient location. Figure 11 is a cross section of the beam port with the moderator inserted near the outer end. Aluminum and sulfur (3/4 Al + 1/4 S) were the materials chosen because they are effective in removing fast neutrons while passing intermediate ones. Aluminum and sulfur are relatively inexpensive and can be inserted in the port with little difficulty. The aluminum and sulfur are followed by lead to reduce the gamma ray flux from core and induced gamma rays. A chamber which can be flooded with water and which can be used as a beam shutter will be installed in front of the moderator.

Figures 12 and 13 show the dose rates calculated for this beam. In Figure 12 it is seen that the neutron dose behaves more like that from the 35 keV beam than from the lower energy neutron beams. This moderator has not removed as many of the high energy neutrons as would be desirable. Still the advantage depth is 6 cm and the ratio of doses is about 2. The dose rate at 2 cm from the front face of the phantom is 0.9 rad/min, which is high enough for sample and animal tests.

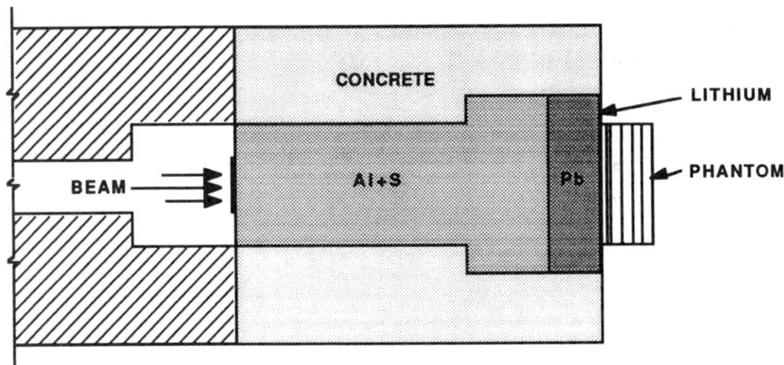

Figure 11.   Diagram of the NCT beam that could be inserted into MURR port F. The moderator consists of 60 cm of Al-S and 10 cm of Pb.

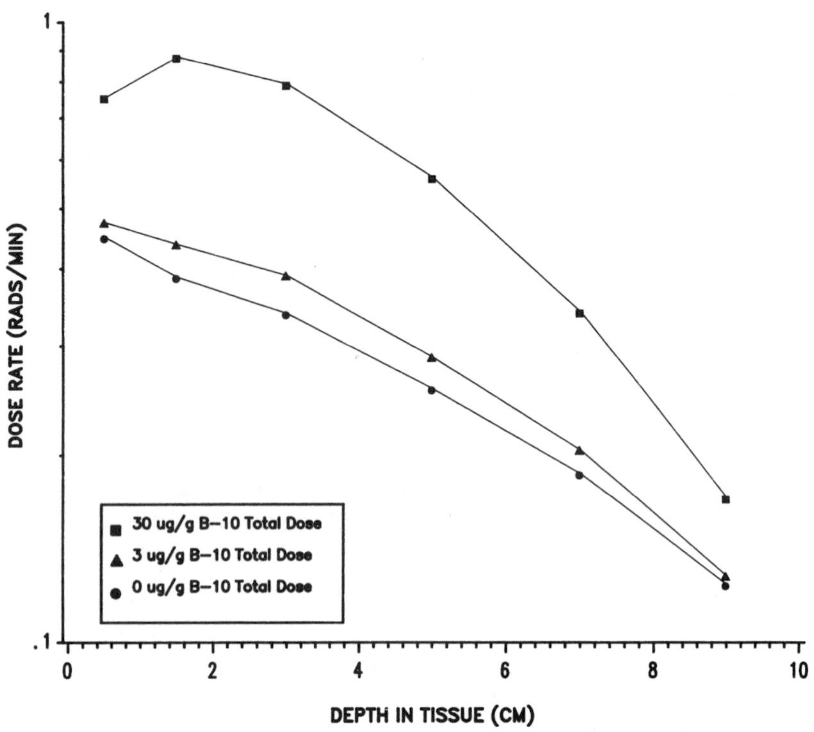

Figure 12.   The total doses from the MURR port F beam.

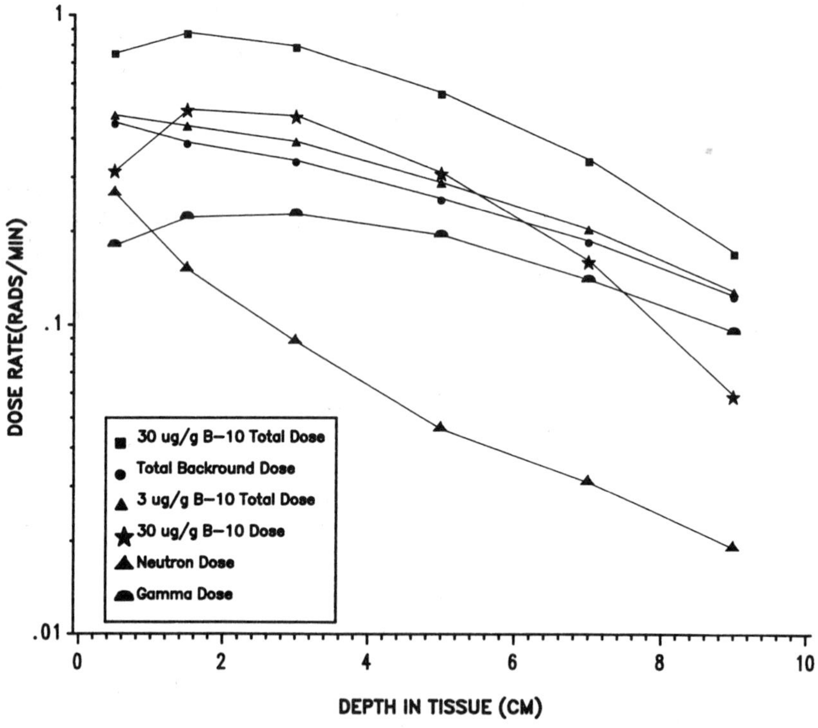

Figure 13.   The contributing parts to the total dose from the MURR port F beam.

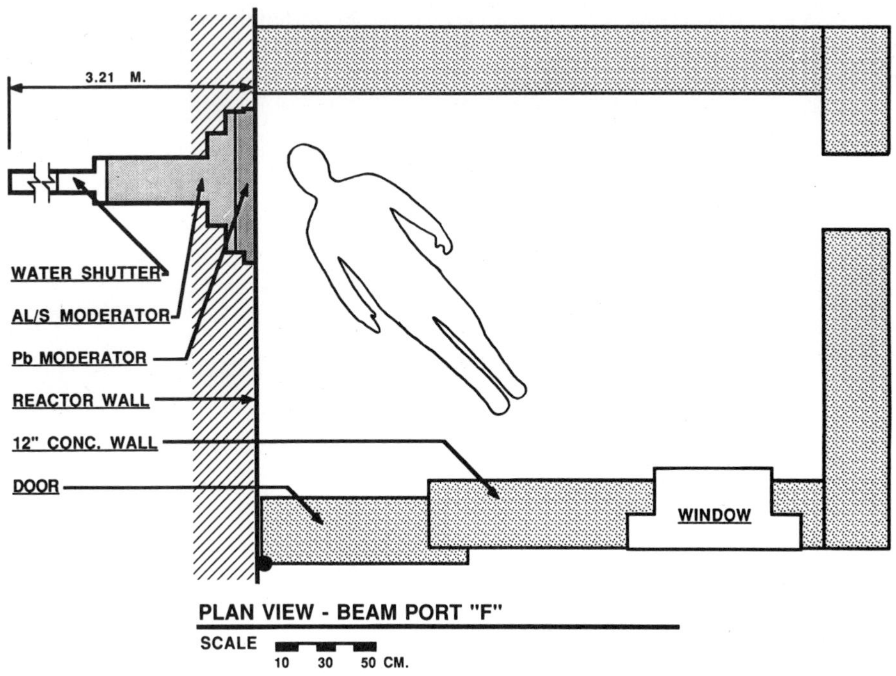

Figure 14.   Diagram of the therapy irradiation room that could be assembled at the MURR port F. The walls are already in place.

A separate computer simulation was performed to evaluate the contribution of the core gamma rays. It was found that, for this design, the core gamma rays would add about 20% to the gamma contribution. The total intensity of this beam design could be increased by a factor of 2.5 by building the moderator in ports B or E of the MURR instead of port F. This increase of a factor of 2.5 is realized at these two ports because these ports are 15 cm in diameter instead of 10 cm and because, for these ports, the distance from the source to the outer edge of the biological shield is less.

Modest improvements in this beam may be realized in the future by adjustments of the amounts of aluminum, sulfur, and lead. A savings can be realized by putting the beam at MURR port F because part of the shielding for a treatment room is already in place at this port. Figure 14 shows the layout of the treatment room that could be realized.

CONCLUSION

The work presented in this paper uses the neutron/photon code, MCNP, to evaluate the design of two different beams at the MURR. The neutron currents and doses delivered to a phantom from these beams are compared to several "ideal" neutron beams of intermediate energy. The first beam, the one designed for the thermal column, is nearly ideal, delivering a beam of intermediate energy neutrons with which NCT could be completed within a few minutes. The background doses from other sources of radiation are relatively small. The second beam, the one designed for MURR port F, will be easier to build, but will be less intense and will have poorer background ratios. However, this beam will be both intense enough and sufficiently low in background so that it will be effective for sample and animal tests and preliminary patient trials.

ACKNOWLEDGMENTS

The authors gratefully acknowledge important assistance from several sources. Dr.

Gene Moum of MURR was most helpful in assisting in getting the MCNP running on the Micro-VAX. Mr. Steve Clement of MIT generously shared his advice and experience in how to run MCNP. Dr. John Russell of Theragenics gave helpful council on the design of the beams. This research was partially supported through a grant from the National Science Foundation.

REFERENCES

1. Proc. First Int. Symp. on Neutron Capture Therapy, Cambridge, MA, 1983, R. G. Fairchild and G. L. Brownell, eds., Brookhaven National Laboratory, BNL-51730 (1984).

2. Proc. Second Int. Symp. on Neutron Capture Therapy, Tokyo, 1985, H. Hatanaka, ed., Nishimura Co., Ltd., Niigata, Japan (1986).

3. Proc. U.S. Dept. of Energy 1986 Workshop on Neutron Capture Therapy, R. G. Fairchild and V. P. Bond, eds., Brookhaven National Laboratory, BNL-51994 (1987).

4. Proc. Third Int. Symp. on Neutron Capture Therapy, Strahlenther. Onkol., D. Gabel, ed., 165(2/3):5-257 (1989).

5. "Nuclear Reactors Built, Being Built, or Planned in the United States," National Technical Information Service, U.S. Department of Commerce, PB82-903002.

6. T. J. Less and R. M. Brugger, "Reactor Moderated Intermediate Energy Neutron Beams for Neutron Capture Therapy," Strahlenther. Onkol., 165(2/3):87 (1989).

7. T. J. Less, "Reactor Moderated Intermediate Energy Neutron Beams for Neutron Capture Therapy," Ph.D. Dissertation, Nuclear Engineering Department, University of Missouri-Columbia, Columbia, MO (1987).

REACTOR-BASED NEUTRON BEAMS

PROGRESS TOWARDS BORON NEUTRON CAPTURE THERAPY AT THE HIGH FLUX REACTOR PETTEN

R. L. Moss

Commission of the European Communities
Joint Research Centre
Institute of Advanced Materials
Petten Establishment
The Netherlands

ABSTRACT

During 1988 the first positive steps were taken to proceed with the design and construction of a neutron capture therapy facility on the High Flux Reactor (HFR) at Petten. The immediate aim is to realise within a short time (summer 1989), an epithermal neutron beam for radiobiological and filter optimisation studies on one of the 10 small aperture horizontal beam tubes. The following summer, a much larger neutron beam, i.e., in cross section and neutron fluence rate, will be constructed on one of the two large beam tubes that replaced the old thermal column in 1984. This latter beam tube faces one whole side of the reactor vessel, extending from a 50 x 40 cm input aperture to a 35 x 35 cm exit hole. The radiotherapeutic facility will be housed here, with the intention to start clinical trials at the beginning of 1991.

This paper describes the present status of the project and includes: a general description of the pertinent characteristics with respect to NCT of the HFR; results of the recently completed preliminary neutron metrology and computer modeling at the input end of the candidate beam tube; the structure and planning of the proposed Work Programme; and the respective direct and indirect participation and collaboration with the Netherlands Cancer Institute and the European Collaboration Group on BNCT.

INTRODUCTION

The development of a neutron capture therapy facility on one of the beam tubes at the High Flux Reactor at Petten is currently underway. Although the projected neutron beam will be suited for NCT in general, the present project is aiming directly at the implementation of the ^{10}B(n,alpha)^{7}Li reaction. It is intended that the first patient treatments could begin as early as Spring 1991. Prior to this, it is also planned to install a relatively smaller epithermal neutron beam facility on one of the other available beam tubes around the reactor. The research and therapy activities on both facilities will be coordinated by the European Collaboration Group on BNCT.

In the paper presented here, a brief description of the HFR, along with its unique organisational structure within Europe, will give an indication of why the HFR is considered to be the most suitable materials testing reactor in Europe for the development of Europe's first NCT facility. The results of the initial neutron metrology work and computer modeling to confirm the nuclear characteristics of the candidate beam tube are given.

To realise such a facility however, it is recognised, as elsewhere [1], that a multi-disciplinary team of experts is required, i.e., clinicians, radiotherapists, radiobiologists, chemists, and clinical and nuclear physicists. At Petten, and indeed throughout Europe, a research or medical centre providing all these disciplines does not exist. Hence, collaboration on a multi-national scale has been imperative. In this respect the BNCT group at Petten sought collaboration on two fronts. Firstly, a local or Dutch NCT Group was formed and presently consists of members from the HFR Division of JRC Petten, the Physics Department at the Netherlands Energy Research Foundation (ECN) at Petten, and the Radiotherapy Department at the Netherlands Cancer Institute in Amsterdam. This group will be directly involved with the construction and installation, the initial animal and patient irradiations, and the addressing of local issues. The second group is the European Collaboration Group (ECG) on BNCT, lead by Prof. Detlef Gabel from Bremen University. This group will oversee and coordinate the use of the facility. The format and planned tasks of the ECG are presented in more detail in a later section.

Finally, some concluding remarks are given on the intended facility. It is realised that with respect to this conference, reporting of detailed design work is somewhat premature. It is nevertheless important that the NCT community at large be made aware of the activities at Petten and within Europe.

THE HIGH FLUX REACTOR (HFR) PETTEN

Since 1962, following an agreement between the Dutch Government and the Commission of the European Communities (CEC), the HFR, while being on Netherlands soil, is owned and financed by the CEC . By special contract, the reactor is operated by ECN (Dutch employees), but the bulk of the experimental work at the pool- and core-irradiation facilities, is carried out by CEC employees at the Joint Research Centre (JRC) at Petten. Hence, all reactor and experimental development and upgrading is authorised and financed by the JRC. In this respect, for example, the reactor vessel was replaced in 1984 and last year, the primary heat exchangers were renewed. Such upgrading actions are an astute policy that is unable to be followed elsewhere in Europe. The HFR thus offers a reliable and guaranteed operation for the future.

<u>General Characteristics</u>

The HFR is a 45 MW(t) materials testing reactor (ORR type), cooled and moderated by light water [2]. The core of the reactor is an 8 x 9 rectangular array (see Figure 1), consisting of 33 fuel assemblies of the MTR type (93% ^{235}U enriched), 6 control rods, 16 beryllium reflector elements along 3 sides of the core, and 17 free positions into which experimental facilities may be placed. In addition, the reactor has the so-called poolside facility where up to 10 experiments can be placed to simulate power transients by movement of the experiment towards the reactor vessel, while maintaining the reactor power constant.

<u>Characteristics with Respect to NCT</u>

More pertinently with respect to BNCT, the reactor is equipped with 12 horizontal beam tubes located around three sides of the reactor vessel (see Figure 1). The majority of activities on the beam tubes are solid state and nuclear physics experiments carried out and financed by ECN. In addition, a dry thermal neutron radiography facility is available. Ten of the beam tubes are cylindrical with an average diameter of 18.5 cm. The output fluence rates (thermal) vary, depending on the individual beam configuration, from 10^6 to 3 x 10^7 neutrons cm^{-2}s^{-1}.

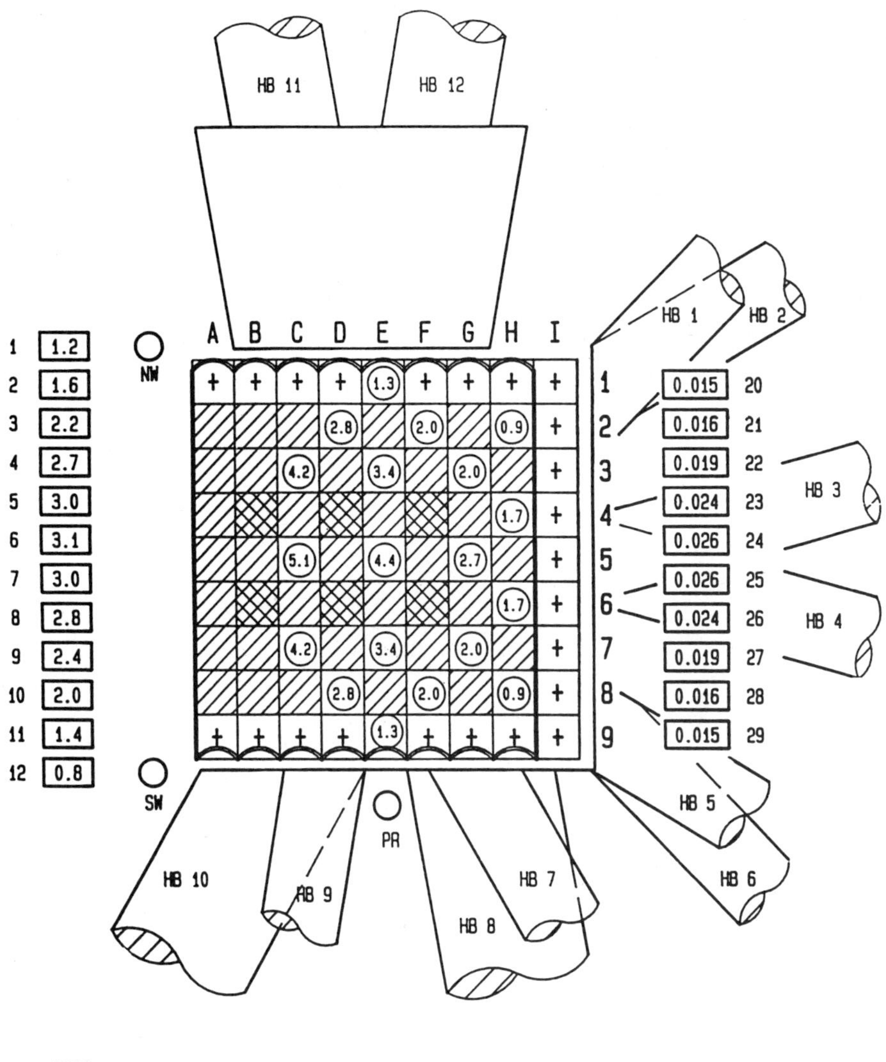

Figure 1.   Cross-Section Through Standard HFR Core Configuration.

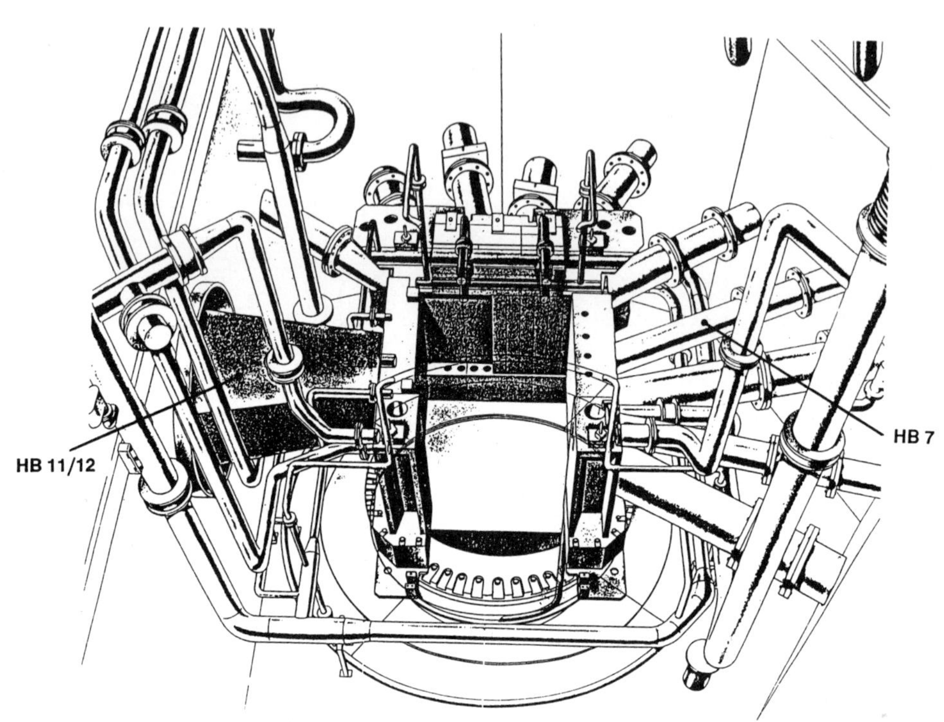

Figure 2.   In-Pool Schematic of Core-Box, Showing Configuration and Distribution of Beam Tubes.

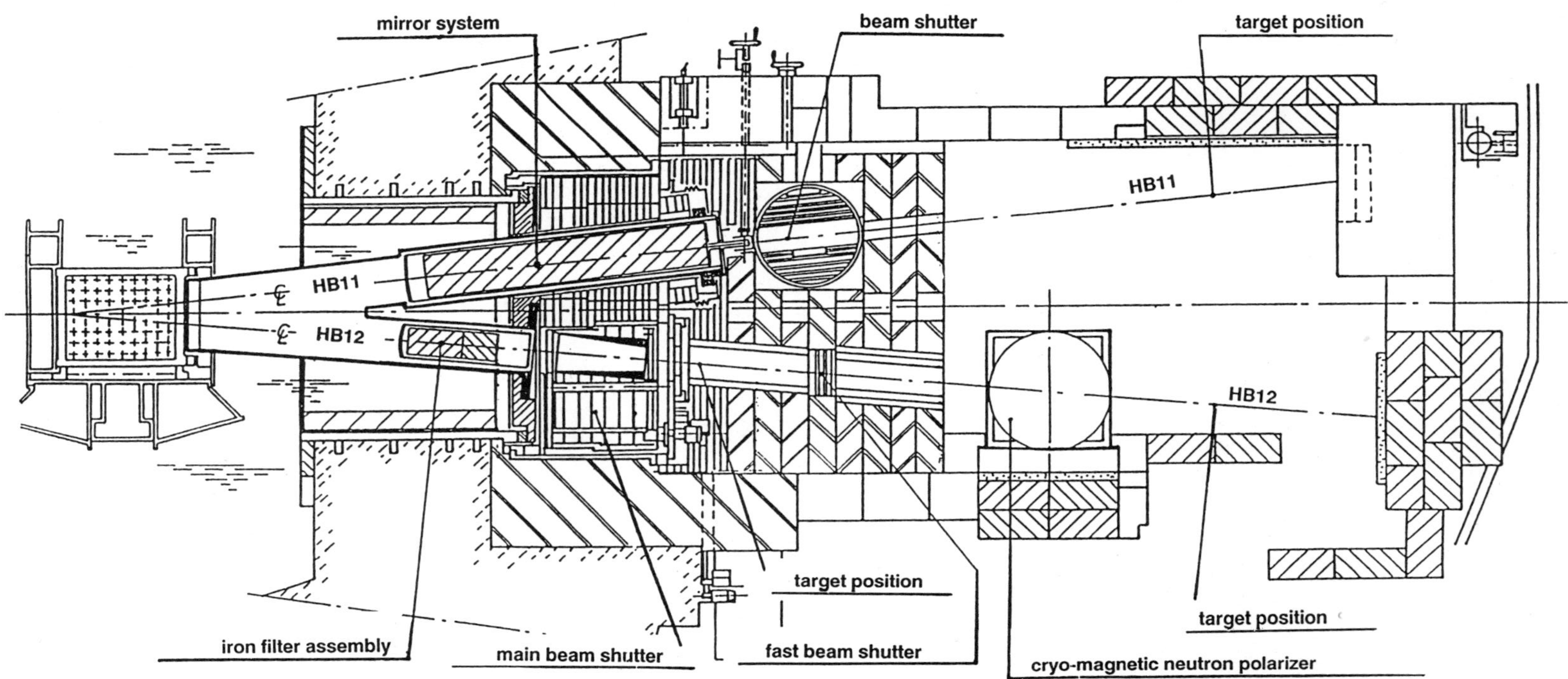

Figure 3.   Present Set-Up at Beam Tubes HB11/12.

On the north side of the reactor however, a specially designed large (50 x 40 cm) beam tube was installed during the vessel replacement procedure in 1984 [3]. The relative size of this tube is shown clearly in the in-pool schematic shown in Figure 2. The distance or water gap between the beam tube and the reactor vessel is 0.5 cm. The tube splits some 1.3 m from the vessel into 2 relatively large beam tubes, called HB11 and HB12 (see Figure 3). Figure 3 also indicates the present set-up of the beam tubes and the operational environment. The output end of beam tube HB11 is 35 x 35 cm and is located almost 4 m from the reactor vessel. Present usage, fundamental nuclear physics, can take advantage of a 3×10^9 $cm^{-2}s^{-1}$ thermal fluence rate from a focusing mirror system at HB11 and up to 10^8 $cm^{-2}s^{-1}$ 24 keV fluence rate through an iron filter assembly in HB12.

The present set-up, as shown in Figure 3, in addition to the respective mirror and iron filter systems, indicates the potential space in front of the beams where a therapy room could be housed. The large size, i.e., nuclear and physical, of HB11 offers a unique possibility to develop a filter assembly yielding a fluence rate of approximately 10^{10} $cm^{-2}s^{-1}$, with energies in the range of a few eV to a few keV, i.e., a high flux epithermal neutron beam. Much care will be taken in the future design work to minimise the fast neutron and gamma-ray components in the beam, aiming at a low proton- and gamma-load of the irradiated biological structures.

The installation of the necessary filter assembly and accompanying shielding evidently would require several weeks work, ideally when the reactor is shut down. Due to the reactor's strict operation and planning schedules, this could only be executed during the annual long summer outage period (6-8 weeks). Taking into account the status of the design work at present, this implies the 1990 summer period as the earliest possible date for implementation. Nevertheless, to gain experience and optimisation of parameters for epithermal neutron beam design, it is also planned to configure as early as this summer (1989) one of the smaller beam tubes (HB7). A programme of work, presented below, has therefore been formulated to utilise this beam tube with the principal aim to achieve as best as possible all the necessary data, such that extrapolation of results to the larger beam tube, HB11, can be confidently carried out.

<u>Neutron Metrology on HB11</u>

To confirm the spectrum and fluence rate at the input end of the beam tube, an extensive exercise of measurements and calculations has been carried out [4]. The work was jointly executed by JRC Petten and the ECN Metrology group. A wide range of detector sets and wires were used to determine the thermal and fast neutron fluence rates, as well as the neutron spectrum. The most convenient place near HB11, gained from previous experience, was to place the detector sets in the wedge-shaped vertical channels between the two beryllium assemblies and the northern innerside of the reactor vessel (see Figure 4). The results of the measurements were used to adjust the calculated neutron spectrum at the same position. From these calculations, the angular fluence rate spectrum at the entrance of the beam tube was then determined.

The positions of the various detector sets and wires as placed in the reactor are shown in Figure 4. Seven detector holders were used, consisting of nickel and cobalt wires. Three of the sets included extra foils to determine the fluence rate of neutrons with intermediate energies in the neutron spectrum. In addition, the central holder contained two foil sets encapsulated in cadmium boxes. The various activation/fission reactions of the extended detector sets are shown in Table One. The measurements took place at a reactor power of 532 kW.

The adjusted fluence rate spectrum per unit lethargy in the water gap between the beryllium reflector elements and the reactor vessel, for the central measuring position 6, is shown in Figure 5. The activation energy ranges are also indicated. The respective fast

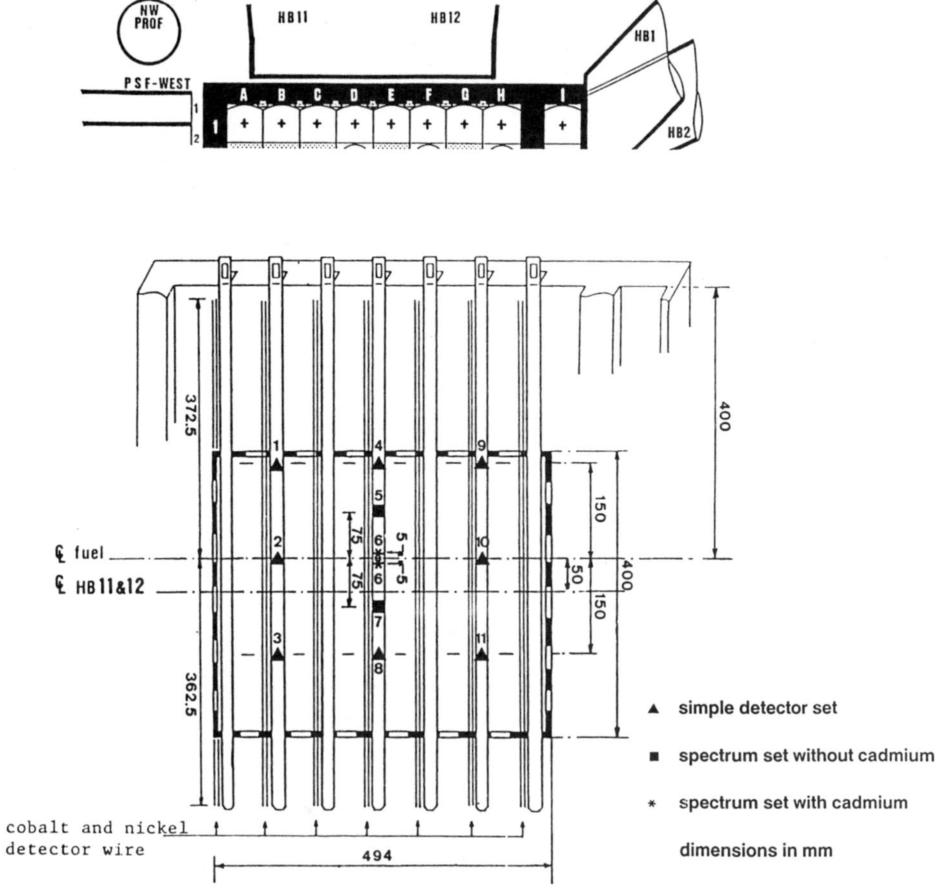

Figure 4.   Positions of the Detector Holders Including the Detector Sets and Wires.

175

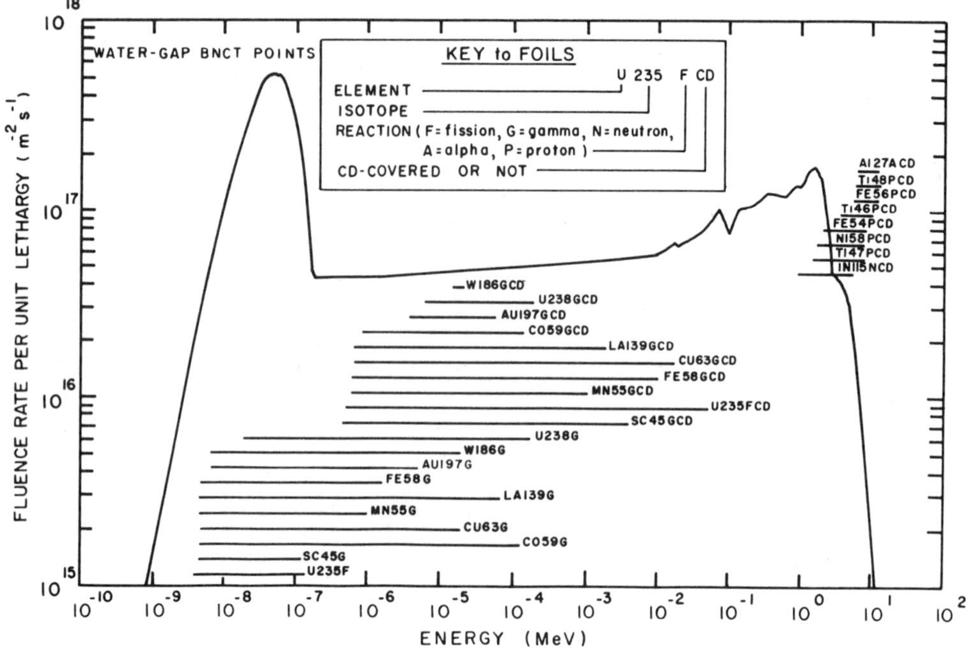

Figure 5.   Adjusted Fluence Rate Spectrum Per Unit Lethargy, Valid for Position 6A/B.

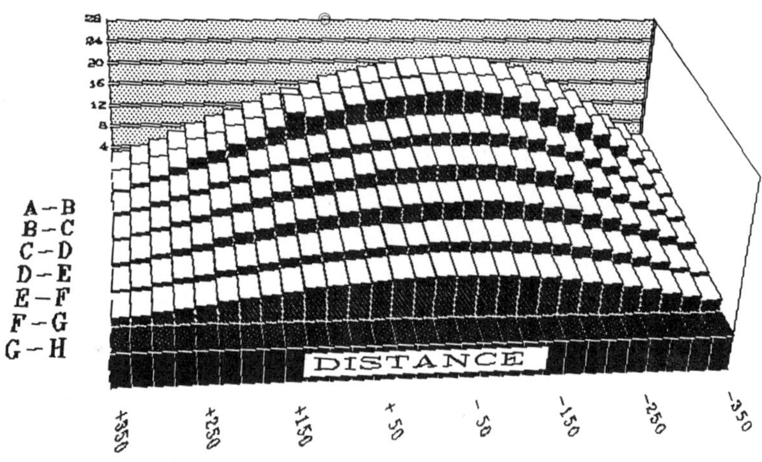

Figure 6.   Vertical Fast Fluence Rate Distribution in Front of Beam Tube HB11/12.

and thermal fluence rate distributions in front of the beam tube entrance are shown in Figures 6 and 7. The respective peak fluence rates are 2.1×10^9 cm^{-2}s^{-1} and 9.2×10^9 cm^{-2} s^{-1}.

Calculations using the transport code DOT [5] with 27 neutron broad energy groups were performed to determine the angular fluence rate spectrum for a discrete angle directed along the axis of the beam tube. The resulting spectrum is shown in Figure 8. For BNCT applications, it was calculated that for an epithermal beam (20 eV - 70 keV), an angular fluence rate per unit lethargy of 4.1×10^{11} cm^{-2}s^{-1}sr^{-1}, just inside the beam tube can be obtained for the present configuration. The following exercise will be to determine the effect of replacing one or more of the beryllium elements by aluminium plugs. With the integrated fluence rate over the beam tube cross section, it is expected that an input fluence rate of approximately 10^{12} cm^{-2}s^{-1} can be readily achieved, which should suffice following correct filter optimisation in producing an output fluence rate of the order of 10^{10} cm^{-2}s^{-1}.

Table One. Activation/Fission Reactions of the Extended Detector Sets.

^{45}Sc (n,γ) ^{46}Sc
^{55}Mn (n,γ) ^{56}Mn
^{58}Fe (n,γ) ^{59}Fe
^{59}Co (n,γ) ^{60}Co
^{63}Cu (n,γ) ^{64}Cu
^{139}La (n,γ) ^{140}La
^{186}W (n,γ) ^{187}W
^{197}Au (n,γ) ^{198}Au
^{235}U (n,f) f.p.
^{238}U (n,γ) ^{239}U

Thermal and Intermediate Neutrons

^{27}Al (n,α) ^{24}Na
^{46}Ti (n,p) ^{46}Sc
^{47}Ti (n,p) ^{47}Sc
^{48}Ti (n,p) ^{48}Sc
^{54}Fe (n,p) ^{54}Mn
^{56}Fe (n,p) ^{56}Mn
^{58}Ni (n,p) ^{58}Co
^{115}In (n,n') ^{115}In^m

Fast Neutrons

The measurements and calculations therefore confirmed that the beam tube HB11 offers the opportunity to develop a large cross section, high flux epithermal neutron beam. In addition, as seen later, the decision to proceed with the design and construction was more than prudent.

THE BNCT PROJECT AT PETTEN

To realise the (boron) neutron capture therapy facility on HB11, the following Work Programme, highlighting the essential tasks, is planned as shown in Table Two.

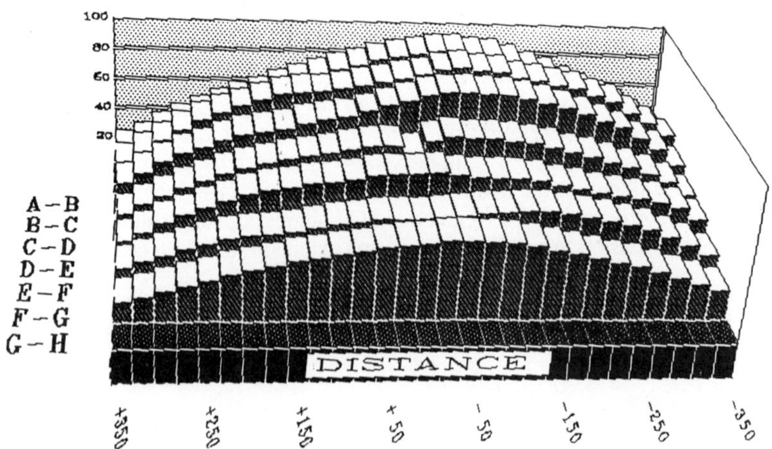

Figure 7. Vertical Thermal Fluence Rate Distribution in Front of Beam Tube HB11/12.

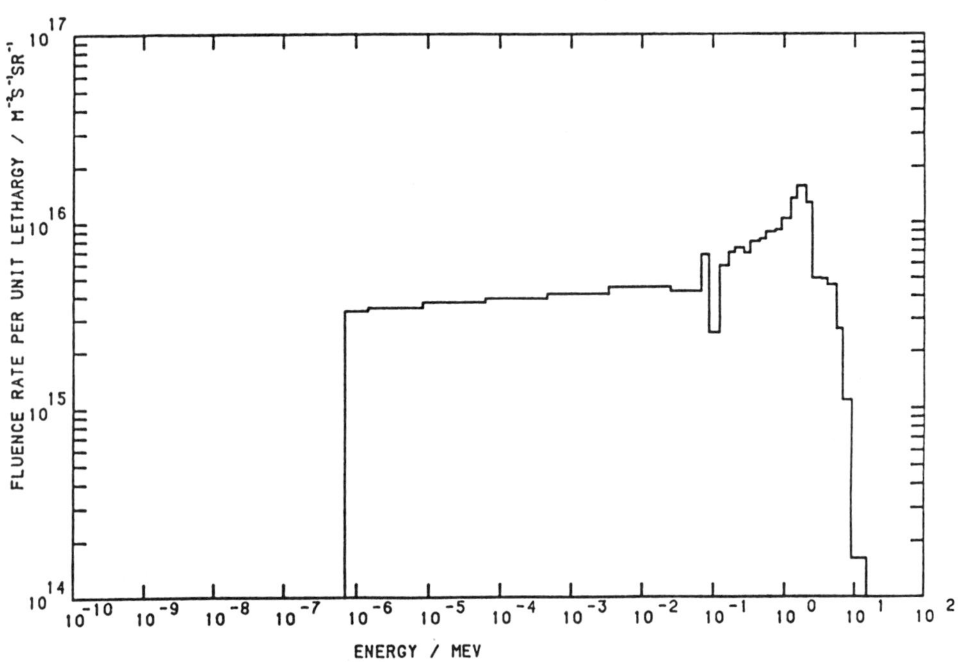

Figure 8. Angular Fluence Rate Spectrum for a Discrete Angle Directed
Along the Axis of Beam Tube HB11.

Table Two. Planning Schedule for the Petten BNCT Facility.

```
|________________\ Year  || ___1989___ || ___1990___ || ___1991___ |
|  Item          \(1/4)  || 1 | 2 | 3 | 4 || 1 | 2 | 3 | 4 || 1 | 2 | 3 | 4 |
|                        ||   |   |   |   ||   |   |   |   ||   |   |   |   |
|design calcn (HB7)      || ======|   |   ||   |   |   |   ||   |   |   |   |
|construction (HB7)      ||   |===|   |   ||   |   |   |   ||   |   |   |   |
|installation (HB7)      ||   |   === |   ||   |   |   |   ||   |   |   |   |
|metrology    (HB7)      || === |   ========== |   |   |   ||   |   |   |   |
|animal irrad (HB7)      ||   |   |   |   || ====================>|   |   |   |
|dosimetry    (HB7)      ||   |   |   | ====================>|   |   |   |   |
|                        ||   |   |   |   ||   |   |   |   ||   |   |   |   |
|design calcn (HB11)     ||   ==================== |   |   ||   |   |   |   |
|construction (HB11)     ||   |   |   |   ||=======|   |   ||   |   |   |   |
|installation (HB11)     ||   |   |   |   ||   |   ===== |   ||   |   |   |   |
|metrology    (HB11)     ||   | ==|   |   ||===|   |   =====================>|
|                        ||   |   |   |   ||   |   |   |   ||   |   |   |   |
|animal irrad (HB11)     ||   |   |   |   ||   |   | ===================>|
|therapy      (HB11)     ||   |   |   |   ||   |   |   | =================>|
|dosimetry    (HB11)     ||   |   |   |   ||   |   | ===================>|
|                        ||   |   |   |   ||   |   |   |   ||   |   |   |   |
|treatment planning      ||   |   | ===================================>|
```

An explanation of the above items, giving related research activities and development work, is as follows :

Beam Tube HB7

Design calculations - computer modeling techniques using Monte Carlo codes, e.g., MCNP, have recently been started in conjunction with the Harwell Laboratories in the UK. The modeling will consider, amongst other things; Be reflector elements or Al plugs at the periphery of the reactor core; an Al spectrum shifter at the input end of the beam tube; compare recommended [7,8] filter combinations Al/D_2O and Al/Ar/S, various gamma and fast neutron filters; and determine the corresponding shielding requirements.

Beam metrology - measurements of the gamma-ray, and the slow and fast neutron components will be carried out and compared to calculations. Apart from comparison purposes, the results will be used for dosimetry and shielding requirements. Methods to be used are multiple activation detector sets, proton-recoil techniques, hydrogen proportional-counter spectrometer, etc.

Construction and installation - optimisation of the above parameters will lead to the construction of the small facility this summer. It is intended to minimize the complexity of the facility as much as possible so that exchange of the filter combinations and access for instrumentation can be readily carried out.

Animal irradiations - cell culture studies as a function of field size, depth in tissue equivalent phantom, and beam filtration; pharmokinetic and distribution studies of several candidate boron compounds using atomic emission spectrometry and track-etch techniques to measure boron distributions within mouse tumours and normal tissues; tumour response studies in animal tumours measured by growth delay and tumour cure; and phase I clinical pharmokinetic and toxicity tests (unirradiated) of promising new compounds.

Dosimetry - performance of a number of experiments to determine the neutron, proton, and gamma-ray characteristics in different phantoms exposed to the beam,

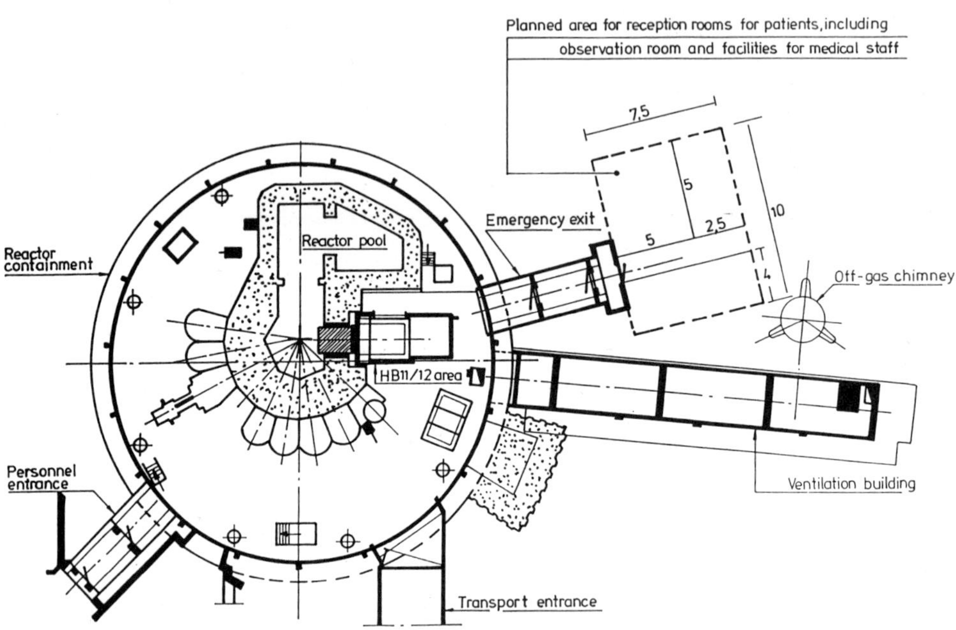

Figure 9. Plan Through Reactor Building Indicating the Convenient Placing of the Emergency Personnel Exit with Respect to the Proposed BNCT Facility.

180

e.g., using BF$_3$ counters, bare and Cd-covered gold foils, Cd-covered indium foils, fission counters, TLDs, etc.

<u>Beam Tube HB11</u>

<u>Design calculations</u> - on the basis of the above exercises, supplemented by filter test experiments at HB11, the neutron beam will be redesigned taking into account the specific geometry of HB11.

<u>Beam metrology</u> - measurements similar to those performed on HB7 will be carried out, with the option to repeat them for anticipated beam configuration changes.

<u>Construction and installation</u> - the complete task is planned to take no longer than two years. After performing the dosimetric characterisation of the beam and testing the different methods for boron concentration determination, patient treatment can be initiated.

<u>Clinical facility</u> - extension of the design phase to the therapy facility will follow in close cooperation with the radiotherapists. Factors such as: accessibility of patient, medical and scientific staff to the reactor building and treatment room; the construction of a medical preparation room inside and adjacent to the reactor; a separate well-ventilated room for experimental animal housing including space for an incubator for storing cells in tissue culture will all be investigated. Additionally, a waiting room for patients outside and adjacent to the entrance of the reactor (see Figure 9). Instrumentation requirements include: monitoring the beam in the treatment room; accurate positioning of patients by means of lasers; visual display units; audio-contact, etc. A treatment couch is required that can be rotated 180° around the beam aperture to facilitate coplanar field irradiation.

<u>Dosimetry and treatment planning</u> - calculations will be performed to derive the neutron, proton, and gamma-ray characteristics of the epithermal neutron beam in phantoms as a function of various parameters such as beam cross section and composition of the phantom. By combining the beam characteristics and information about boron uptake, detailed dose calculations will be made for each individual patient after establishing the treatment technique and fractionation scheme. A number of real-time and integral dose monitoring systems will be installed for measuring the actual dose delivered to the patient.

<u>Therapy (clinical feasibility studies)</u> - patients with glioblastoma will be exposed to fractionated epithermal neutron irradiations after slow infusion of sulphydryl boron compounds. Similarly, melanoma patients will be treated after receiving boron-phenylalanine administration. Boron concentrations in blood and tumour tissue will be performed, using the on-site available prompt gamma-ray spectroscopy and compared with results using quantitative neutron capture radiography. Tumour response and survival time, as well as local and systemic normal tissue toxicity are the main endpoints of the study.

EUROPEAN COLLABORATION GROUP ON BNCT

The European Collaboration Group on BNCT was formulated at Abingdon, UK in the summer of 1987. The intention of the group is to coordinate exchange of information and collaboration programmes between European Research and Medical Centres, with the eventual aim of performing BNCT on optimised facilities throughout Europe. The group presently consists of over 30 members, representing over 10 European countries. Recently a Declaration of Intent (CEC terminology) for a Concerted Action on BNCT has been prepared by Prof. Gabel (Bremen) and Dr. Schofield (Harwell), and submitted to the European Commission's Medical and Health Research Division in Brussels. The Action enables, through financial support from the CEC, coordination of the various BNCT activities at the different institutes in Europe. A successful application is anticipated and would allow the European Collaboration Group to meet several times in the coming three years, including the possibility to exchange staff and material among the various institutes.

The Concerted Action will consist of the Project Leader (Prof. Gabel) and his Project Management Team (Drs. Chiaraviglio, Durrant, Larsson, Moss, and Schofield) who will coordinate general BNCT activities within Europe. The principal aim is to create preconditions for clinical implementation at Petten of BNCT of tumours at the earliest possible date and to create the possibility for additional treatment centres in the next few years. Preconditions include:

- design, construction, physical and dosimetric characterisation of epithermal neutron beams for a European facility at the JRC Petten,

- evaluation of the irradiation effects on animal models, especially late irradiation effects, leading to safe estimates of initial radiation doses in patients,

- toxicological and pharmacological evaluation in patients of presently identified boronated tumour seekers, leading to optimised administration schemes and improved tumour uptake,

- synthesis and pharmacological evaluation in animal models of new and improved boronated tumour seekers for glioma and melanoma,

- development of and screening for tumour seekers for possible use in the extracorporal treatment of liver malignancies, and

- experimental evaluation of the possible merits of extracorporal bone marrow irradiation for removal of malignant cells.

CONCLUDING REMARKS

Within Europe, no single research institute or medical centre has the resources or available multi-disciplinary team to realise a BNCT facility. The Commission of the European Communities is therefore fortunate to have the High Flux Reactor at Petten under its auspices and also to be in a position to support the design and construction of a NCT facility on the reactor and additionally, as anticipated, to support the European Collaboration Group (ECG) on BNCT.

The BNCT project is currently underway. The first epithermal neutron beam will be available after this summer. The experience to be gained on this relatively smaller facility will be exploited and utilised for the planned full-scale therapeutic facility on beam tube HB11, the following year. The results of the design details, with various parameter studies, are unfortunately still in the early stages and therefore cannot be discussed at this Workshop. Nevertheless, it is considered by the ECG that the intended plans at Petten and the progress to-date should be presented. It is expected that at the 4th International Symposium on BNCT in 1990, a more detailed report on the design of the facility will be available. In addition, the results on the first animal irradiations and related activities will be presented.

REFERENCES

1. R. G. Zamenhof, H. Madoc-Jones, O. K. Harling, and J. A. Bernard, Jr., "A Multidisciplinary Program Leading to a Clinical Trial of Neutron Capture Therapy at Tufts – New England Medical Center and the Massachusetts Institute of Technology," <u>Strahlenther. Onkol.</u>, 165 (2/3):254 (1989).

2.   H. Roettger et al., "High Flux Materials Testing Reactor Petten, Characteristics of Facilities and Standard Irradiation Devices," <u>EUR 5700 EN</u> (1986).

3.   F. Stecher-Rasmussen, "Capture Gamma-Ray Spectroscopy and Related Topics," in <u>Proc. AIP Conf.</u>, No. 125, Knoxville, TN (1984).

4.   R. W. A. Kraakman et al., "Neutron Fluence Rate and Neutron Spectrum Measurements in Front of HB11 and HB12. (Part of a Feasibility Study for a BNCT Facility)," <u>ECN-89-06</u> (1989).

5.   "The DOT 3.5 Two-Dimensional Discrete Ordinates Radiation Transport Code," Radiation Shielding Information Center, Oak Ridge National Laboratory (1976).

6.   J. F. Breismeister, ed., "MCNP – A General Monte Carlo Code for neutron and photon transport, Version 3A" Los Alamos National Laboratory, <u>LA-7396-M</u>, Rev. 2 (1986).

7.   R. G. Fairchild, J. A. Kalef-Ezra, S. Fiarman, L. Wielopolski, J. Hanz, S. Mussolino, and F. Wheeler, "Optimization of an Epithermal Beam for NCT at the Brookhaven Medical Research Reactor (BMRR)," <u>Strahlenther. Onkol.</u>, 165(2/3):84 (1989).

8.   G. Constantine, J. A. B. Gibson, K. G. Harrison, and R. Schofield, "Harwell Research on Boron Neutron Capture Therapy," <u>Strahlenther. Onkol.</u>, 165(2/3):92 (1989).

INSTALLATION AND TESTING OF AN OPTIMIZED EPITHERMAL NEUTRON BEAM AT THE BROOKHAVEN MEDICAL RESEARCH REACTOR (BMRR)

R. G. Fairchild,[1] J. Kalef-Ezra,[1,2] S. K. Saraf,[1] S. Fiarman,[1]
E. Ramsey,[3] L. Wielopolski,[3] B. H. Laster,[1] and F. J. Wheeler[4]

[1] Medical Department, Brookhaven National Laboratory, Upton, NY
[2] University of Ioannina, Ioannina, Greece
[3] Health Sciences Center, State University of New York, Stony Brook, NY
[4] Idaho National Engineering Laboratory, Idaho Falls, ID

INTRODUCTION

Initial clinical trials of Neutron Capture Therapy (NCT) in the United States were unsuccessful. Lack of success has been attributed to two causes: (1) absence of selective localization of boron in tumor cells, and (2) poor penetration in tissue of the thermal-neutron beams used. Since then, improved compounds have been developed which can be selectively targeted to tumor [1-3]. In addition, improvements have been made in neutron delivery. At a workshop on neutron sources for NCT held in 1986, it was recommended that current technology be utilized to produce pure epithermal-neutron beams for NCT. These would provide the increased penetration in tissue required for improved therapy. The study group on neutron beams recommended that these beams should have an epithermal-neutron flux density of $\sim 1 \times 10^9$ n/cm^2-s (or more) to enable application of therapy within ~1 hour (or less) [4].

While the possibility exists that various filter configurations can be designed which would produce monoenergetic neutron beams at various energies, such beams tend to have intensities which are insufficient for therapeutic application. In an effort to maximize intensity, we have chosen to utilize the entire reactor core as a source of neutrons (i.e., the complete core as viewed from the point of irradiation) and to use the broad epithermal energy region (1 to 10,000 eV) for the production of thermal neutrons at depth in tissue for NCT. Concomitantly, appropriate moderators, or "spectrum shifters," are used to selectively suppress the undesirable fast neutrons (E>10 keV).

Various calculations indicate that an optimized epithermal-neutron beam can be produced by moderating fission neutrons either with a combination of Al and D_2O, or with Al_2O_3 [4]. We have designed, installed, and tested an Al_2O_3-moderated epithermal-neutron beam at the Brookhaven Medical Research Reactor (BMRR). The epithermal-neutron fluence rate of 1.8×10^9 n/cm^2-s produces a peak thermal-neutron fluence rate of $\sim 2 \times 10^9$ n/cm^2-s in a tissue-equivalent (TE) phantom head, depending on the configuration. Thus a single therapy treatment of 5×10^{12} n/cm^2 can be delivered in 30-45 minutes.*

---

* All irradiation times are given for a BMRR power of 3 MW, which is the highest power which can be delivered continuously.

Neutron Beam Design, Development, and Performance for Neutron Capture Therapy
Edited by O. K. Harling *et al.*
Plenum Press, New York, 1990

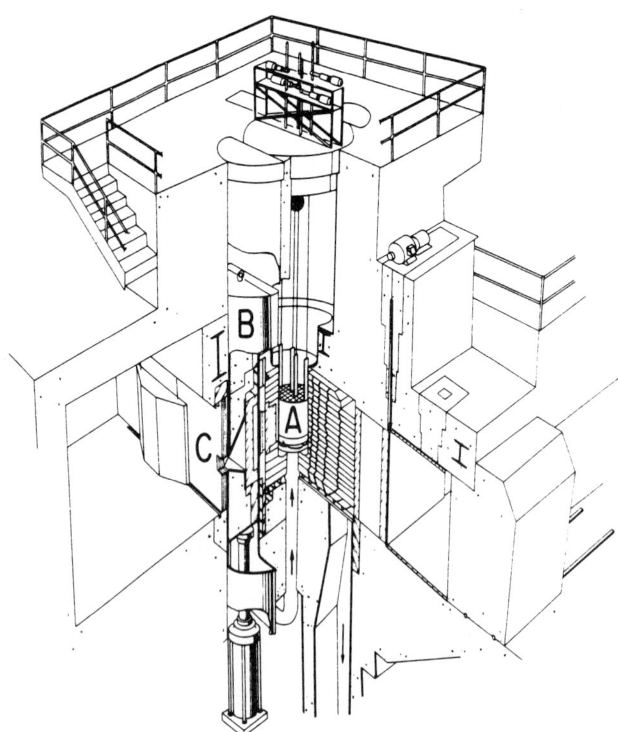

Figure 1. Cross-sectional view of the Brookhaven Medical Research Reactor, showing the core (A), removable shutter (B), and one of two identical patient irradiation facilities (C). The current configuration has an epithermal-neutron beam in the East Irradiation Facility and a thermal beam in the West Facility. Maximum reactor power is 3 MW.

MATERIALS AND METHODS

The design, construction, installation, and testing of this epithermal-neutron beam was done in a collaborative project between Brookhaven National Laboratory (BNL) and Idaho National Engineering Laboratory (INEL). Installation of the Al_2O_3 filter arrangement was done at the East Irradiation Facility of the BMRR (see Figure 1). The 5-MW (3-MW continuous power) reactor was designed and built in 1959 primarily for use as a neutron source for medical and biological experiments [5]. A cross section of the irradiation facility is shown in Figure 2. Regions A and B are housed in a 20-ton shutter which was designed so that it could be easily removed for the installation of various filters and/or moderators. Removal of the shutter (in two parts) is accomplished with an overhead crane in approximately one hour. This flexibility has been fully utilized in these experiments, as a number of permutations have been evaluated in arriving at the "current" configuration [6-7]. Region C has two empty aluminum tanks, which can be filled with liquids, such as D_2O, or solid "microspheres", such as Al_2O_3.

An effort has been made to compare calculated values of beam parameters with experimental measurements of the same parameters at each step in the filter installation. Calculations were made at INEL with one-dimensional (cylindrical) models for the SCAMP and the ANISN discrete ordinate codes. This combination of codes couples the cross-section library of SCAMP (ENDF/B-V) with the high-order scattering and secondary gamma production of the ANISN model. In addition, final design and "as-built" analyses were carried out with a two-dimensional model using the DOT 4.3 code and the Bugle-80 ENDF/B-IV cross-section library. Cylindrical (r-z) geometry was used, with the z-axis coinciding with the beam axis [8].

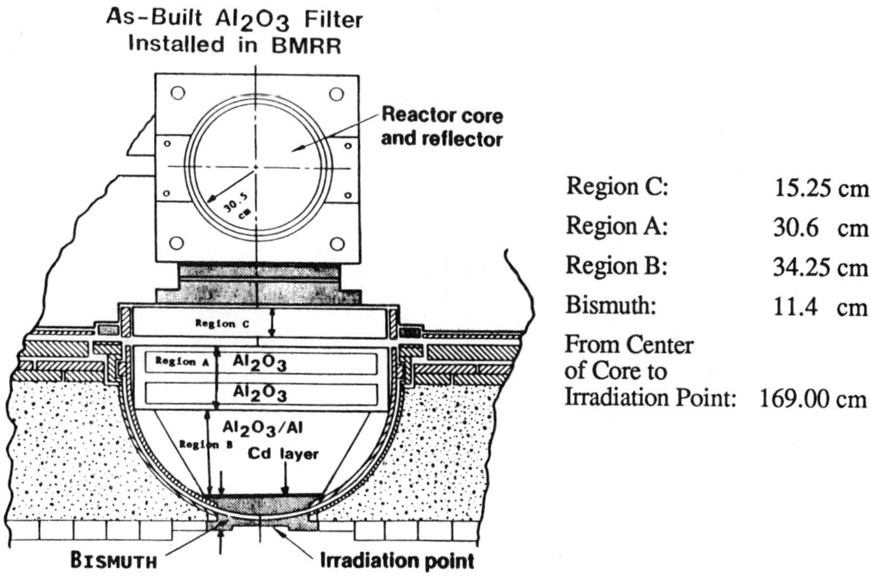

Figure 2. Cross section of epithermal-neutron beam facility showing reactor core and filter/moderator arrangement in beam shutter for the "current" configuration in Table One.

Various shutter configurations were evaluated. These included a completely empty shutter, the reference case "R" in which 18 cm of D_2O served as the moderator, and the "current" configuration of 45.7-cm Al_2O_3, 19.7-cm Al, 11.4 cm Bi, and 0.051-cm Cd (see also [8]). It should be noted that the reference case 'R' had been installed in 1965 to produce a Cd-filtered epithermal-neutron beam [9,10].

Measurements of total dose to soft tissue, fast-neutron dose, and γ-dose, as well as the thermal, epithermal, and fast-neutron fluence rates were made at each stage in the filter installation. These were compared to calculated values. Paired-ionization chamber measurements (tissue-equivalent [TE] and graphite-CO_2 chambers) were used to evaluate the total dose and both the fast neutron and γ-components of the mixed radiation fields. Threshold and fission foils were also used to evaluate the fast-neutron dose and ^7LiF thermoluminescent dosimeters were used to verify γ-dose measurements. Gold, sodium, and copper foils were used to measure thermal and resonance neutrons. Thermal neutron depth-flux curves were measured in a 16.6 cm x 23 cm cylinder filled with TE fluid [9,10]. Details of the dosimetric techniques are given in [11]. Fast-neutron dose distributions in the phantom were obtained from the measured absorbed dose (free in air), and attenuated as a function of depth as calculated [8]. The gamma doses from photons generated in the phantom were obtained from values measured previously with similar thermal-neutron distributions [10], and normalized to the peak thermal-neutron flux density in the present head phantom. Normalized values of the gamma dose (used in this paper) were somewhat higher than theoretical calculations of the γ-dose from the present beam [8].

RESULTS

The results of eight shutter configurations are summarized in Table One, varying from a completely empty shutter (configuration 1) to the "current" geometry in which a total

Table One

Calculated and Experimental Values of Beam Parameters for Various Shutter Configurations (BMRR, 1 MW).

Configuration	1	2	3	4	R	8	BNL - Inter. Configuration	Current Configuration
Region C Region A Region B	0 0 0 7.6-cm Bi	0 16.5-cm Al 0 0 7.6-cm Bi	0 17.8-cm Al 9.6-cm D_2O 0 7.6-cm Bi	12-cm D_2O 16.5-cm Al 0 0 7.6-cm Bi	12-cm D_2O 2.5-cm Al 6-cm D_2O 15.6-cm Bi	12-cm D_2O 17.8-cm Al 9.6-cm D_2O 0 7.6-cm Bi	0-cm D_2O 17.8-cm Al 9.6-cm D_2O 11.42-cm Al 22.86-cm Al_2O_3 11.4-cm Bi 0.051-cm Cd	0-cm D_2O 8.25-cm Al 22.86-cm Al_2O_3 11.42-cm Al 22.86-cm Al_2O_3 11.4-cm Bi 0.051-cm Cd
Fast n-KERMA [rad/min]								
ANISN (BNL) ANISN (INEL) DOT (INEL) Experim. (BNL) Experim. (INEL)	3272 1410 1070	1166 553 360	98.8 59 55.1[†] 52	86.1 – 49	48.9 45 27	10.4 – 9	1.03 – 2.3	0.60 – 1.03* 1.75 1.68
Gamma-KERMA [rad/min]								
ANISN (BNL) DOT (INEL) Experim. (BNL)	30.12 43	28 42	24.3 19.9[†] 31	26.7 30	7.76 9	20.2 22	0.733 0.81	0.46 0.12* 0.4
Epithermal Flux 0.4 eV – 10 keV $[10^{10}\ cm^{-2}s^{-1}]$								
ANISN (BNL) ANISN (INEL) DOT (INEL) Experim. (BNL) Experim. (INEL)	2.76 0.82 0.474	1.68 0.53 0.26	1.05 0.29 0.34[†] 0.14	0.871 – 0.116	0.503 0.25 0.126	0.264 – 0.122	0.0824 0.066	0.078 0.066* 0.060 0.062

Configuration	1	2	3	4	R	8	BNL-Inter. Configuration	Current Configuration
Thermal Flux 0.0 – 0.4 eV [10^{10} cm^{-2}s^{-1}]								
ANISN (BNL)	4.69	1.62	1.39	1.77	3.2	1.08	0.00057	0.00056
Experim. (BNL)	2.9	1.2	0.61	0.81	2.0	0.4	0.0031	0.0024
Thermal Peak in Phantom [10^{10} cm^{-2}s^{-1}]								
ANISN (BNL)	11.4	6.83	2.24	1.9	0.98	0.50	0.23	0.21
Experim. (BNL)	–	1.5	0.64	0.41	0.28	0.12	–	0.085
Fast n-KERMA / Epithermal Flux [10^{-11}rad/(n-cm^{-2})]								
ANISN (BNL)	198	116	15.7	16.5	16.2	6.57	2.08	1.28
ANISN (INEL)	287	174	33.9	30.3	–			
DOT (INEL)								2.59*
Experim. (BNL)	376	231	64.3	73.3	35	11.3	5.71	4.87
Experim. (INEL)								4.51
Gamma-Kerma / Epithermal Flux [10^{-11}rad/(n-cm^{-2})]								
ANISN (BNL)	1.82	2.78	3.86	5.11	2.57	12.75	1.48	0.99
DOT (INEL)								0.492*
Experim. (BNL)	14.3	26.2	34.2	39.5	11.5	28.5	2.06	1.12

* DOT calculations are for 15 cm Bi.   † DOT calculations are for 11.4 cm Bi.

Table Two

Summary of Beam Parameters for Current Optimized Epithermal-Neutron Beam
(65.4 cm Al_2O_3 and Al).

Power	3 MW
Epithermal-neutron flux density* (n/cm^2-s)	1.8×10^9
Thermal-neutron flux density	2.8×10^9 (no added filtration)
(peak flux at ~2-cm depth	2.5×10^9 (0.5-mm Cd added)
in phantom; n/cm^2-s)	1.9×10^9 (1.0-mm ^{6}Li added)
Absorbed dose from fast neutrons free in air*	5.3 rad/min
Absorbed dose from gammas, free in air*	1.2 rad/min
Fast-neutron dose per epithermal neutron	4.87×10^{-11} rad/(n-cm^2)
Gamma dose per epithermal neutron	1.12×10^{-11} rad/(n-cm^2)
Fast-neutron dose per thermal neutron (no added filtration)	3.15×10^{-11} rad/(n-cm^2)

* Measured at center of irradiation port face.

of 65.4 cm of Al_2O_3 and Al was used to moderate the beam. The relative fast neutron and γ-contaminations in the new beam (per epithermal neutron) have been reduced to ~10% of the contaminations present in the old Cd-filtered epithermal-neutron beam developed in 1965 (configuration R). The resultant thermal-neutron fluence rate of ~2×10^9 n/cm^2-s is ~3 times less than that generated with the old beam (configuration R).

The various beam parameters for the current configuration are summarized in Table Two.[†] Assuming a one-to-one correspondence between incident epithermal-neutron flux and thermal-neutron flux at depth in tissue (a conservative estimate), 45 min would be needed to deliver a therapeutic fluence of 5×10^{12} thermal neutrons/cm^2. This is within the suggested time limit of 60 min as recommended by the Physics Committee at the 1986 Workshop on NCT [4].

Three different configurations were evaluated for irradiating a TE phantom head in the 25 cm x 25 cm epithermal-neutron beam (irradiation point, Figure 2). These were: no added filtration at the phantom, 0.5-mm Cd filter added, and 1-mm ^{6}Li added. Results of the thermal-neutron fluence rate measurements are given in Figure 3 for a reactor power of 1 MW. It can be seen that by increasing the filtration at the point of irradiation, the peak/surface (P/S) ratio increases from 2.8 to 4.2 to 4.9 for the three geometries respectively, while at the same time the peak intensity is reduced from 2.8 to 1.9×10^9 n/cm^2-s. Depending on experimental conditions, increased peak/surface ratios may be useful for providing increased skin sparing for situations in which it is desirable to leave the skin and skull intact. As described in the discussion, it is anticipated that these flux densities will reduce irradiation times to $\leq$30 min.

For the purpose of this Workshop, effective dose rate (rad x RBE) curves have been plotted on semi-log scales for the $^{14}N(n,p)^{14}C$ and $^{10}B(n,\alpha)^7Li$ reactions, as well as

---

[†] Parameters are given in Table Two for a reactor power of 1 MW, as measured. Results can be extrapolated linearly to the maximum power of 3 MW.

for the fast-neutron (H-recoil) and γ-dose (from the reactor and H(n,γ)D reactions). As prescribed by the workshop organizers, RBEs of 1.6 have been used for the nitrogen and fast-neutron (N and H) dose, and 2.3 for the ^{10}B dose. Also, a nitrogen concentration of 1.84%, was assumed for brain, as opposed to the value of 2.6% reported for "standard man" for whole body nitrogen.

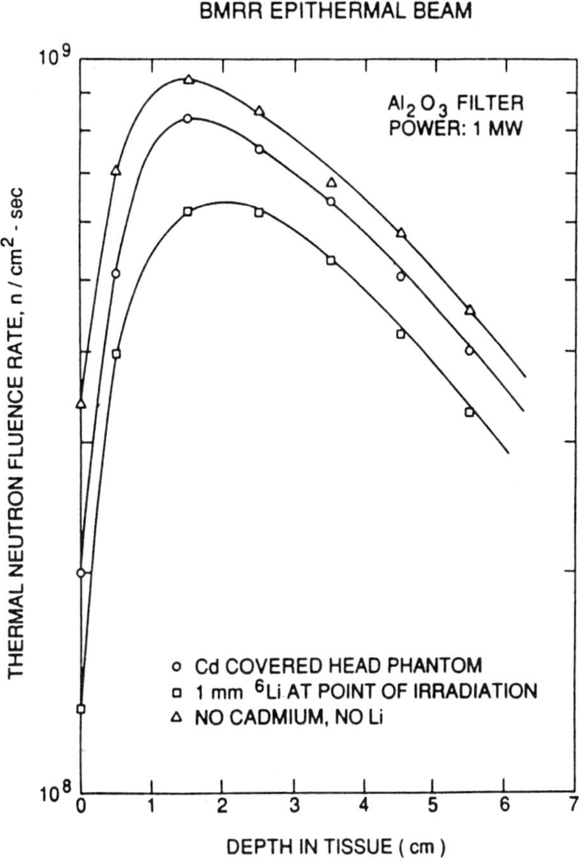

Figure 3. Thermal-neutron flux densities generated in tissue-equivalent head phantom (16.6 cm x 23 cm cylinder) using an incident epithermal-neutron beam (current configuration; BMRR power 1 MW).

Results are given in Figure 4 for the current filter configuration (shown in Figure 1) and summarized in Table Two (65.4-cm Al$_2$O$_3$ and Al). The flux distribution obtained with 0.5 mm Cd-added filtration was used to generate Figure 4. It is assumed that a similar curve can be obtained with a thin (<1-mm) ^{6}Li filter [9,10]. In addition, the same parameters have been graphed for the depth-flux curves obtained with 1-mm ^{6}Li filtration (Figure 5). Here, the large P/S value of 4.9 and the peak flux density at 2.0-cm depth may prove useful under certain situations. Parameters requested for comparison at this Workshop (advantage depth, advantage depth dose rate, and advantage depth dose ratio) are summarized in Table Three for the two beams.

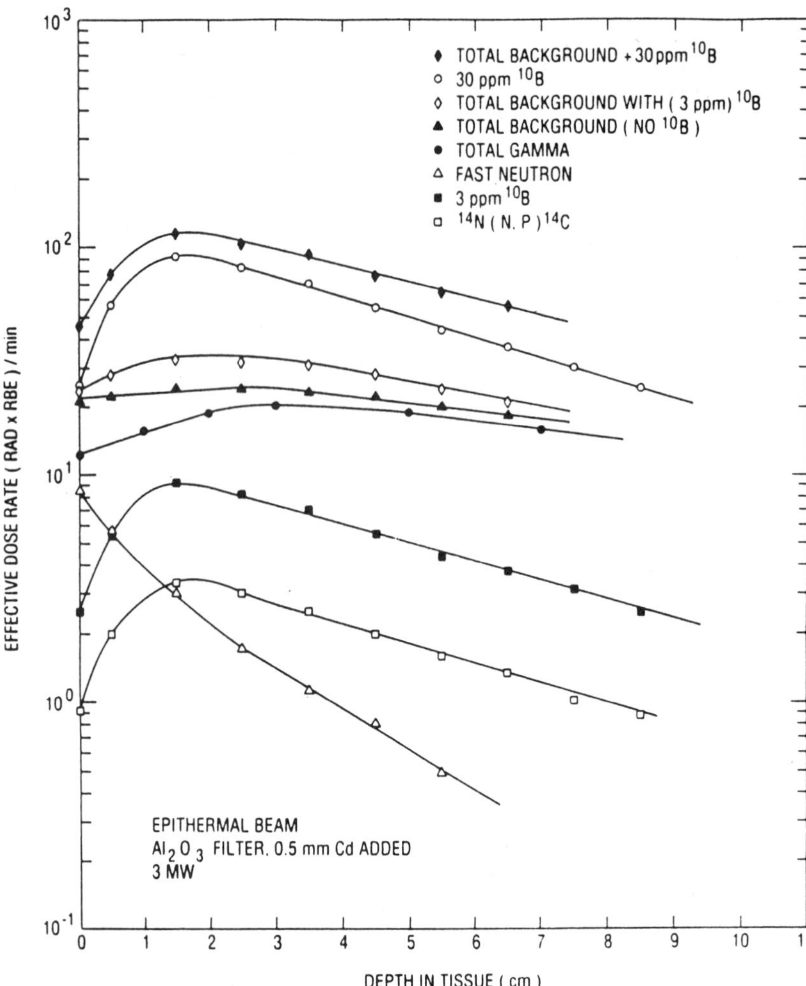

Figure 4. Biologically-effective dose rate for the "current" epithermal-neutron beam configuration at the BMRR (power = 3 MW). The flux distribution obtained with a Cd filter 0.5-mm thick at the point of irradiation was used. Values for both RBE, and the ^{14}N and ^{10}B content were as prescribed for this Workshop.

Calculations of the advantage depths by Wheeler et al. in this Workshop are somewhat lower than those shown in Table Three, for approximately the same beam and geometry (18 cm x 23 cm phantom, 20-cm diameter beam) [8]. This may be due to the use of a 5-cm thick collimator in the latter calculations or to the use of a relatively higher γ-dose rate in Figures 4 and 5 of this paper.

DISCUSSION

For the purpose of beam comparison for this Workshop, it was requested that plots of beam components be made on semi-log scales as in Figures 4 and 5. However, biological response is more readily conceptualized with a linear scale than with a logarithmic ordinate scale of physical radiation dose. Thus, for the purpose of evaluating

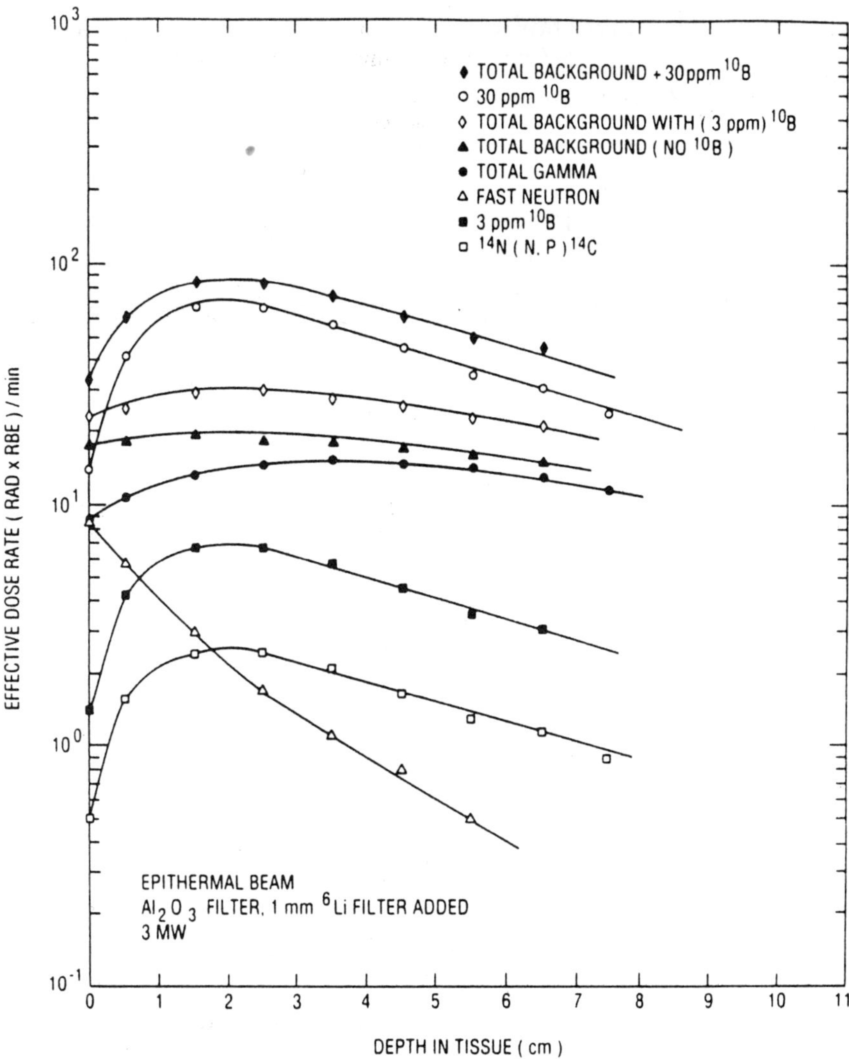

Figure 5. Same as Figure 4, but with a 1-mm thick ^{6}Li-filter added at the point of irradiation.

the significance of depth-dose curves in this discussion, data have been developed on a linear scale. This has been done for the case of 1-mm ^{6}Li added-filtration, shown in Figure 6. Here, the adventitious radiation components (N, H, and γ) are plotted along with the total (N+H+γ), as well as a separate curve showing the distribution from 3-ppm ^{10}B. From Table Two, it can be seen that the reactor-produced γ is negligible compared to that from the H(n,γ)D reaction. Hence, the N and γ curves represent unavoidable contributions to normal tissue dose produced by the thermal-neutron distribution. The contribution from fast neutron dose (H) could be reduced by further moderation, but at the cost of reduced beam intensity. Calculations indicate that the fast-neutron dose (H) could be reduced relative to epithermal neutrons by ~1/2 (to ~2.4x10^{-11} rad/epithermal neutron) by utilization of the now empty "C" region, with a concomitant reduction of epithermal-neutron fluence rate by ~ 1/2. Such a reduction in fast-neutron dose is graphed in Figure 6, where it can be

Table Three

Beam Parameters at 3 MW for the 65.4-cm (Al$_2$O$_3$ + Al) Moderated Beam,
for Comparison at this Workshop.

		0.5-mm Cd-added filtration	1.0-mm ^{6}Li-added filtration
i)	Minimum advantage depth	9.2 cm	8.2 cm
ii)	Maximum advantage depth	11.1 cm	10.4 cm
iii)	Advantage depth dose rate	33. (rads x RBE)/min*	30. (rads x RBE)/min*
iv)	Advantage depth dose ratio	3.6*	3.2*
v)	% low-LET dose	24.2%*	23.3%
vi)	% high-LET dose	7.0%*	6.8%*
vii)	% ^{10}B (30µg/g) dose	65.2%*	66.1%*

* With 3 µg/g ^{10}B in healthy tissue.

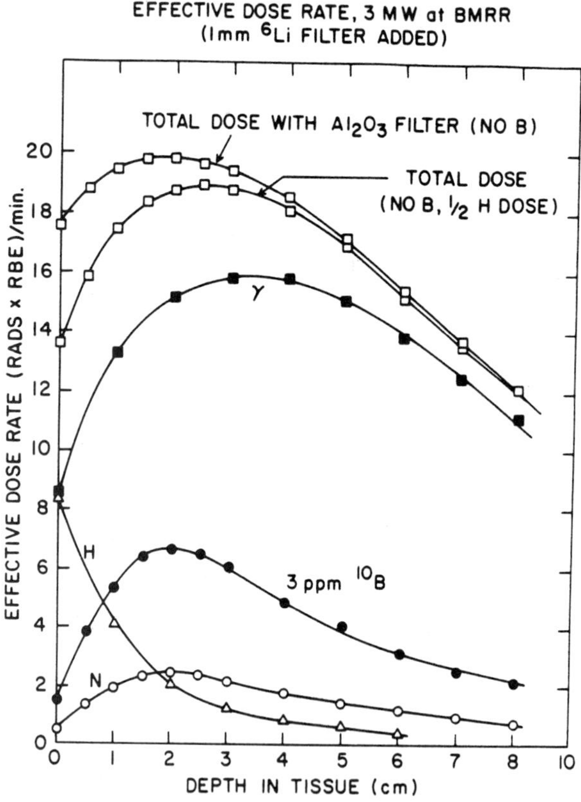

Figure 6.   Same parameters as Figure 5, but plotted on a linear scale.   In addition, the effects of reducing the fast-neutron dose (H) by a factor of 2, to 2.4x10^{-11} rad/epithermal neutron, is shown in the "total dose" curves.

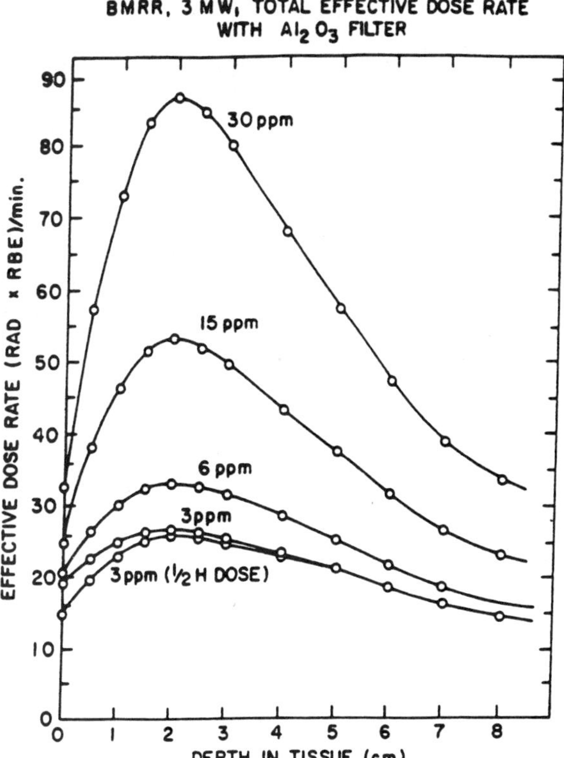

Figure 7. Total effective dose rate curves are shown for the same parameters as Figures 5 and 6 but with 30, 15, 6, and 3 ppm ^{10}B added to tissue.

seen that the peak dose to normal (boron free) tissue (at ~2 cm) would then be reduced by ~5%. The net reduction in total biologically effective dose to normal tissue due to the 50% reduction in fast-neutron dose is insignificantly small, and becomes increasingly so as the presence of boron is introduced in normal tissues.

Figure 7 illustrates the situation in which boron is present in tissue. Total biologically effective dose curves are shown for 30, 15, 6, and 3-ppm ^{10}B. If it is assumed that 30 ppm is in tumor and 3 ppm is in normal tissue (i.e., a tumor-to-normal tissue or "T/N" concentration ratio = 10), reducing the fast-neutron dose component by a factor of 2 (to 2.4×10^{-11} rad/epithermal neutron) would reduce the maximum dose to normal tissue by less than 4%. The effect of reducing the fast-neutron dose is included in the 3-ppm curve, where the effect would be maximum, but has been ignored for the higher B concentrations, because the significance is minimal. Given that a change in Therapeutic Gain (tumor dose/maximum normal tissue dose) of at least 10% would be necessary to produce significant changes in local control, a reduction of 4% or less in the maximum dose to normal tissue would not be worthwhile. Reducing the fast-neutron dose would have even less significance for T/N ratios <10. With present compounds, T/N ratios in excess of 5 are not expected.

It should be emphasized here that we believe the parameter of importance is the Therapeutic Gain (TG). The consensus is that for the effective treatment of brain tumors, a tumor dose 50% in excess of normal tissue dose is needed to approach curative levels (i.e., TG = 1.5). Because TGs in excess of 1 are not available with conventional therapy (as tumor dose is limited to the tolerance of normal tissues supporting the tumor), the potential of NCT to deliver TGs of from 2 to 3, as indicated in Figure 7, becomes important.

Table Four

Depth and Thermal Flux at Peak for Mono-Directional Neutron Beams
with 1.0 n/cm^2 (from [19]).

Source Energy	Beam Diameter (cm)	Medium	Depth at Peak (cm)	Thermal Flux at Peak (n/cm^2)
29 keV	16.6	Tissue-Equiv.	3.4	1.94
1.4 eV			1.9	3.29
1/E			2.6	2.56
29 keV	10.0	Tissue-Equiv.	3.4	1.11
1.4 eV			1.9	2.26
1/E			2.4	1.74

Radiation therapy is based on the premise that the dose to normal tissue will be raised to the tolerance levels, in the hope of achieving toxic levels in tumor. The dose to normal tissue, as well as therapeutic gain, can be determined from Figure 7 where tumor and normal tissue dose can be evaluated based on depth in tissue and boron content. It appears unlikely that radiation oncologists would deliver whole-brain irradiation in a single effective dose exceeding 1000 (rad x RBE). Assuming 30 ppm in tumor and a T/N of 5, therapy in a single dose would take ~30 min, as evaluated from Figure 7. With fractionated therapy, effects of edema would be reduced and total dose could be increased. A point of major importance in NCT is that the ability to target ^{10}B selectively to tumor potentially provides beam localization on a cellular level. In view of the local recurrence characteristic of malignant brain tumors, it is anticipated that large irradiation fields will be used (i.e., >10-cm diameter fields), in order to include areas of potential recurrence in the treatment volume. Protection of normal tissue should come from clearance or restriction of ^{10}B from these tissues (i.e., T/N ratios $\geq$5).

It has been suggested that 24-keV neutrons could be useful for clinical applications of NCT [12]. The fast-neutron dose at the surface would be ~24x10^{-11} rad/incident neutron [13], as opposed to the value of 4.9x10^{-11} rad/incident neutron for an Al_2O_3-moderated beam with a 1/E spectral distribution. Given the dose distribution shown in Figure 6, it is clear that if the fast-neutron (H) dose is increased by a factor of ~5 (as it would be for the 24-keV beam), the therapeutic gain would be reduced significantly. Such a reduction in TG may be unwarranted because intense beams of 1/E-neutrons are now available, as described in this paper. Further, the distribution of thermal neutrons generated by 1/E, 2-keV, and 24-keV neutrons in water has been reported to be similar, in that the location of peak thermal-neutron flux density and depth of penetration does not vary significantly. Thus, the increased surface dose from 24-keV neutrons may not be offset by increased depth of penetration [14]. The above analysis is based on the assumption of a one-to-one correspondence between incident neutron intensity and thermal-neutron flux generated at depth in a head phantom. Table Four shows that this assumption is expected to be valid for 24-keV neutrons. It is also assumed that the RBEs are similar for the beams in question.

Three important points should be noted from the above discussion:

1. The maximum dose to normal tissue occurs at a depth of 2 cm for all conditions (i. e., with or without boron). Thus, one would expect brain tissues at ~2-cm depth to be the "critical" organ.

2. The biological half-life for BSH in humans has been found to be from 6-10 hours [15] to a few days [16], when administered slowly. These data are supported by

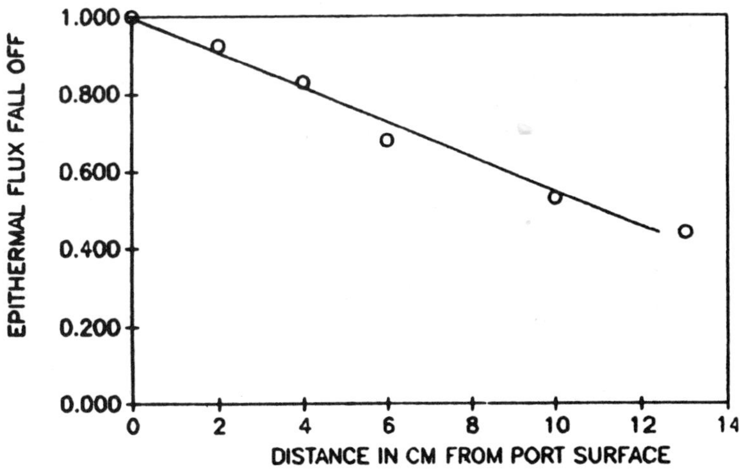

Figure 8. Fall-off in air of the epithermal-neutron flux density along the beam axis.

similar findings obtained following intravenous infusion in rodents [17]. Therefore, it is unlikely that tumor-boron concentrations will be reduced during therapy by amounts in excess of ~5%. In addition, the dimer form of BSH (BSSB) has been shown to have a longer biological half-life and to produce tumor-boron concentrations of about twice those found with BSH [17]. Thus, it is quite likely that BSSB will be found advantageous for new U.S. clinical trials of BNCT for malignant gliomas. In any case, because it is anticipated that therapy will be delivered in multiple fractions, as recommended by the recent International Workshop on Clinical Aspects of NCT [3], time per fraction should be <30 min.

3.   For tumors within 1-2 cm of the midline of the head (~7-cm depth), it is expected that bilateral (opposed port) irradiations will be used. Previous studies have shown that with an optimized epithermal-neutron beam, 35-ppm ^{10}B and a T/N of 10, Therapeutic Gain (TG) is equal to 3.6 over the central 10 cm of a phantom head [18]. With the current example in which 30 ppm is assumed in tumor and a T/N of 5, the use of bilateral irradiations would raise the TG at the midline (7 cm) from ~1.2 to 2. At the BMRR, bilateral irradiations would be carried out with a 2.5-cm thick collimator at the point of irradiation. The fall-off in air of the beam intensity at 2.5 cm from the port face is about 10%, as shown in Figure 8.

SUMMARY

NCT is a binary system, in which ^{10}B is physiologically targeted to tumor and then allowed to interact with thermal neutrons generated in the treatment volume by an externally applied neutron beam. Consequently, an unusually large number of parameters are obtained, which bear on the resultant Therapeutic Gain (TG). However, a perusal of these data, as illustrated in Figure 7, indicates that the TG would increase significantly beyond values projected in this paper if the absolute amount of ^{10}B could be increased above 30 ppm. For example, increasing ^{10}B concentration in tumor to 45 ppm would increase TG by ~33% (with a T/N of 5). A similar increase in TG would follow an increase in T/N from 5 to 10. Those associated with the development of boron compounds for NCT feel that such developments are within reach.

ACKNOWLEDGMENTS

This research was supported in part by the U.S. Department of Energy under Contract No. DE-AC02-76CH00016. Accordingly, the U.S. Government retains a nonexclusive, royalty-free license to publish or reproduce the published form of this contribution, or allow others to do so, for U.S. Government purposes.

REFERENCES

1. Proc. Workshop on Boron Compounds Suitable for Neutron Capture Therapy for the Treatment of Cancer, Sponsored by Rad. Res. Program, Div. of Cancer Treatment, NCI, Bethesda, MD; Held at Annapolis, MD (May 1988).

2. Proc. Third Int. Symp. on Neutron Capture Therapy, Strahlenther. Onkol., D. Gabel, ed., 165(2/3):5-257 (1989).

3. Proc. 1988 Workshop on Clinical Aspects of Neutron Capture Therapy, R. G. Fairchild, V. P. Bond, and A. D. Woodhead, eds., Basic Life Sciences Series, Vol. 50, Plenum Press, New York (1989).

4. Proc. U.S. Dept. of Energy 1986 Workshop on Neutron Capture Therapy, R. G. Fairchild and V. P. Bond, eds., Brookhaven National Laboratory, BNL-51994 (1987).

5. J. B. Godel, "Description of Facilities and Mechanical Components, Medical Research Reactor (MRR)," Brookhaven National Laboratory, BNL-600 (T-173) (1960).

6. R. G. Fairchild, J. Kalef-Ezra, S. Fiarman, and F. Wheeler, "Physics Aspects of Boron Neutron Capture Therapy: Epithermal Neutron Beam Optimization," in Proc. 1988 ANS Int. Reactor Physics Conf., Jackson Hole, WY, Vol. II, p. 2-423 (Sept. 1988).

7. R. G. Fairchild, J. A. Kalef-Ezra, S. Fiarman, L. Wielopolski, J. Hanz, S. Mussolino, and F. Wheeler, "Optimization of an Epithermal Neutron Beam for NCT at the Brookhaven Medical Research Reactor (BMRR)," Strahlenther. Onkol., 165(2/3):84 (1989).

8. F. J. Wheeler, D. K. Parsons, D. W. Nigg, D. E. Wessol, L. G. Miller and R. G. Fairchild, "Physics Design for the Brookhaven Medical Research Reactor Epithermal Neutron Source." (These Proceedings.)

9. R. G. Fairchild, "Development and Dosimetry of an 'Epithermal' Neutron Beam for Possible Use in Neutron Capture Therapy. I: 'Epithermal' Neutron Beam Development," Phys. Med. Biol., 10(4):491 (1965).

10. R. G. Fairchild and J. L. Goodman, "Development and Dosimetry of an 'Epithermal' Neutron Beam for Possible Use in Neutron Capture Therapy. II: Absorbed Dose Measurements in a Phantom Man," Phys. Med. Biol., 11:15 (1966).

11. S. K. Saraf, J. Kalef-Ezra, R. G. Fairchild, B. H. Laster, S. Fiarman, and E. Ramsey, "Epithermal Beam Development at the BMRR: Dosimetric Evaluation." (These Proceedings.)

12. C. A. Perks, A. J. Mill, G. Constantine, K. C. Harrison, and J. A. B. Gibson, "A Review of Boron Neutron Capture Therapy (BNCT) and the Design and Dosimetry of a High Intensity, 24 keV, Neutron Beam for BNCT Research," Brit. J. Radiol., 61:1115 (1988).

13. "Protection Against Neutron Radiation," <u>U.S.N.B.S. Handbook 63</u> (1965).

14. R. C. Greenwood, "The Design of Filtered Epithermal Neutron Beams for BNCT," in <u>Proc. U.S. Dept. of Energy 1986 Workshop on Neutron Capture Therapy</u>, R. G. Fairchild and V. P. Bond, eds., Brookhaven National Laboratory, <u>BNL-51994</u>, p. 123 (1987).

15. G. C. Finkel, C. E. Poletti, R. G. Fairchild, D. N. Slatkin, and W. H. Sweet, "Distribution of ^{10}B after Infusion of $Na_2{}^{10}B_{12}H_{11}SH$ into a Patient with Malignant Astrocytoma: Implications for Boron Neutron Capture Therapy," <u>Neurosurgery</u>, 24:6 (1989).

16. W. H. Sweet, J. R. Messer, and H. Hatanaka, "Supplementary Pharmacological Study Between 1972 and 1977 on Purified Mercaptoundecahydrododecaborate," in <u>Boron-Neutron Capture Therapy for Tumors</u>, H. Hatanaka, ed., Nishimura Co., Ltd., Niigata, Japan, p. 59 (1986).

17. D. N. Slatkin, D. D. Joel, R. G. Fairchild, P. L. Micca, M. M. Nawrocky, B. H. Laster, J. A. Coderre, G. C. Finkel, C. E. Poletti, and W. H. Sweet, "Distributions of Sulfhydryl Borane Monomer and Dimer in Rodents and Monomer in Humans: Boron Neutron Capture Therapy of Melanoma and Glioma in Boronated Rodents," in <u>Proc. 1988 Workshop on Clinical Aspects of Neutron Capture Therapy</u>, R. G. Fairchild, V. P. Bond, and A. D. Woodhead, eds., Basic Life Sciences Series, Vol. 50, Plenum Press, New York, p. 179 (1989).

18. R. G. Fairchild, D. N. Slatkin, J. Coderre, P. Micca, B. Laster, S. B. Kahl, P. Som, I. Fand, and F. Wheeler, "Optimization of Boron and Neutron Delivery for Neutron Capture Therapy (NCT)," <u>Pigment Cell Research</u>, 2:309 (1989).

19. F. J. Wheeler, "Neutron Flux Calculations for Head Phantom," Internal Memo, <u>Whlr-22-87</u> (1987).

NEUTRON CAPTURE THERAPY BEAMS AT THE MIT RESEARCH REACTOR

J. R. Choi,[†] S. D. Clement,[†] O. K. Harling,[†] and R. G. Zamenhof[*]

[†] Nuclear Reactor Laboratory
Massachusetts Institute of Technology
Cambridge, MA

[*] Department of Radiation Oncology
Tufts-New England Medical Center
Boston, MA

ABSTRACT

Several neutron beams that could be used for neutron capture therapy at MITR-II are dosimetrically characterized and their suitability for the treatment of glioblastoma multiforme and other types of tumors are described. The types of neutron beams studied are: 1) those filtered by various thicknesses of cadmium, D_2O, ^6Li, and bismuth; and 2) epithermal beams achieved by filtration with aluminum, sulfur, cadmium, ^6Li, and bismuth. Measured dose vs. depth data are presented in polyethylene phantom with references to what can be expected in brain. The results indicate that both types of neutron beams are useful for neutron capture therapy. The first type of neutron beams have good therapeutic advantage depths (approximately 5 cm) and excellent in-phantom ratios of therapeutic dose to background dose. Such beams would be useful for treating tumors located at relatively shallow depths in the brain. On the other hand, the second type of neutron beams have superior therapeutic advantage depths (greater than 6 cm) and good in-phantom therapeutic advantage ratios. Such beams, when used along with bilateral irradiation schemes, would be able to treat tumors at any depth in the brain. Numerical examples of what could be achieved with these beams, using RBEs, fractionated-dose delivery, unilateral, and bilateral irradiation are presented in the paper. Finally, additional plans for further neutron beam development at MITR-II are discussed.

INTRODUCTION

In recent years, there has been renewed interest in using neutron beams along with boron-containing chemical carriers to treat a form of brain tumor, called glioblastoma. This approach, called neutron capture therapy or NCT, offers the potential for a new treatment modality which could cure this otherwise fatal disease. Because it is desired to be able to treat this tumor at any depth in the brain, there have been significant efforts to develop a neutron beam of epithermal energy that can provide useful neutron fluxes at up to 7-cm depth in tissue. Seven centimeters is about one half the brain diameter viewed from the side of a typical human head.

In this paper, we report on the design and performance of several neutron beams at the Massachusetts Institute of Technology's Research Reactor (MITR-II). MITR-II is a heavy-water reflected, light-water cooled and moderated nuclear reactor which utilizes flat,

Neutron Beam Design, Development, and Performance for Neutron Capture Therapy
Edited by O. K. Harling *et al.*
Plenum Press, New York, 1990

plate-type, finned, aluminum clad fuel elements highly enriched in U-235. The high power density core design enables the reactor to deliver high neutron fluxes up to 10^{14} n/cm^2-s, fast or slow, at the licensed power level of five megawatts thermal. This high neutron flux can be accessed through the medical therapy beam aperture located directly below the core. Through the use of various filters, including D_2O, cadmium, ^{6}Li, bismuth, aluminum, and sulfur, the average beam energy can be shifted to either increase intensity or increase penetration. Several different filter configurations have been studied to experimentally determine neutron beam performance for NCT of tumors at various depths in the brain. With relatively minor modifications to the existing beam port, a range of filtered beam performance is obtainable at MITR-II. Useful neutron beams for therapy range from thermal to epithermal in energy and approach the best possible ideal beams. Test results for several of these beams are discussed in this paper.

EXPERIMENTAL PROCEDURES

NCT Dosimetry

Dosimetry in NCT is made complicated by the fact that there are three major radiation dose components that need to be measured simultaneously and separately. These are doses due to thermal neutrons, fast neutrons, and gamma rays. In addition, the dosimetry is further complicated by the induced gamma rays produced from various materials in the tissue through the capture of thermal neutrons. Therefore, it is not sufficient to merely make measurements in air, but rather measurements must be made in phantom to simulate the neutron moderation and capture in tissue.

To simulate the moderation of neutrons in the human brain, cylindrical polyethylene phantoms, 20.3-cm high and 17.8 cm in diameter, were used in all measurements of beam performance. The polyethylene phantoms differ from the actual human brain tissue in several different ways. First, because the chemical composition of the brain tissue and the polyethylene plastic are significantly different, the hydrogen and carbon fractions and densities are different (see Table One). Second, the polyethylene lacks any oxygen or nitrogen.

Table One. Polyethylene Head Phantom Specification and Composition.

Overall Specifications					
Shape:	Right-circular cylinder				
Dimensions:	17.8 cm in diameter, 20.3 cm in height				
Composition:	Polyethylene, 0.143 wt. fraction H, 0.857 wt. fraction C				
Density:	0.913 g/cc				
Hydrogen Density:	7.68 E+22 atoms/cc				
Material	Density	Hydrogen	Nitrogen	Carbon	Oxygen
Brain	1.05 g/cc	10.57%	1.84%	13.97%	72.59%
Polyethylene	0.91 g/cc	14.29%	N/A	85.71%	N/A

The hydrogen density difference is important for two reasons. First, because hydrogen is such an excellent moderator of neutrons, the different hydrogen density will result in a different thermal neutron flux at depth. A higher hydrogen density will result in faster moderation and in more capture of neutrons, which will produce a lower neutron flux at depth. Monte Carlo calculations by Clement [1] of neutron moderation in polyethylene and in brain-equivalent material show about ten percent higher neutron flux at depth (5 to 8 cm) in brain-equivalent material. One can therefore expect from that, because the thermal neutron capture in brain-equivalent material is less, the induced gamma rays produced will also be less. Indeed, calculations also show lower induced gamma-ray doses in brain-equivalent material compared to polyethylene [1].

The lack of oxygen and nitrogen in the phantom seems at first to be an important omission. However, oxygen behaves in a similar manner to carbon in neutron interactions [2]. Therefore, carbon is often substituted for oxygen in tissue dosimetry with little loss in accuracy. The principal nitrogen interaction with the thermal neutrons is the ^{14}N(n,p)^{14}C reaction [3]. Because the range of the resulting proton is only a few microns, all the energy from the ^{14}N(n,p)^{14}C reaction is deposited locally, where the neutron is absorbed. Therefore, rather than measuring the dose, it can be calculated using the ^{14}N(n,p)^{14}C reaction cross section and the appropriate KERMA factors provided that the correct thermal neutron fluxes are known.

Despite the differences in the neutron interaction properties of polyethylene and brain, phantom measurements using the former are very useful in evaluating the performance of NCT beams because they provide relative comparisons between different designs. If desired, accurate beam performance in actual brain tissue can be calculated using the Monte Carlo technique discussed by Clement et al. in another paper in these proceedings [1]. For final dosimetric evaluation of neutron beams, tissue-equivalent head phantoms of realistic head geometry will be used.

<u>Dose Measurements</u>

Thermal neutron flux measurements were made by activation of pure gold foils as described in the ASTM protocol for determining thermal neutron fluxes [4]. The gold foils used in these experiments were small disks 0.35 cm in diameter and 0.002 cm in thickness and weighed approximately 4 mg. At this diameter and thickness, the self shielding and the flux depression that often go along with gold foil measurement are not significant [5]. The activated gold foils were counted in a computerized Ge-Li detector connected to an MCA and an ND9900 controller. The counter was calibrated with an NBS standard source and the efficiency of the detector in the desired energy range was determined with suitable standards. The estimated accuracy of our thermal neutron flux measurements is $\pm$ 5%.

The depth-dose distribution of thermal neutrons along the central axis was obtained by activating gold foil detectors in the center of the polyethylene disks, making up the phantom, at various depths. Appropriate KERMA factors were then used with these fluxes to obtain the ^{10}B(n,α)^{7}Li dose and the ^{14}N(n,p)^{14}C dose as a function of depth in the phantom.

The fast neutron and gamma-ray dose rates in brain were determined by utilizing the paired-ion chamber technique in phantom [6,7]. Usually a neutron sensitive chamber, such as a tissue-equivalent ion chamber, is paired with a neutron insensitive chamber, such as a graphite-walled ion chamber. The ion chambers respond to mixed radiation fields differently and as follows:

$$R = kD_N + hD_\gamma + R_T \tag{1}$$

where:

$$R = \frac{\text{chamber response to total mixed field}}{\text{chamber sensitivity to calibration source}}$$

$$k = \frac{\text{chamber sensitivity to neutrons in the field}}{\text{chamber sensitivity to calibration source}}$$

$$h = \frac{\text{chamber sensitivity to gamma rays in the field}}{\text{chamber sensitivity to calibration source}}$$

R_T = total response from thermal neutrons

D_N = dose from fast neutrons

D_γ = dose from gamma rays

Thermal neutron response is a linear function of the thermal neutron flux and has been determined by experiment [8]. Once the thermal neutron response is subtracted from equation (1) above, the resulting equation becomes a function of fast-neutron doses, gamma-ray doses, neutron sensitivity, and gamma-ray sensitivity. Because these ion chambers have similar sensitivity to gamma rays whether it be from mixed fields or calibration sources, gamma-ray sensitivity, h, can be set equal to 1. Then equation (1) can be re-written for each chamber, where the subscripts S and U stand for tissue-equivalent (TE) chamber and the graphite chamber respectively. Solving these two equations simultaneously for D_N and D_γ results in equations (2) and (3), where R_S and R_U are the responses of the TE and the graphite-walled ion chambers after correcting for the thermal neutron responses:

$$D_N = \frac{R_S - R_U}{k_S - k_U} \tag{2}$$

$$D_\gamma = \frac{k_S R_U - k_U R_S}{k_S - k_U} \tag{3}$$

Neutron sensitivity factors, k_S and k_U, are functions weakly dependent on neutron energy. They can be found in the literature for many common detector materials. Moreover, because they can be numerically integrated to fit any given neutron spectrum, the measured responses of the two chambers, R_S and R_U, can be used to determine the doses from fast neutrons and gamma rays (D_N and D_γ).

Ion chambers, used for these measurements, were manufactured by EG&G and utilize continuous gas flow. The A-150, tissue-equivalent, plastic-walled ion chamber was 0.1 cc in volume and was continuously flushed with methane-based tissue-equivalent gas. The graphite-walled ion chamber was 0.2 cc in volume and was continuously flushed with carbon dioxide gas. The current and the charge from the ion chambers were collected with a model 614 Keithley digital electrometer. The chambers were used with a continuous gas flow rate of 20 cc/min and a collection voltage of 250 V.

A polyethylene head phantom identical in composition and overall dimensions to the one use d for gold-foil activation was used for the dual ion chamber measurements. The phantom had predrilled holes at several depths throughout its length where ion chambers could be placed. The holes not being used for measurements were filled with plugs made out of the same polyethylene material. Each ion chamber was used to make dose measurements along the central axis at various depths in the phantom. Combining the results with the thermal neutron correction factor as described in the previous section and making use of the thermal neutron flux depth profile allowed the construction of separate depth-dose profiles for fast-neutron dose, total gamma-ray dose, thermal-neutron dose (^{14}N(n,p)^{14}C), and B-10 dose (^{10}B(n,α)^{7}Li) along the central axis.

THE MITR-II MEDICAL IRRADIATION FACILITY

General Description

The medical irradiation facility at MITR-II includes a surgical operating room that is located directly beneath the reactor core (see Figure 1). An opening in the room's ceiling with a multi-layered shutter system allows the neutron treatment beam to be admitted. Additional absorbers, filters, and special collimators can be inserted in the beam-line. The room is outfitted with a scrub sink, X-ray viewer, and fluorescent lights. There also exists a hydraulically elevated operating table.

The currently available, unmodified MITR-II medical therapy beam shown in Figure 2 is a high intensity, D_2O-filtered neutron beam with a bismuth shield in its path to reduce the incident gamma radiation. The beam is controlled by three shutters. These shutters are the H_2O shutter tank, the boral shutter, and the lead shutter. The beam can be moderated or shifted in energy by the D_2O filter/moderator located above the H_2O shutter tank.

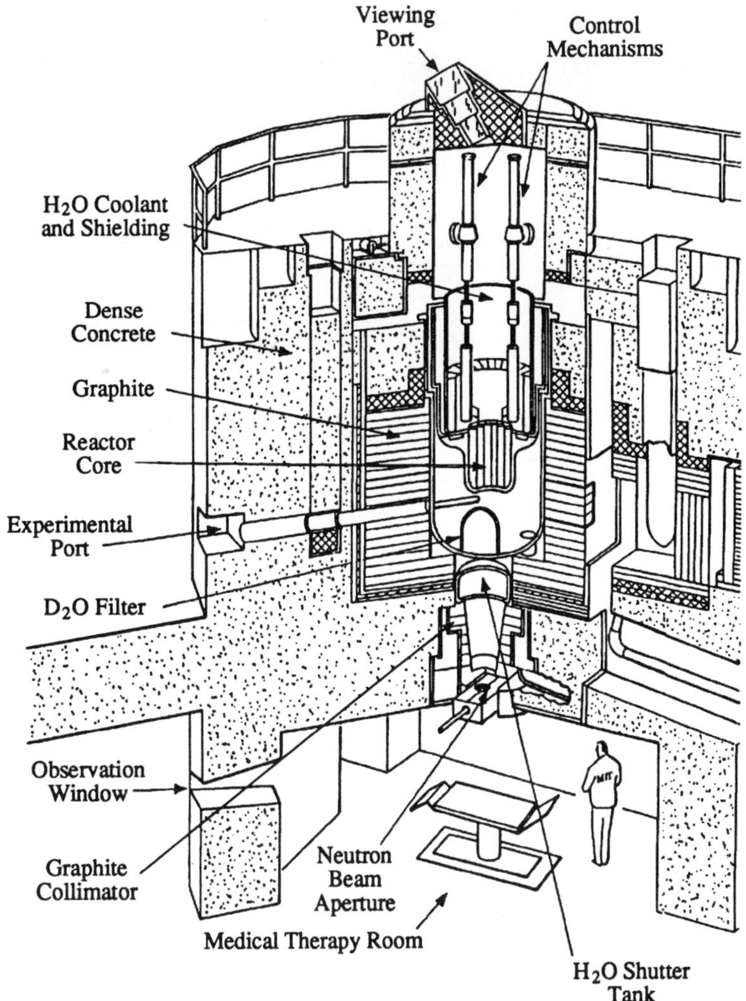

Figure 1. Isometric view of the medical irradiation facility at MITR-II showing the relative location of the core, the D_2O filter, H_2O shutter tank, and the graphite collimator.

The D_2O tank level, which can be adjusted infinitely within its two extreme settings, varies the energy spectrum of the beam. In its full open position with only the 15.2 cm of bismuth filter, a thermal neutron flux of greater than 3.5×10^{10} n/cm^2-s is available at the patient position. The cadmium ratio, as determined from gold foils, is greater than 50. In its fully closed position, with a similar bismuth filter, the thermal neutron flux decreases to about 2.3×10^{10} n/cm^2-s and the cadmium ratio rises to greater than 400. The H_2O shutter tank, the boral shutter, and the lead shutter are located in the ceiling of the medical therapy room and comprise independently operable shutter systems which are used to turn the beam on and off.

Between the lead shutter and the H_2O shutter tank, there is approximately 60 liters of empty volume in the tapered graphite collimator, in which various filter materials may be placed. This space can be filled with aluminum and/or sulfur to create an epithermal neutron beam or it can be left void to provide a lower energy neutron beam. Because of the various shutter systems, modifications can be made to the MITR-II beam without affecting normal reactor operation. The medical beam can also be used without affecting other experiments in the reactor.

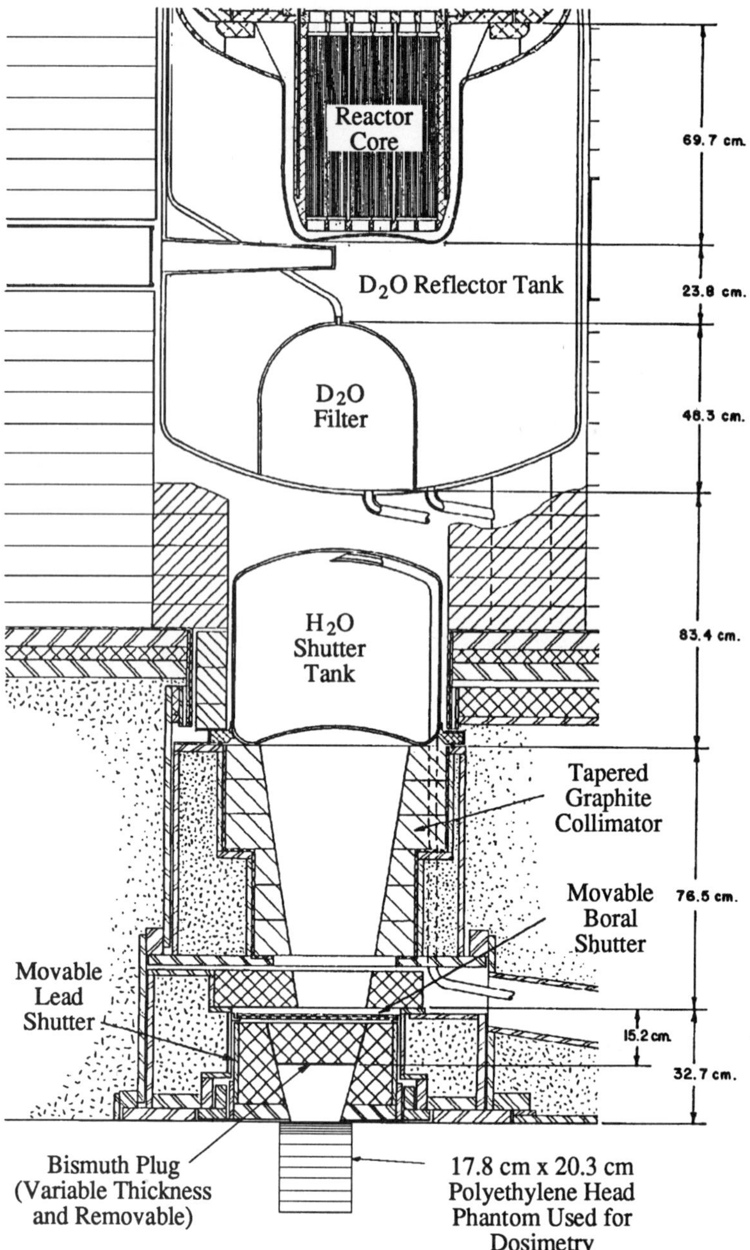

Figure 2. Cross-sectional view of the MITR-II medical therapy beam with its system of shutters.

<u>Performance of the D$_2$O/Bismuth-Filtered Beams</u>

During the course of this study, two major parameters were changed in the beam. The first was the amount of D$_2$O present in the D$_2$O filter/moderator. The second was the presence of a very thin sheet of cadmium right above the 15.2-cm bismuth filter. Both the D$_2$O filter and the sheet of cadmium function to shift the energy spectrum of the neutron beam.

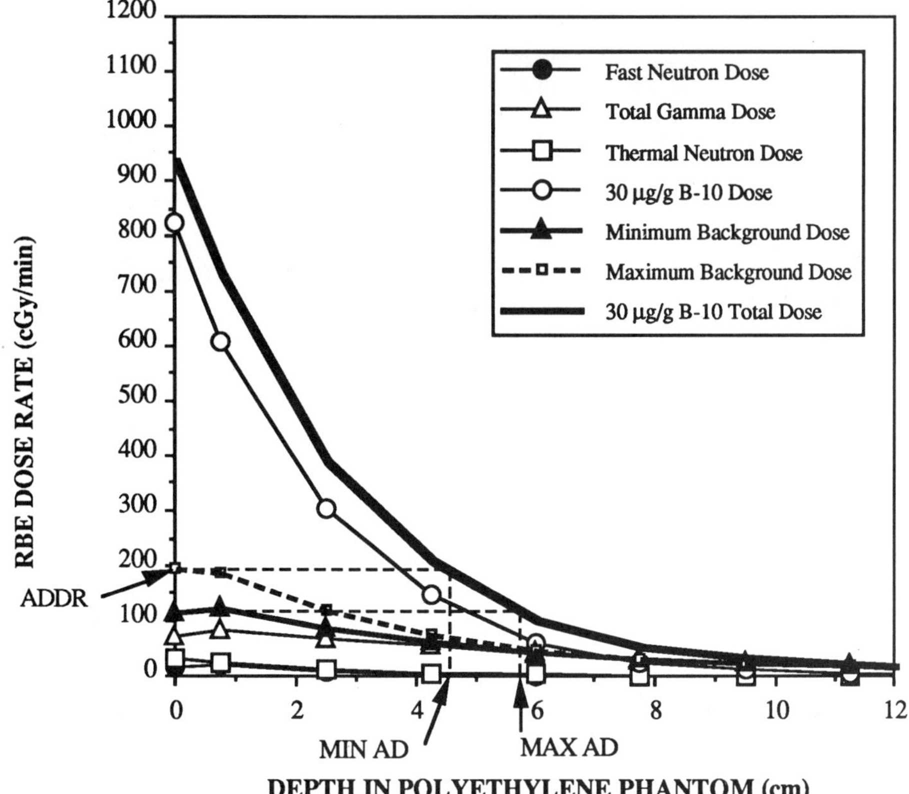

Figure 3. Typical comprehensive dose vs. depth curve along the central axis of the polyethylene phantom for the D_2O/bismuth-filtered beam at MITR-II (M11); D_2O at maximum level.

The desired thickness of the cadmium, for an epithermal beam, was determined by the following criteria. The filter should block better than 90% of the 0.025-eV neutrons but should allow more than 90% of the >0.5-eV neutrons to pass through. A cadmium thickness of 0.018 cm (0.007 in) was chosen based on these criteria. It will block more than 90% of the 0.025-eV neutrons, but will allow better than 90% of the >0.5-eV neutrons to pass through.

Comprehensive dose vs. depth profile measurements were made in the polyethylene phantom with the D_2O filter empty, 1/2 full, 3/4 full, and completely full. The D_2O filter level was adjusted in the range of completely full to completely empty with an accuracy better than 0.5 cm. With the 0.018-cm (0.007 in) cadmium filter, the dose profile measurements were made with the D_2O filter empty and 5/8 full. A representative dose vs. depth profile along the central axis of the phantom from these studies is shown in Figure 3. For the most part, there is little buildup in the thermal neutron flux. The slopes of the B-10 dose and the thermal-neutron dose show the exponential decay with depth that is characteristic of highly-thermalized neutron beams. The total background dose shows rapid decay in intensity with depth. Results summarized in Table Two show various neutron beam characteristics such as the neutron flux, the cadmium ratio, and the depth at which the maximum neutron flux occurred.

There are three figures of merit used to described the various NCT beams. These are described more fully in the companion paper by Clement et al. [1]. Briefly, they are the advantage depth (AD), the advantage ratio (AR), and the advantage depth dose rate (ADDR). The advantage depth (AD) is defined as the depth where the total therapeutic dose is equal to the maximum background dose. The maximum advantage depth is for the

Table Two. Thermal-neutron flux and cadmium ratio at the surface of the polyethylene phantom for D_2O, bismuth, and cadmium-filtered beams at MITR-II. Measurements were made along the central axis of the polyethylene phantom.

Filter Number	D_2O Filter Level	Cadmium Presence	Maximum Thermal Neutron Flux	Cadmium Ratio
M10	Empty	None	3.5 E+10 n/cm^2-s @ 0.0 cm	60
M13	1/2	None	2.5 E+10 n/cm^2-s @ 0.0 cm	60
M14	3/4	None	2.4 E+10 n/cm^2-s @ 0.0 cm	250
M11	Full	None	2.3 E+10 n/cm^2-s @ 0.0 cm	440
M16	Empty	0.018 cm	5.4 E+09 n/cm^2-s @ 0.6 cm	7
M15	5/8	0.018 cm	2.2 E+09 n/cm^2-s @ 0.0 cm	40

hypothetical situation where the tumor-to-blood boron dose ratio is equal to infinity. The minimum advantage depth is where the tumor-to-blood boron dose ratio is set equal to 10 to 1. An effective 10 to 1 ratio of B-10 dose in tumor to blood may be realistically achievable in BNCT applications using the currently available boron compounds. The second important figure of merit is the advantage ratio (AR). This is defined as the quotient of the integral of the total therapeutic dose and the integral of the total background dose over a given depth. For this paper, that depth is chosen to be the maximum advantage depth. The third figure of merit is the advantage depth dose rate (ADDR). This is the total therapeutic dose at the minimum advantage depth. Table Three summarizes the beam characteristics in terms of these figures of merit, advantage depths, advantage ratio, advantage depth dose rate, and the dose fractions of the total integral therapeutic dose due to high-LET radiation dose, low-LET radiation dose, and B-10 dose.

As expected, a major effect of changing the level of the D_2O filter was the shift in the neutron spectrum as indicated by the variations in the cadmium ratios. With increased levels of D_2O in the filter, the mean neutron beam energy decreased and the fast neutron doses decreased as indicated by the relative decrease in the high-LET dose percentages in Table Three. With the D_2O filter more than 3/4 full, the resulting neutron beam energy was very close to thermal. Therefore, the fast neutron doses became inconsequential

Table Three. AD, AR, ADDR, and dose fractions of the total integral therapeutic dose for the D_2O, bismuth, and cadmium-filtered beams at MITR-II. The ADDR is determined at the minimum AD. Measurements were made along the central axis of the polyethylene phantom.

Filter Design Number	D_2O Filter Level	Max/Min RBE Advantage Depth (cm)	RBE Advantage Ratio	RBE Adv. Depth Dose Rate (RBE-cGy/min)	High-LET Dose (%)	Low-LET Dose (%)	30μg/g B-10 Dose (%)
M10	Empty	4.8/3.9	3.5	390	10.7	17.9	71.4
M13	1/2	4.5/3.6	4.0	270	9.8	15.4	74.8
M14	3/4	4.8/3.8	4.7	200	5.9	15.2	78.9
M11	Full	5.7/4.5	5.1	190	4.9	14.6	80.5
M16	Empty†	4.6/4.2	2.2	115	26.1	19.2	54.7
M15	5/8†	4.6/3.8	2.8	31	10.9	24.8	64.3

† 0.018-cm sheet of cadmium present on top of the bismuth plug.

compared to the combined boron and total gamma dose. The background dose in these configurations was almost entirely due to the gamma rays induced in the polyethylene phantom by the H(n,γ)D reaction. Accordingly, the advantage ratio is very favorable, being around 5. This is better than 80% of what is possible from an ideal neutron beam of thermal energy [1].

The D_2O, bismuth, and cadmium-filtered beams at MITR-II are very high in intensity and are well suited for the treatment of shallow tumors due to their high advantage ratio and relatively low background doses. The total treatment time for most of these beams, given a maximum background dose to healthy tissue of 2000 RBE-cGy, is less than 20 minutes. The maximum advantage depths in polyethylene, close to 5 cm for some of these beams, allow them to be used for tumors up to intermediate depths.

<u>Engineering Design of the Aluminum/Sulfur-Filtered Epithermal Beam</u>

Exhaustive Monte Carlo studies have indicated that an MITR-II neutron beam filtered by approximately 35 cm of pure sulfur and 10 cm of bismuth would give favorable results for neutron capture therapy [1]. These studies also revealed that this filter should be placed far away from the patient in order to reduce the background dose from filter-induced gamma rays. Although a filter made of solid sulfur is desirable for its neutron characteristics, sulfur is not a good material to work with from an engineering point of view. It is very fragile, forms toxic vapors if heated to over 100 °C, and readily reacts with hydrogenous material to form many different toxic or corrosive compounds. However, if the sulfur can be isolated from the surroundings in an inert medium that can also provide structural strength and integrity, then such a filter can readily be inserted in a neutron beam. Aluminum is an excellent material for this purpose. It is readily available and can be machined easily. Furthermore, it also possesses a useful neutron transmission window for epithermal neutrons much like that of sulfur [9]. Indeed, subsequent Monte Carlo calculations of a filter made of sulfur together with the necessary aluminum structure predicted dose vs. depth characteristics similar to that of a pure sulfur filter.

Based on these considerations, a solid cylindrical crystalline sulfur filter 35 cm in height and 15.2 cm in diameter, encased in an aluminum can, was designed and fabricated for installation close to the bottom of the H_2O shutter tank. A thin sheet of cadmium, 0.018 cm in thickness, was also utilized with the aluminum sulfur filter. The cadmium sheet was placed on top of the aluminum can to reduce the thermal-neutron induced gamma rays that would be produced by the sulfur. The outer cone region was filled with aluminum shot doped with 1000 ppm by weight of natural lithium followed by bismuth shot. The lithium in this specially formulated Al/Li alloy was designed to reduce the amount of induced gamma rays from the filter structure. Specifically, it was designed to absorb one half of the thermal neutrons that would otherwise be absorbed in aluminum. According to Monte Carlo studies, this ratio provided the best compromise between neutron transmission and production of aluminum-induced gamma rays. The central filter assembly containing the sulfur was held in position 30 cm above the top of the bismuth collimator by an aluminum spacer. Monte Carlo studies also showed that about 8 cm of solid bismuth was required to attenuate the incident core gamma rays. Ideally, this filter also should be placed high in the beam line to provide better collimation. However, because of the complexity and expense associated with placing the bismuth filter close to the core, we decided to place the bismuth filter immediately underneath the sulfur filter. Also, an additional 4.9 cm of bismuth was placed in the newly designed bismuth collimator to reduce any structure-induced gamma rays. Figure 4 shows the configuration of the test sulfur filter with aluminum/lithium shot, and bismuth shot as installed in the graphite collimator region of the MITR-II medical beam.

The entire epithermal filter was designed so that it could be quickly installed and removed. The filter is modular so that the central section containing sulfur, the aluminum spacer support, bismuth, and the outer cone region containing aluminum and bismuth shot are not permanently joined. With the central sections removed, the therapy beam essentially returns to the D_2O/bismuth-filtered beam. The versatile, modular design of the filter allows the complete changeover from sulfur-filtered epithermal beams to D_2O/bismuth-filtered beams in a matter of hours.

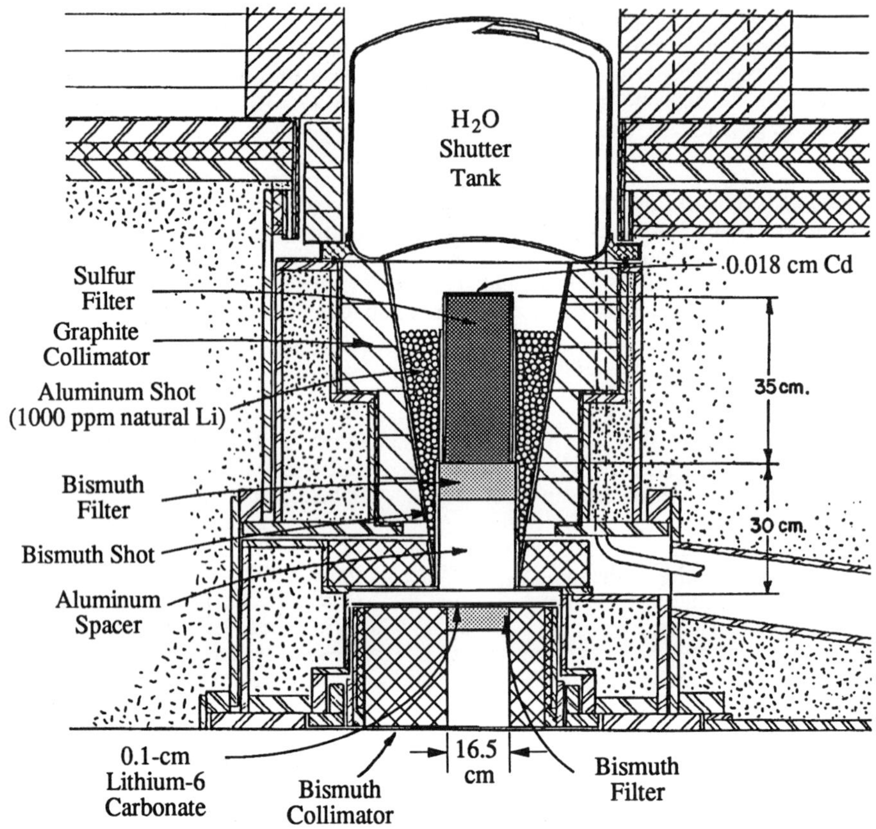

Figure 4. Cross-sectional view of the sulfur and aluminum filters as installed in the tapered graphite collimator region of the medical therapy beam at MITR-II.

Performance of the Sulfur/Aluminum-Filtered Epithermal Beam

With the current aluminum and sulfur filters, the MITR-II therapy beam is transformed to an epithermal beam with acceptably low background radiation. The typical measured dose vs. depth profile along the central axis shows the characteristic epithermal shape with the peak flux occurring at about 1 to 2 cm in depth in polyethylene (see Figure 5). Note that the major background components, fast neutrons and gamma rays, all have intensities on the order of a few RBE-centigrays per minute; while the therapeutic dose is on the order of tens of RBE-centigrays per minute. This order of magnitude difference between the therapeutic dose and the background dose gives an acceptable therapeutic advantage for this beam. The sum of these background doses determines the ADDR and the total treatment time. With these intensities, therapeutic dose to the tumor can be delivered in several fractions or in less than two hours of total irradiation time.

Tables Four and Five summarize the major characteristics of the various measured aluminum/sulfur-filtered beams at MITR-II. The collimation length in Table Five refers to the minimum distance between the lowest point of the epithermal filter to the top of the measuring phantom. One can immediately see the importance of collimation for the advantage depth. All of the filters with collimation lengths greater than 30 cm have advantage depths of 6 cm or more. The drawback to this is that, in all cases, the advantage ratios are not as high as those filters with smaller advantage depths. Note that there is no

 Measured thermal neutron flux and cadmium ratio at the surface of the polyethylene phantom for several sulfur/aluminum-filtered beams at MITR-II. Measurements were made along the central axis of the polyethylene phantom. $^6Li_2CO_3$ was placed on top of the bismuth collimator and/or on top of the phantom.

Filter Number	Bismuth Thickness	$^6Li_2CO_3$ Thickness	Maximum Thermal Neutron Flux	Cadmium Ratio
M19	7.6 cm	0.1 cm	4.5 E+09 n/cm^2-s @ 0.6 cm	5.7
M20	7.6 cm	0.1 cm	7.2 E+08 n/cm^2-s @ 1.0 cm	4.7
M23	7.7 cm	0.1 cm	7.3 E+09 n/cm^2-s @ 0.6 cm	7.5
M25	10.1 cm	0.1 cm	4.3 E+09 n/cm^2-s @ 0.6 cm	7.8
M26	10.1 cm	0.1 cm	1.4 E+09 n/cm^2-s @ 1.0 cm	5.5
M27	15.9 cm	0.1 cm	5.0 E+08 n/cm^2-s @ 0.6 cm	5.5
M37	12.5 cm	0.1 cm	2.3 E+09 n/cm^2-s @ 0.6 cm	6.3
M41	12.5 cm	0.2 cm	1.0 E+09 n/cm^2-s @ 1.3 cm	2.4

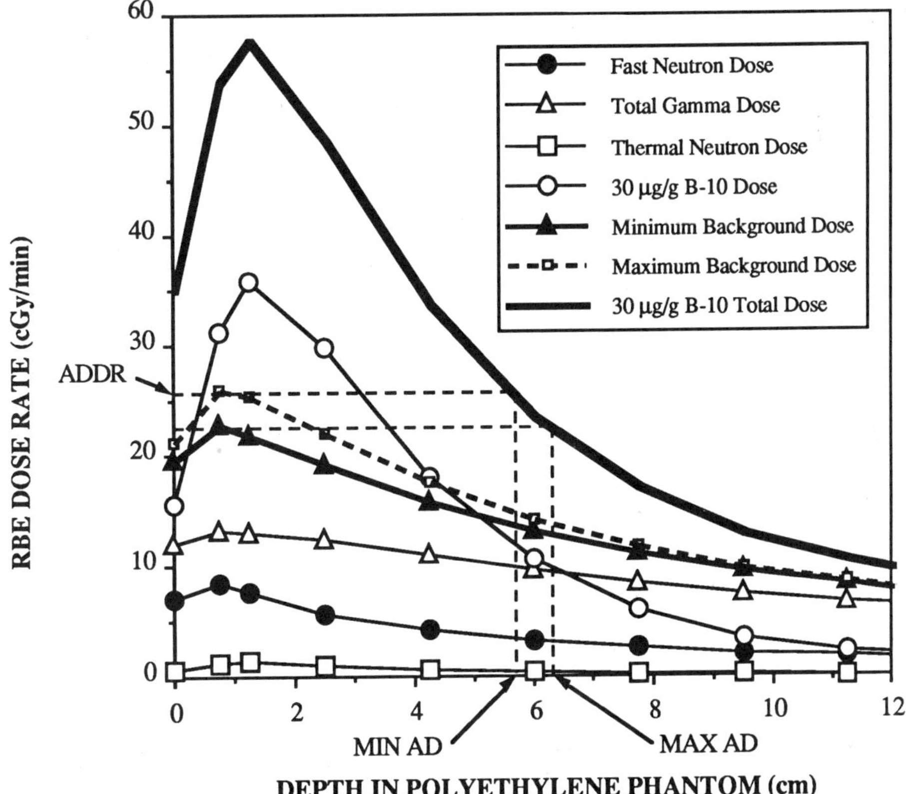

Figure 5. Typical comprehensive dose vs. depth curve along the central axis of the polyethylene phantom for the aluminum/sulfur-filtered beam at MITR-II (M41); D_2O at minimum level.

Table Five.   AD, AR, ADDR, and dose fractions of the total integral therapeutic dose for the sulfur/aluminum-filtered beams at MITR-II. The ADDR is determined at the minimum AD. Measurements were made along the central axis of the polyethylene phantom.

Filter Design Number	Collimation Length (cm)	Max/Min RBE Advantage Depth (cm)	RBE Advantage Ratio	RBE Adv. Depth Dose Rate (RBE-cGy/min)	High-LET Dose (%)	Low-LET Dose (%)	30 µg/g B-10 Dose (%)
M19	20	5.6/4.8	2.58	70	10.1	28.7	61.2
M20	64	7.0/5.9	1.83	17	12.8	41.8	45.4
M23	30	6.0/5.0	2.02	155	3.4	46.1	50.5
M25	10	5.7/4.7	2.68	65	10.3	27.1	62.7
M26	30	5.9/5.0	2.23	25	12.5	32.4	55.2
M27	35	6.2/5.3	3.34	7	14.4	15.5	70.1
M37	23	6.3/5.3	3.21	38	9.9	21.3	68.8
M41	23	6.3/5.7	2.24	26	15.4	29.4	55.2

such thing as the best neutron beam for therapy. Some have better advantage depths, some better advantage ratios, and some better advantage depth dose rates. All are desirable qualities. There is always a tradeoff among these figures of merit. This means that the 'best beam' is a compromise of all three qualities. A beam with good advantage depth such as M20 has a lower advantage ratio. A beam with high intensity such as M23 can deliver a therapeutic dose in less than 20 minutes, but has less advantage depth. A beam with high advantage depth ($\approx$ 75% of ideal beam) and high advantage ratio ($\approx$ 67% of ideal beam) such as M27 has low advantage depth dose rate. However, in summary, all of the aluminum/sulfur-filtered beams can be used in therapy. When minimum treatment time is critical, M23 is the best beam for the job. If maximum advantage depth is deemed important, then M20 can be used. For the highest therapeutic advantage, M27 can be used. The Monte Carlo calculations showed that a maximum advantage depth of 6 cm or more in polyethylene corresponds to 7.5 cm or more in an elliptical brain tissue. This means that all of the neutron beams listed above with an advantage depth of about 6 cm in polyethylene can treat tumors at any depth in the real human brain with therapeutic advantage.

EXAMPLES OF CLINICAL BEAM USE FOR NCT

Clinical Example of Shallow Tumor Treated with D₂O/Bismuth-Filtered Beam

The D_2O/bismuth-filtered beams such as M10 or M11 at MITR-II are well suited for treating shallow tumors through neutron capture therapy. Assuming a patient with a spherical tumor 2 cm in diameter and centered 3 cm below the scalp, the dose profile data for tumor and the surrounding healthy tissue was developed using realistic values for RBE. A tumor boron concentration of 30 µg/g and tumor-to-blood ratio of 3 to 1 were assumed. This implies an effective tumor to blood boron dose ratio of about 10 to 1 as discussed by Clement et al. [1]. The total maximum RBE-dose to healthy tissue was limited to 2000 RBE-cGy. Effective RBEs for fast neutrons, recoil protons, and alpha particles of 1.6, 1.6, and 2.3 were used respectively. Because fractionated therapy was assumed, an RBE of 0.5 was also assumed for all gamma-ray doses. For simplicity, results of neutron moderation and absorption in the elliptical human head were taken from experiments conducted in a cylindrical polyethylene head phantom with the specifications given in Table One.

Figure 6, Figure 7, and Table Six present the results for the described scenario assuming the M11 beam with the characteristics given in Tables Two and Three. Figure 6 shows the dose profile for unilateral irradiation of the tumor. Note the very substantial

Table Six. Summary of total RBE/fractionated doses delivered to tumor lying 3 cm below the surface by a D₂O/bismuth-filtered beam (M11) at MITR-II. A boron concentration of 30 μg/g in tumor and an effective B-10 dose ratio in tumor-to-blood of 10 to 1 is assumed. Maximum RBE-dose to healthy tissue is limited to 2000 RBE-cGy.

Type of Irradiation	Unilateral	Bilateral
Maximum Tumor RBE-Dose:	5500 (RBE-cGy)	5500 (RBE-cGy)
Average Tumor RBE-Dose:	3900 (RBE-cGy)	3950 (RBE-cGy)
Maximum Background RBE-Dose:	2000 (RBE-cGy)	2000 (RBE-cGy)
Average Background RBE-Dose:	720 (RBE-cGy)	1360 (RBE-cGy)

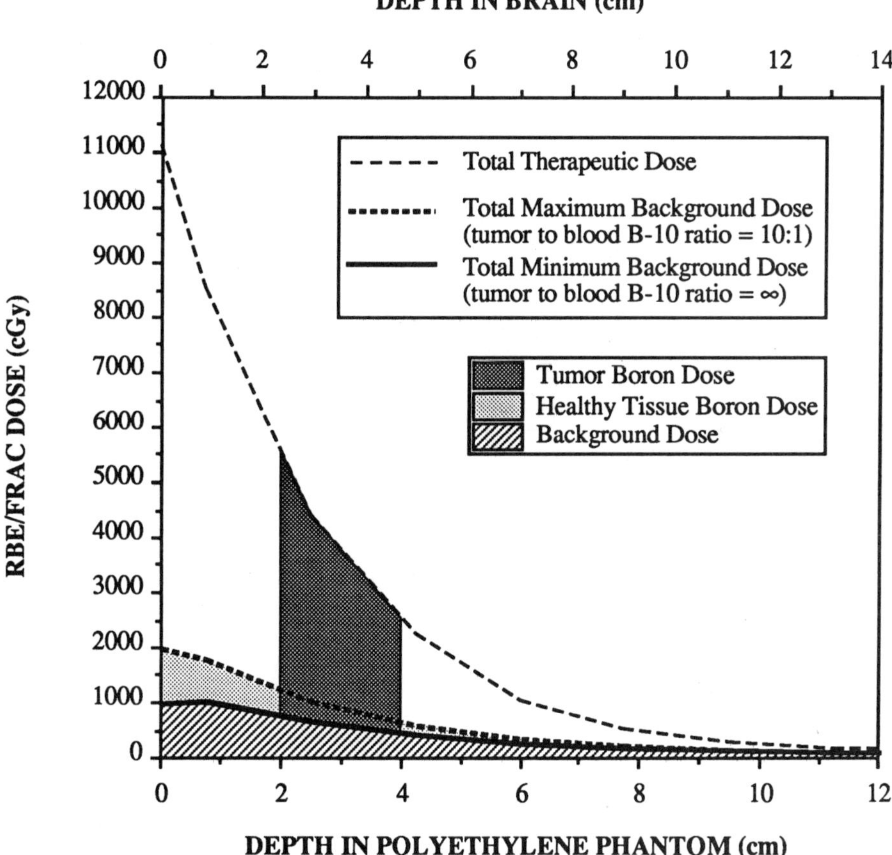

Figure 6. Measured dose along the central axis of the polyethylene phantom for the treatment of a 2-cm diameter tumor located 3 cm below the surface using D₂O/bismuth-filtered beam (M11) at MITR-II based on maximum background RBE-dose of 2000 RBE-cGy. Unilateral fractionated irradiation into the polyethylene head phantom. Equivalent brain depth is based on hydrogen atom density.

increase in dose delivered to the tumor vs. the healthy tissue. The average RBE-dose delivered to the tumor is around 4000 RBE-cGy compared to the average RBE-dose delivered to the healthy tissue of only 720 RBE-cGy. Maximum background RBE-dose to any healthy tissue is 2000 RBE-cGy.

Figure 7, on the other hand, is a dose vs. depth plot of bilateral irradiation of the same tumor. For bilateral irradiation, the diameter of the polyethylene head was chosen to be 12 cm. Because the hydrogen density in polyethylene is 17% greater than in brain, 12 cm of polyethylene has similar neutron moderating properties as 14 cm of brain. Fourteen centimeters is the average side-to-side width of the human head. Therefore, the results from calculations on 12 cm of polyethylene can be applied to the human head with reasonable accuracy.

For bilateral irradiation, the average RBE-dose delivered to the tumor is also around 4000 RBE-cGy. However, the average background RBE-dose is now over 1300 RBE-cGy. The maximum background RBE-dose, because it is the limiting dose to the patient, was kept at 2000 RBE-cGy. There is also a drop in the tumor to background dose ratio.

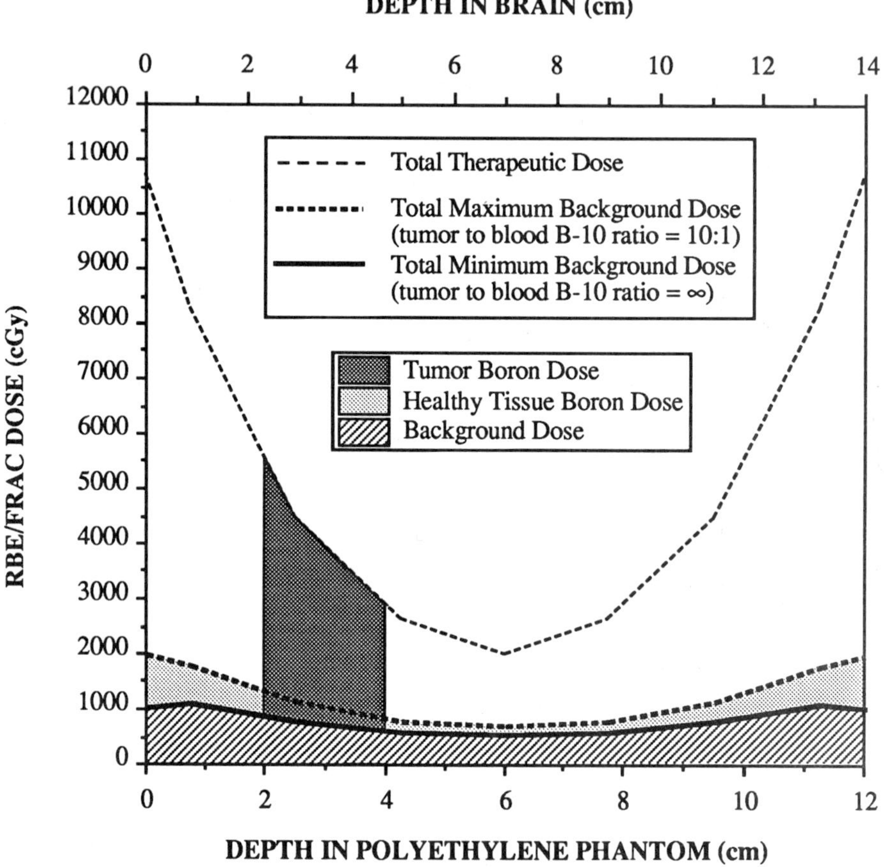

Figure 7. Measured RBE-dose along the central axis of the polyethylene phantom for the treatment of a 2-cm diameter tumor located 3 cm below the surface using D_2O/bismuth-filtered beam (M11) at MITR-II based on maximum background dose of 2000 RBE-cGy. Bilateral fractionated irradiation into the polyethylene head phantom. Equivalent brain depth is based on hydrogen atom density.

Table Seven. Summary of total RBE/fractionated doses delivered to tumor lying 6 cm below the surface by a sulfur/aluminum-filtered beam (M41) at MITR-II. A boron concentration of 30 µg/g in tumor and an effective B-10 dose ratio in tumor to blood of 10 to 1 is assumed. Maximum RBE-dose to healthy tissue is limited to 2000 RBE-cGy.

Type of Irradiation	Unilateral	Bilateral
Maximum Tumor RBE-Dose: Average Tumor RBE-Dose:	2500 (RBE-cGy) 2050 (RBE-cGy)	3230 (RBE-cGy) 3200 (RBE-cGy)
Maximum Background RBE-Dose: Average Background RBE-Dose:	2000 (RBE-cGy) 1100 (RBE-cGy)	2000 (RBE-cGy) 1730 (RBE-cGy)

This occurs because the D_2O/bismuth-filtered beam has a poor advantage ratio at high depth and because there exist high background doses at both edges of the brain. These results indicate that a unilateral irradiation with an M10 or M11 type of beam offers superior performance with good therapeutic advantage for tumors at shallow depth relative to bilateral irradiation.

<u>Clinical Example for Deep-Seated Tumor Treated with the Sulfur/Aluminum-Filtered Epithermal Beam</u>

A sulfur/aluminum-filtered beam such as M41 (see Tables Four and Five) at MITR-II, with its large advantage depth and advantage ratio, is well suited for treating deep-seated tumors through the intact skull with fractionated bilateral irradiation therapy. Assuming a patient with a tumor 2 cm in diameter and centered 6 cm below the scalp, the dose vs. depth data for tumor and the surrounding healthy tissue was developed in a similar manner to that used in the previous section. As discussed earlier, because of the hydrogen density difference between polyethylene and brain, 6 cm in polyethylene will correspond to 7 cm in real brain (i.e., for this example, the tumor is centered 7 cm below the scalp in real brain tissue). Again, a tumor boron concentration of 30 µg/g and an effective B-10 dose ratio between tumor and blood/tissue of 10 to 1 was assumed. Identical RBEs were used as in the previous shallow-tumor example.

The results are shown graphically in Figures 8 and 9 and tabulated in Table Seven. Figure 8 shows the result of unilateral irradiation. Although a therapeutic advantage is achieved, the average tumor RBE-dose is only about 2050 RBE-cGy compared to the average background RBE-dose of 1100 RBE-cGy. On the other hand, the bilateral irradiation, shown in Figure 9, gives a very high average tumor RBE-dose of over 3200 RBE-cGy compared to the average background RBE-dose of 1730 RBE-cGy. The average background dose is not much different from the maximum background dose. This is a very desirable shape for the background dose/depth distribution. The therapeutic dose component is also relatively flat, with somewhat more structure or variation in dose versus depth than the total background component.

For deep-seated tumors, bilateral irradiation with epithermal neutrons delivers superior performance over unilateral irradiation. This is due both to the shape of the therapeutic dose vs. depth curve and to the shape of the total background dose curve. Because the overall background dose is relatively flat over all depths, there is no penalty for irradiating from both sides of the brain as in the case for lower energy beams. With a bilaterally delivered epithermal beam, of the M41 type, it is possible to deliver almost 50% more dose to a deep-seated tumor, compared to unilateral irradiation, while staying within the limits of maximum tolerable dose for healthy tissue.

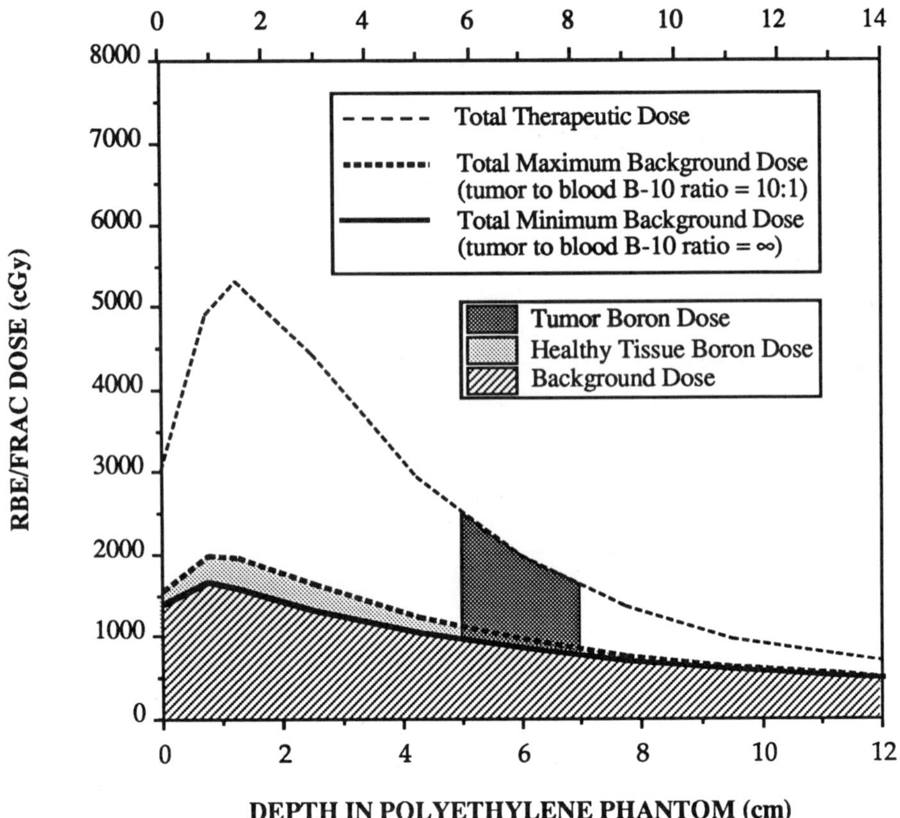

Figure 8. Measured RBE-dose along the central axis of the polyethylene phantom for the treatment of a 2-cm diameter tumor located 6 cm below the surface using aluminum/sulfur-filtered beam (M41) at MITR-II; based on maximum background RBE-dose of 2000 RBE-cGy. Unilateral fractionated irradiation into the polyethylene head phantom. Equivalent brain depth is based on hydrogen atom density.

For any given clinical situation, however, a decision on which neutron beam should be used should be based on comprehensive treatment planning simulations such as discussed by Zamenhof et al. [10]. In order to validate these simulations for a beam filter, comprehensive spatial dose distribution measurements should be made not only in a right-circular cylinder polyethylene head phantom, but also in a phantom that is identical in size, shape, and composition to the human head.

SUMMARY

We have measured the performance of neutron beams, with a wide range in energy and intensity, available at MITR-II for use in neutron capture therapy. Sulfur/aluminum-filtered epithermal-neutron beams, with their larger advantage depths, were shown to be very promising for the treatment of deep-seated tumors. There are a number of different 'best' epithermal beams at MITR-II depending upon the desired figure of merit. For best advantage depth, we have M20. For best advantage ratio, we have M27. When the highest RBE-dose rate is deemed important, we have the M23 beam that is capable of delivering 2000 RBE-cGy to the tumor in less than 15 minutes. An epithermal beam such as M41, used in the previous section, is an example of a good compromise on all three

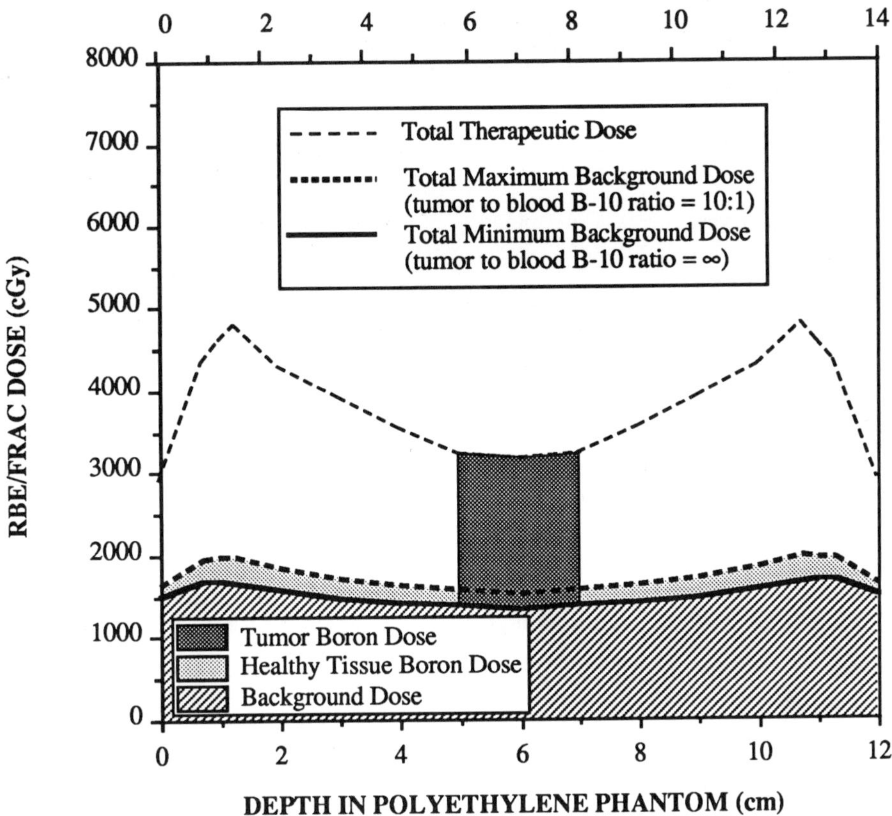

Figure 9. Measured dose along the central axis of the polyethylene phantom for the treatment of a 2 cm diameter tumor located 6 cm below the surface using aluminum/sulfur-filtered beam (M41) at MITR-II; based on maximum background dose of 2000 RBE-cGy. Bilateral fractionated irradiation into the polyethylene head phantom. Equivalent brain depth is based on hydrogen atom density.

figures of merit. However, these are by no means the only desirable neutron beams for NCT. Neutron beams with lower energies, but better advantage ratios, such as the D_2O/bismuth-filtered beams at MITR-II, are well suited for the treatment of shallow tumors including those from malignant melanoma.

ACKNOWLEDGMENTS

The authors wish to thank Dr. John Bernard and the rest of the MITR-II operations group for their support and assistance in all aspects of this work. This research was funded by the U.S. Department of Energy, Office of Health and Human Assessments, under Grant No. DE-FG02-87ER-6060.

REFERENCES

1. S. D. Clement, J. R. Choi, R. G. Zamenhof, J. C. Yanch, and O. K. Harling, "Monte Carlo Methods of Neutron Beam Design for Neutron Capture Therapy at the MIT Research Reactor (MITR-II)." (These Proceedings.)

2. F. H. Attix, <u>Introduction to Radiological Physics and Radiation Dosimetry</u>, John Wiley and Sons, Inc., New York, p. 479 (1986).

3. R. S. Caswell, J. J. Coyne, and M. L. Randolph, "KERMA Factors of Elements and Compounds for Neutron Energies Below 30 MeV," <u>Int. J. Appl. Radiat. Isot.</u>, 33:1227 (1982).

4. <u>ASTM E262-77</u>, Standard Method for Measuring Thermal Neutron Flux by Radioactivation Techniques.

5. W. Zobel, "Experimental Determination of Corrections to the Neutron Activation of Gold Foils Exposed in Water," <u>ORNL-3407</u>, Topical Report (1963).

6. M. Ashtari, <u>Biological and Physical Studies of Boron Neutron Capture Therapy</u>, Ph.D. Thesis, Massachusetts Institute of Technology, pp. 211-235 (1982).

7. Attix, op. cit., pp. 477-500.

8. Ashtari, op. cit., p. 240.

9. S. F. Mughabghab, M. Divadeenam, and N. E. Holden, <u>Neutron Cross Sections: Vol. 1, Neutron Resonance Parameters and Thermal Cross Sections, Part A: Z = 1 - 60</u>, Academic Press, Orlando, FL (1981).

10. R. G. Zamenhof, S. D. Clement, J. C. Yanch, O. K. Harling, J. F. Brenner, H. Madoc-Jones, and J. C. Yanch, "Monte Carlo Based Dosimetry and Treatment Planning for Neutron Capture Therapy of Brain Tumors." (These Proceedings.)

GEORGIA TECH RESEARCH REACTOR EPITHERMAL BEAM

J. L. Russell, Jr.,[1] W. H. Miller,[2] R. M. Brugger,[2]
and W. H. Herleth[2]

[1] Theragenics Corporation
Norcross, GA

[2] University of Missouri
Columbia, MO

INTRODUCTION

This report describes the design, construction, neutron spectrum measurements, and gamma dose measurements for an epithermal-neutron beam at the Georgia Tech Research Reactor (GTRR). The GTRR facility, the filter geometry, and the materials are described. Measurements and computations of beam parameters and their variation with filter configurations are also presented. An optimal configuration is specified. The beam is also partially characterized by calculating the dose distribution in the target phantom that Prof. Otto K. Harling of the Massachusetts Institute of Technology recently suggested be used as a standard for comparing BNCT beams.

Measurements (Bonner Sphere and gamma only) were made on this same filter and port in January 1988 and results reported [1]. That configuration differed from the current one because position V-3, located at the tip of the reentrant beam port H-1 did not contain a fuel element, but now does. The principal difference is that the presence of the fuel element increased epithermal beam intensity without significantly degrading the fast-neutron or gamma contamination.

GEORGIA TECH RESEARCH REACTOR

The GTRR is located in the Frank H. Neely Nuclear Research Center of the Georgia Institute of Technology in Atlanta, Georgia. It is a heterogeneous, heavy-water moderated and cooled reactor, fueled with plates of aluminum-uranium alloy. It is designed to produce a thermal flux of more than 10^{14} n/cm^2-s at a power of 5 MW and an exit moderator temperature of 137 °F. A horizontal cross section of the GTRR is shown in Figure 1. (Note: Only one of the beam ports is shown.)

The reactor core is approximately two feet in diameter, two feet high and, when fully loaded, contains provisions for up to nineteen fuel assemblies spaced six inches apart in a triangular array. Each assembly contains sixteen fuel plates. The total uranium-235 content of a full loading is 3.2 kg. The fuel is centrally located in a six foot diameter aluminum reactor vessel which provides a two foot thick D$_2$O reflector completely surrounding the core. Surrounding the D$_2$O tank is a twenty foot diameter concrete shield with penetrations to allow extraction of the neutron beams.

Neutron Beam Design, Development, and Performance for Neutron Capture Therapy
Edited by O. K. Harling *et al.*
Plenum Press, New York, 1990

The reactor is housed in a sixty foot diameter containment shell which provides a twenty foot wide experiment hall around the outside of the reactor shield. The beams exit into this hall, and this is where patients would receive their therapeutic radiation.

The reactor contains twenty-two horizontal openings, a thermal column, and a bio-medical irradiation facility. Stations H-1 through H-10 are horizontal beam ports, all of which lie in the horizontal plane passing through the center of the reactor. All ten ports are provided with rotating shutters so that the beam intensity may be reduced without shutting down the reactor. H-1 is a six inch I.D. beam port which extends into the D_2O region to a point sixteen inches from the core center. It is this port into which a filter was incorporated to produce an epithermal beam.

FILTER CONFIGURATION

The aluminum-sulfur-cadmium-lead filter of Figure 2 is installed in port H-1. The cadmium removes the thermal neutrons from the beam, effectively removing all neutrons below ~0.6 eV. The aluminum and sulfur preferentially pass epithermal neutrons below ~30 keV, and block neutrons above that energy. The lead reduces the gamma flux. The 7-inch iris serves to trim the edges of the beam so that the area of patient exposure can be precisely controlled. The purpose of the borated polyethylene surrounding the filter is to absorb scattered neutrons before they can reenter the beam. The rotating shutter is shown in the open position in Figure 2. The iris is located flush with the outside surface of the concrete shield at H-1, which places it ten feet from the reactor center [2].

THEORY

The theory of operation of the filter is that it removes neutrons from the beam by either absorption or scattering of neutrons (the "good geometry" assumption). Beam transmission at a particular energy is simply:

$$T(E) \ = \ \exp[\Sigma(E)t]$$

where: $T(E)$ is the filter transmission at neutron energy E,

            $\Sigma(E)$ is the total neutron cross section of filter material at Energy E,

            E is the neutron energy, and

            t is the filter thickness.

The neutron flux at the target location is:

$$F_f(E) \ = \ F_{nf}(E)T(E)$$

where: $F_{nf}(E)$ is the neutron flux at energy E at the target position, with no filter in the beam tube, and

        $F_f(E)$ is the neutron flux at energy E at the target position, with a filter in the beam tube.

This model neglects those neutrons which scatter only through a small angle and still strike the target or which, after a first scattering, undergo multiple scatterings, reenter the beam at a lower energy, and strike the target. Neglecting these in-scatter neutrons is what is meant by the "good geometry" assumption. The terminology originated with the early work on measurements of total neutron cross sections. The first of the neglected items above is about a 1% effect for this filter. The effect of neglecting the second item is more difficult to assess, but based on the small size of the first should not qualitatively affect the spectrum.

A test calculation using the Monte Carlo code (MNCP) on a similar filter but with

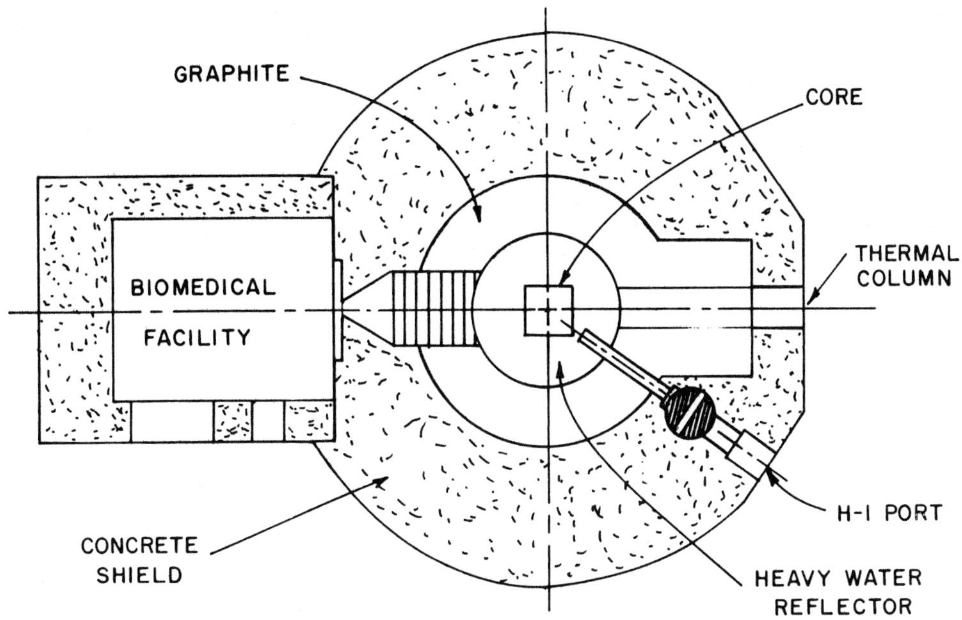

Figure 1.   Horizontal Section of GTRR at the Core Midplane.
(Beam ports H-2 to H-10 not shown.)

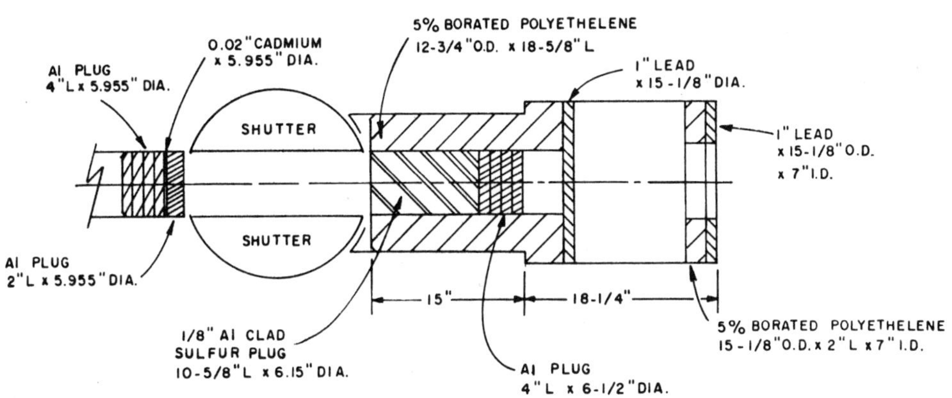

Figure 2.   Epithermal Filter.

Table One

Beam Parameters for 86 W Reactor Power Level

Distance from Shield Face (inches)	Neutrons 5-inch Sphere (n/cm²-s)	Neutrons 2-inch Sphere (n/cm²-s)	Gammas (mrad/hr)
12	4,458±386	4,295±372	7.2±.6

greater attenuation reduced the neutron flux (0.4 eV to 15 MeV) in H-1 by a factor of 168. Applying the simple attenuation model with ten energy groups per decade, and using a completely different cross-section set, yielded a reduction by a factor of 137. The 20% difference, in the wrong direction to be accounted for by multiple scattering, could be explained by a 4% difference in cross-section sets, density, or dimensional inputs to the two calculations.

The MCNP calculation of the open hole (no filter) neutron spectrum at the H-1 target point indicates that approximately half the neutrons in the 0.4-eV to 15-MeV range are above 10 keV. Adjusting a 1/E tail plus a Watt spectrum to match that ratio gives a physically reasonable spectrum to use for relating the data from the three different neutron detector systems. Using this simple model, Figures 3 and 4 show the assumed open hole and the calculated filtered spectrum for comparison as flux per unit lethargy vs. a logarithmic energy scale. For this type of plot, a square centimeter anywhere under the curve corresponds to the same number of neutrons. Both curves are relative and are not normalized to a particular power level, or to each other. The resulting filtered spectrum shown in Figure 4, by comparison with Figure 3, illustrates the dramatic reduction of the high energy neutrons above 30 keV and the elimination of the very low energy neutrons below 0.6 eV.

MEASUREMENTS AND COMPUTATIONS

The neutron intensity of the beam was measured with Ludlum Measurements Inc. Model 42-6 Neutron Detector (Bonner Spheres), [3]. The efficiency of the neutron detector was measured with two sources, a californium-252 spontaneous fission source (Amersham No. CVN6, Serial 2080NC) of 0.70E7 n/s, and a plutonium-beryllium source (Mound Laboratory No. M-1233) of 1.27E7 n/s. Count rates were measured with the sources located at one meter from the detector, with each of the standard sets of moderating spheres (2", 5", 8", 10", 12") fitted on the detector. The efficiency was determined using the two largest spheres, as in [2]. The standard deviation of the neutron detector calibration is estimated to be 5%. To determine an absolute intensity of the neutron beam, the neutron energy spectrum is integrated over the detector efficiency [3]. The shape of the energy spectrum used for this calculation is that of Figure 4. The gamma intensity of the beam was measured with a portable air ionization chamber, Victoreen Panoramic 470A. The gamma detector was calibrated with a cobalt source for which the strength was determined with an NBS traceable Victoreen Model 500 electrometer, Serial 763, with Model 550-4 ionization chamber, Serial 1392. The standard deviation of the gamma detector calibration is estimated to be 6%. The neutron response of the gamma detector was qualitatively investigated using a one inch thick polyethylene sheet to block beam neutrons. The effect appeared to be small.

Measurements were taken with the reactor power level at 86 W and the result was extrapolated to 5 MW. This extrapolation relied on the linearity of the reactor instrumentation over this range of power level and is estimated to be valid to within 7%.

Neutron flux was measured with both the 2-inch and 5-inch spheres. Because the beam is only 7 inches in diameter, the larger spheres could not be fully illuminated by the beam and their data would have been of questionable interpretation. Refer to Table One.

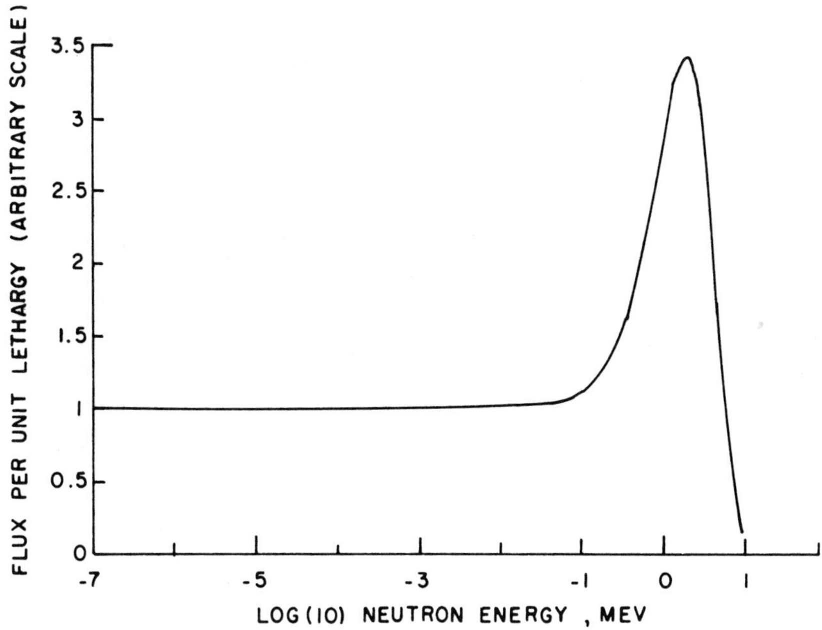

Figure 3. Neutron Spectrum for Empty H-1 Port, Assumed
Shape for Filter Calculations.

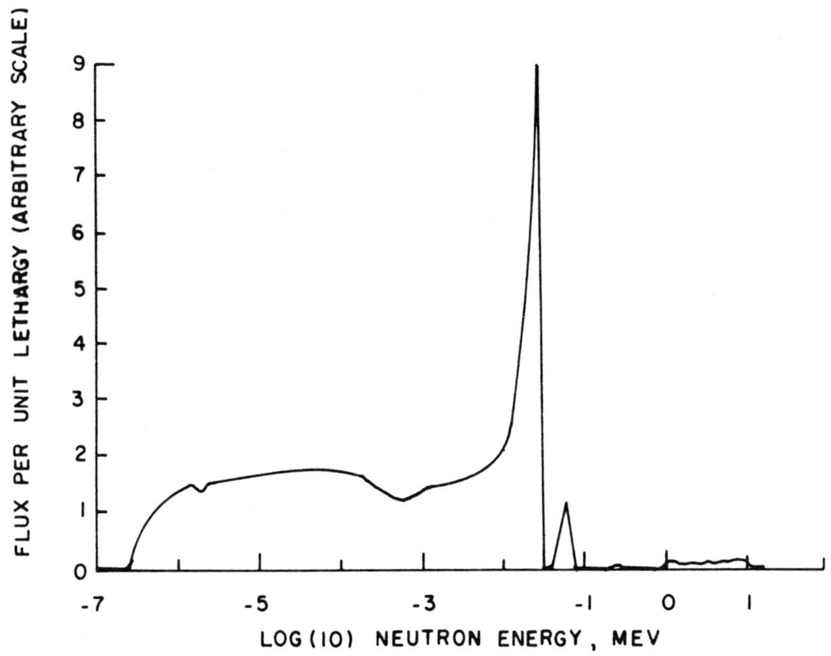

Figure 4. Calculated Shape of Neutron Spectrum for H-1 Port
with Aluminum-Sulfur-Cadmium-Lead Filter.

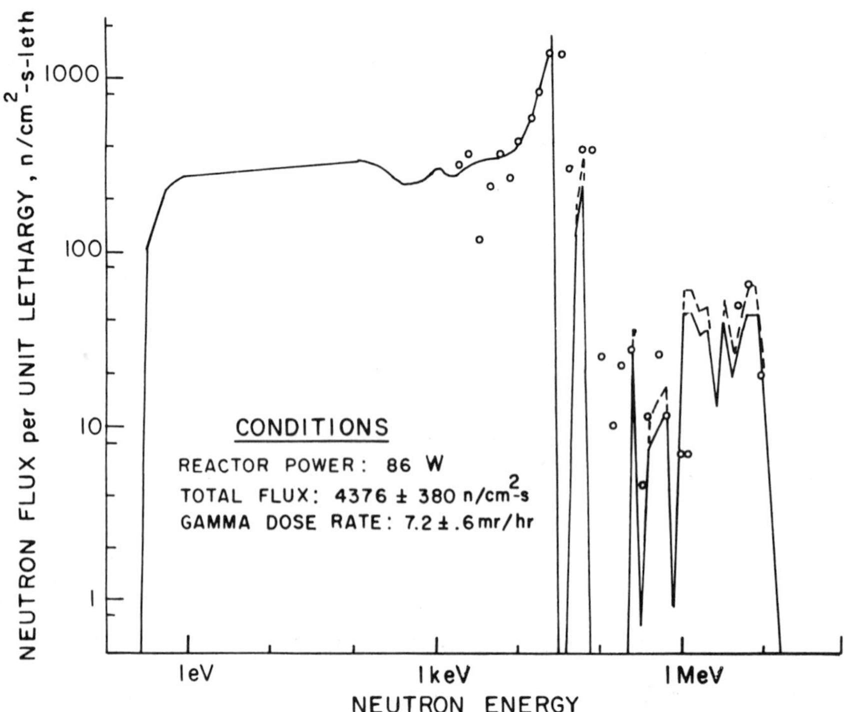

Figure 5. Neutron Spectrum for GTRR Epithermal Beam, Port H-1, Solid Line (Calculated Spectrum), Open Circles (Data), Dashed Line (Spectrum Normalized) to Data Above 100 keV.

NEUTRON SPECTRUM MEASUREMENTS

The neutron spectrum was measured in the range 1.6 keV to 10.5 MeV using two spectrometer systems. The lower energy range, 1.6 keV to 1.3 MeV, was measured with a 1-inch radius, hydrogen gas, recoil spectrometer, Series #270 Spherical Detector, LND Inc., Oceanside, NY. The experimental methods used are those described by Miller [4]. The high range, 4 MeV to 10.5 MeV, was measured with a liquid scintillation proton recoil fast-neutron spectrometer system described by Meyer and Miller [5] and Korsah and Miller [6].

The key technical problem to be overcome by both systems was rejection of the gamma background. This was accomplished for both by pulse shape discrimination. The high ratio of gammas to neutrons in the MeV range limited the lower end of the liquid scintillator to 4 MeV. This left an unmeasured gap from 1.3 MeV (the upper end of the range for the gas spectrometer) to 4 MeV (the lower end of the liquid spectrometer).

The absolute efficiency of the two spectrometers was not measured, but rather was derived from first principles, i.e., the hydrogen cross section and the number of hydrogen atoms in the active volumes of the detectors.

The spectra were measured in the beam center at twelve inches from the shield face, the same as for the gamma and total flux measurements. One gas filling for the gas spectrometer was sufficient for the range 1.6 keV to 1.3 MeV. However, the dynamic range of the electronic system could not span that range. Gas spectra were taken for three

Table Two

Five Megawatt Beam Parameters

	Experimental Filter	Optimal Filter
Configuration	0.02" Cd 10.6" S 10" Al 1" Pb	0.02" Cd 10.6" S 12" Al 0.5" Bi
Treatment Point	12"	18"
Epithermal Neutron (n/cm^2-s)	$(2.5 \pm .30)$E8	$(2.5 \pm .30)$E8
Fast neutron (n/cm^2-s) 30 keV <n< 100 keV	$(.07 \pm .03)$ E8	$(.02 \pm .01)$E8
100 keV <n	$(.08 \pm .02)$E8	$(.04 \pm .01)$E8
Incident Gamma Flux (rad/hr)	420 ± 45	290 ± 35

ranges of electronic settings to span the range. Measurements were made at 86 W because that was the upper limit of the total count rate that the systems could handle. The results are plotted in Figure 5 where they are compared with the computed spectrum which has been normalized with the Bonner Sphere data.

Because the spectrometers have an unmeasured gap from 1.3 to 4 MeV, the computational model was used to quantify the spectrum over the whole range. The following procedure was used:

(1)   It was assumed that the Bonner Sphere measurements provided the best estimate of the total flux.

(2)   It was assumed that the spectrum below 30 keV was best represented by the calculated spectrum normalized to the Bonner Sphere measurements.

(3)   The measured spectrum was summed from 100 keV to 1.3 MeV and from 4 MeV to 10.5 MeV, and divided by the sum over the same energy ranges of the above calculated spectrum to give a ratio of 1.34.

(4)   It was assumed that the best estimate of the spectrum between 100 keV and 15 MeV was the calculated spectrum in that range multiplied by 1.34.

(5)   It is believed that the measured data in the vicinity of the 60-keV neutron spike was strongly influenced by the presence of the intense spike at 25 keV so that it is difficult to extract a quantitative measurement of the intensity of the 60-keV spike.

(6)   It was assumed that the best estimate of the spectrum in the range from 30 keV to 100 keV was the calculated spectrum also multiplied by 1.34.

The dashed curve in Figure 5 is based on the above assumptions.

Table Two summarizes the properties of the epithermal beam extrapolated to the 5 MW reactor power level. The fast neutrons are listed for two energy ranges because the damage per neutron for the lower energy range is about one-tenth that of the higher energy

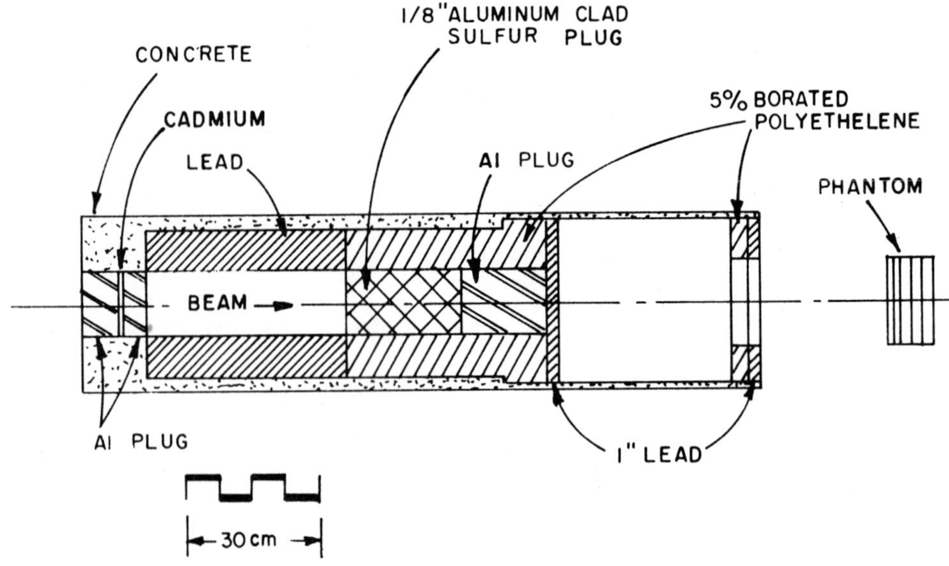

Figure 6.   Geometry Assumed for MCNP Calculation of Phantom Dose.

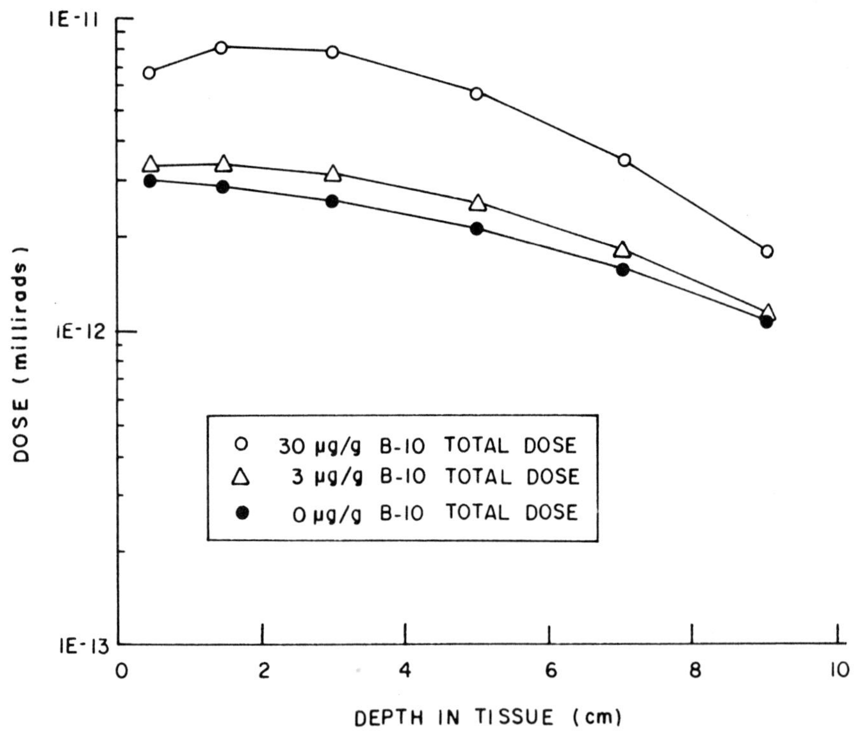

Figure 7.   Calculated Dose Distribution in Phantom Irradiation by Epithermal Beam.

range. For comparison, an optimal beam design (small modification from the existing filter) is presented for comparison. The rationale for the optimal beam parameters is given by Russell [1].

PHANTOM DOSE

The dose in a six-inch diameter cylindrical water phantom was calculated using MCNP. The computational geometry is shown in Figure 6. The calculation was normalized to correspond to one neutron impinging on the reactor side of the filter and the results are shown in Figure 7. The calculation did not include the primary beam gamma dose. Addition of the beam gamma dose as well as renormalization are required for a complete understanding of the phantom dose.

CONCLUSIONS

The neutron spectrum measurements on the epithermal beam from port H-1 of the GTRR confirm the premise that a useful therapeutic epithermal beam can be produced by a "broad band" beam filter. In addition, the simple transmission computational model provides an adequate description of filter performance for the geometry reported. However, if either the source or the patient is near the filter, a more complex computational model would be required.

REFERENCES

1.   J. L. Russell, Jr., "Epithermal Neutron Beam for BNCT," <u>Third Int. Symp. on Neutron Capture Therapy</u>, Bremen, FRG, 31 May - 3 June 1988.

2.   D. J. Noonan, J. L. Russell, Jr., and R. M. Brugger, "A Prototype Epithermal Neutron Beam for Boron Neutron Capture Therapy," <u>Med. Phys.</u>, 13(2):211 (1986).

3.   N. E. Hertel and J. W. Davidson, "The Response of Bonner Spheres to Neutrons from Thermal Energies to 17.3 MeV," <u>Nucl. Instruments and Methods in Phys. Res.</u>, A238:509 (1985).

4.   W. H. Miller, "Neutron Spectroscopy in the 1 to 30 keV Energy Range," Submitted in Dec. 1988, <u>Nucl. Instruments and Methods in Phys. Res.</u>

5.   W. Meyer and W. H. Miller, "Development and Standardization of the UMC NE-213 Liquid Scintillation Proton Recoil Fast Neutron Spectrometer System," National Science Foundation, <u>NSF-GK-40728-UMC-2</u> (Feb. 1975).

6.   K. Korsah, "Development of a Neutron/Gamma Ray Radiation Monitor," Dissertation at the University of Missouri-Columbia, May 1983 (Dissertation Supervisor, W. H. Miller).

NEUTRON BEAM DESIGN AND PERFORMANCE FOR BNCT IN CZECHOSLOVAKIA

J. Burian and J. Rataj

Nuclear Research Institute
Rez near Prague, Czechoslovakia

INTRODUCTION

Conditions for forming an appropriate neutron spectrum on the experimental reactor VVR-S at the Nuclear Research Institute at Rez near Prague were investigated. Each configuration of moderating and shielding layers is referred to as a 'variant.' That designated as Variant C was designed using multigroup transport codes and the results were verified by measuring neutron fluxes, neutron spectra, and the gamma-ray background. In 1988, VVR-S equipment was shut down for reconstruction of the LVR-15 reactor. New conditions for neutron beam performance therefore arose and the calculational results for Variant D were obtained. Experimental verification of the Variant D design will be realized as soon as the reconstruction is finished.

DESIGN AND PERFORMANCE OF THE VVR-S REACTOR

Different configurations of moderating and shielding layers (graphite, heavy water, lead, and bismuth) were placed in the empty region of the thermal column [1,2]. The combination with cavity, designated as Variant C, is presented in Figure 1. It is composed of layers of nuclear graphite, heavy water, lead, and one or two blocks of bismuth of purity 99.99%. The configuration present during experimental verification is shown in Figure 2.

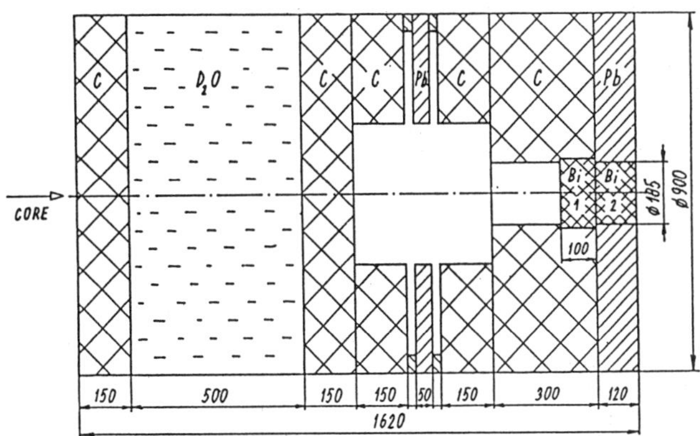

Figure 1. The Configuration of Moderating and Shielding Layers, Variant C. (Dimensions are in mm.)

Neutron Beam Design, Development, and Performance for Neutron Capture Therapy
Edited by O. K. Harling *et al.*
Plenum Press, New York, 1990

<u>Calculational and Experimental Methods</u>

The two-dimensional code DOT was primarily used for the determination of neutron and gamma space-energy distributions in the moderating and shielding layers. The coupled neutron-gamma data library collections of VITAMIN type (DLC-36) made it possible to obtain precise information.

The thermal and epithermal flux densities were measured by gold (Au) and gold/cadmium (Au/Cd) foil activation. Neutron spectra over wide energy ranges were determined with a Bonner moderation spectrometer. A scintillation spectrometer with pulse shape discrimination was used for the mapping of fast neutrons and the gamma-ray background. The dosimetric characteristics were obtained from thermoluminescent and solid state nuclear track detectors.

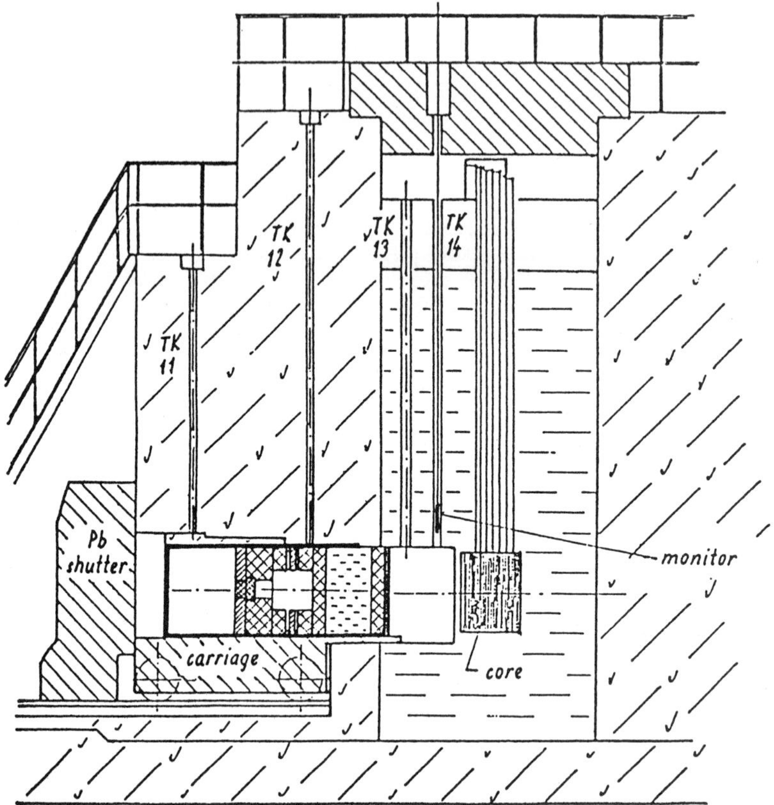

Figure 2. The Configuration Present during Experimental Verification of Neutron Fluxes in the Empty Column of the VVR-S Reactor.

<u>Results for VVR-S Reactor</u>

Very good agreement of the experimental thermal flux density at points within the cavity with calculated results was obtained. However, differences did occur in the region behind the bismuth block because we did not have appropriate multigroup data for bismuth. Instead, data for lead were used. The mapping of neutron and gamma distributions following the last layer shows the same form for neutrons and gammas. This means that impurities in the bismuth were the source of secondary gamma-rays and that the purity of the material was not sufficient.

230

The main measured parameters of the beam are:

Reactor power	1 MW
Thermal neutron tissue dose following Bi	2.5 cGy/min
Gamma dose	0.23 cGy/min
Fast neutron dose	0.02 cGy/min

CONFIGURATION FOR LVR-15 REACTOR

The VVR-S reactor was under reconstruction to create the multi-purpose experimental reactor LVR-15 with a maximum power of 15 MW. New conditions arose in the empty space of the thermal column. The detailed description is shown in Figures 3 and 4. The major difference is the core geometry. Also, a special shutter must be designed between the core periphery and the steel vessel.

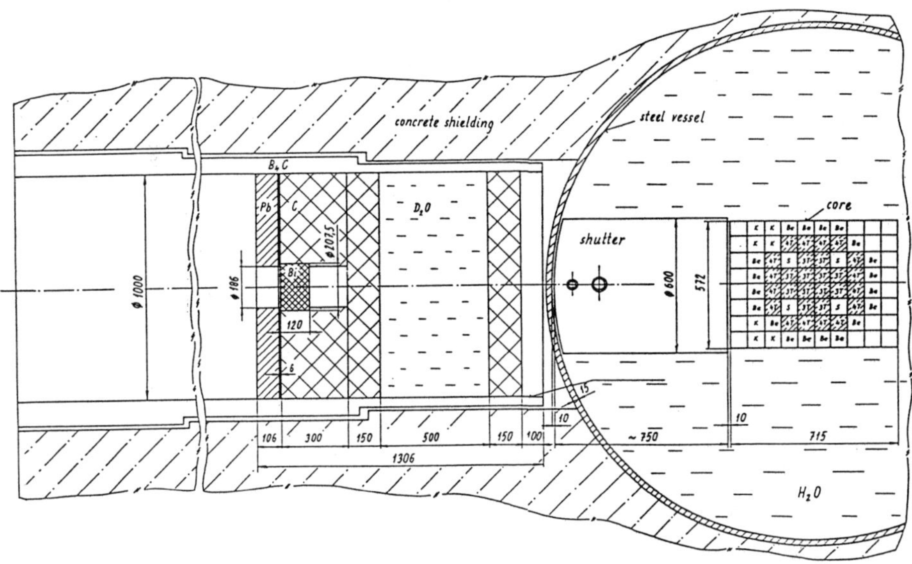

Figure 3. The LVR-15 Beam Configuration, Top View.

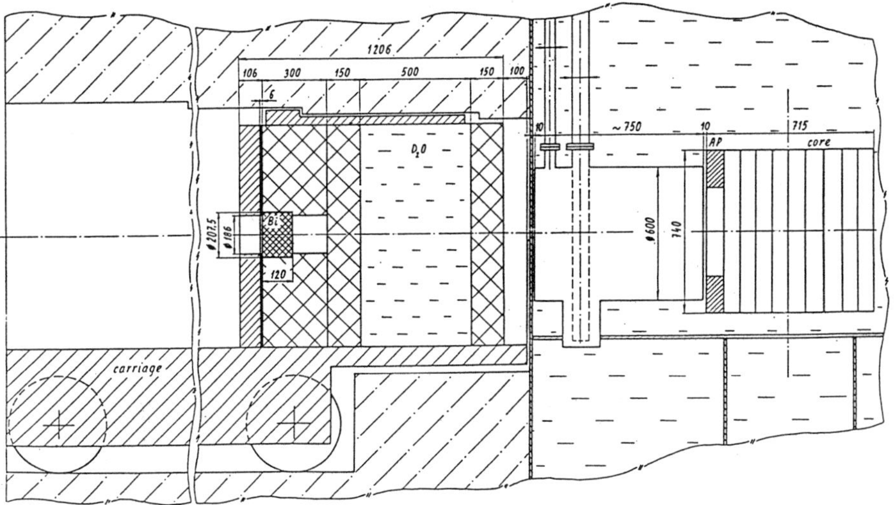

Figure 4. The LVR-15 Beam Configuration, Lateral View.

231

A new configuration of materials, Variant D, is being prepared to be placed on the carriage for experimental verification. Major changes are that the big cavity was left out, the bismuth block is of higher purity (99.999%), and activation and gamma sources in lead were reduced by using a thin layer of boron carbide.

<u>Results for LVR-15 Reactor</u>

Only calculational results are currently available for Variant D. The space-energy distributions of the neutrons and gamma-rays were determined in all moderating and shielding layers. If KERMA-dose factors for brain-equivalent tissue from Figure 5 are assumed [3,4], the course of the response along the axis is as shown in Figure 6.

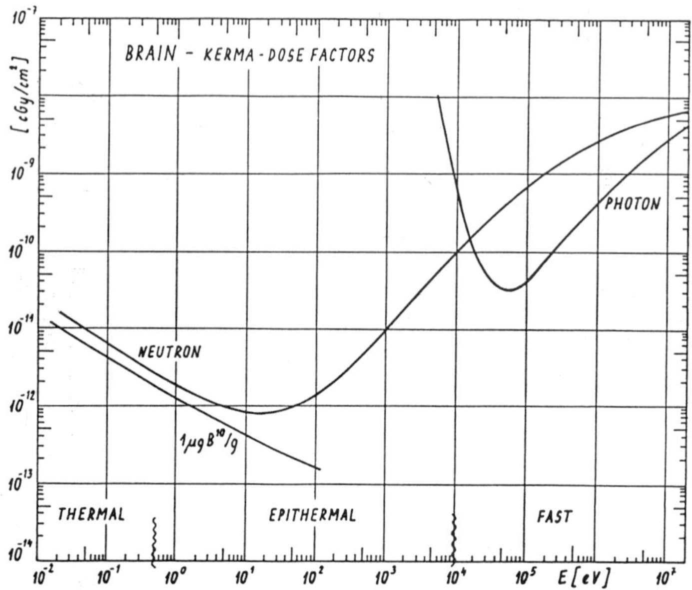

Figure 5. KERMA-dose Factors for Brain-Equivalent Tissue.

Unfortunately, the use of nuclear data for lead instead of bismuth influenced results at the end of the configuration. It is expected that the following values will be obtained from experimental verification for the actual configuration with bismuth:

Reactor power	1 MW
30 µg/g B-10 total dose	25 cGy/min
3 µg/g B-10 total dose	2.5 cGy/min
Total gamma dose	3 cGy/min
Neutron dose	2 cGy/min

CONCLUSIONS

The conditions required for realization of BNCT principles may be attained on the LVR-15 reactor. During 1989 the new conditions will be verified and the distribution of neutrons and gamma-rays in polyethylene head phantoms will be measured.

The design of an epithermal neutron beam is the center of our attention, naturally. Remodeling of the LVR-15 facility is planned and the combination of Al-D$_2$O layers will be investigated. In order to realize the concept in a relatively short time, we hope to benefit from the experience and cooperation of the BNCT international community.

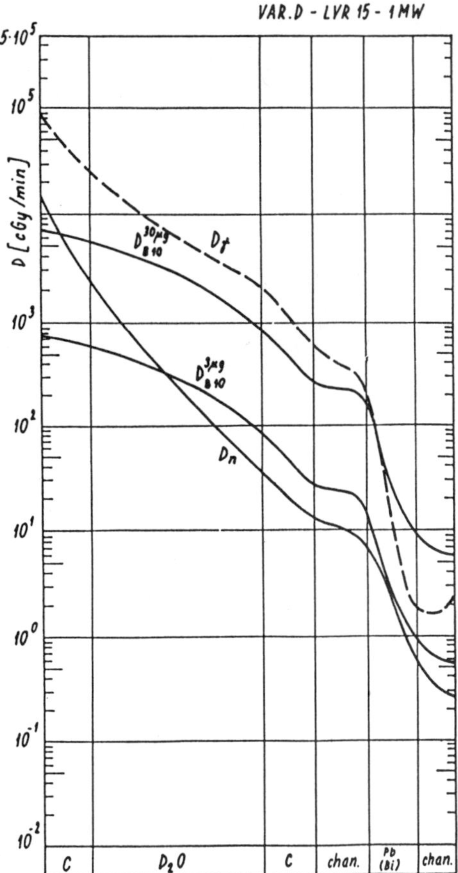

Figure 6. The Response for a 'Small' Piece of Brain-Equivalent Tissue Without the Perturbing Effects of the Surrounding Tissue Along the Axis for Variant D.

REFERENCES

1. J. Burian, P. Karas, E. Fymanová, B. Janský, M. Marek (ÚJV Řež), Z. Prouza, and J. Heřmanská (Ústav biofyziky a nukleární medicíny UK), F. Spurný (ÚDZ ČSAV), "The Verification of Conditions for BNCT through Measurements in the Thermal Column of the VVR-S Reactor," (in Czech), ÚJV 7940 (1987).

2. J. Burian, J. Rataj, M. Marek (ÚJV), Z. Prouza, and J. Heřmanská (Ústav biofyziky a nukleární medicíny UK), F. Spurný (ÚDZ ČSAV), "Verification of Conditions for BNCT on the VVR-S Reactor," (in Czech), ÚJV 8761 (1989).

3. J. H. Hubbel, "Photon Mass Attenuation and Energy-Absorption Coefficients from 1 keV to 20 MeV," Int. J. Appl. Radiat. Isot., 33:1269 (1982).

4. R. S. Caswell, J. J. Coyne, and M. L. Randolph, "KERMA Factors of Elements and Compounds for Neutron Energies Below 30 MeV," Int. J. Appl. Radiat. Isot., 33:1227 (1982).

NEUTRON SPECTRUM MEASUREMENTS IN THE ALUMINUM OXIDE FILTERED BEAM FACILITY AT THE BROOKHAVEN MEDICAL RESEARCH REACTOR

G. K. Becker, Y. D. Harker, L. G. Miller,
R. A. Anderl, and F. J. Wheeler

Idaho National Engineering Laboratory
EG&G Idaho, Inc.
Idaho Falls, ID

ABSTRACT

Neutron spectrum measurements were performed on the aluminum oxide filter installed in the Brookhaven Medical Research Reactor (BMRR). For these measurements, activation foils were irradiated at the exit port of the beam facility. A technique based on dominant resonances in selected activation reactions was used to measure the epithermal neutron spectrum. The fast and intermediate-energy ranges of the neutron spectrum were measured by threshold reactions and ^{10}B-shielded ^{235}U fission reactions. Neutron spectral data were derived from the activation data by two approaches: (1) a short analysis which yields neutron flux values at the energies of the dominant or primary resonances in the epithermal activation reactions and integral flux data for neutrons above corresponding threshold or pseudo-threshold energies, and (2) the longer analysis which utilized all the activation data in a full-spectrum, unfolding process using the FERRET spectrum adjustment code.

This paper gives a brief description of the measurement techniques, analysis methods, and the results obtained.

INTRODUCTION

The Power Burst Facility/Boron Neutron Capture Therapy (PBF/BNCT) development program schedule required the use of an epithermal neutron beam before the PBF beam facility would be available. The beam was needed to carry out the acute dose tolerance study on healthy dogs and the treatment protocol on spontaneous tumor dogs. Calculations on other United States test reactors confirmed that the Brookhaven Medical Research Reactor (BMRR) could provide the highest epithermal neutron beam with an acceptable fast-neutron and gamma component for the canine program.

A joint Idaho National Engineering Laboratory/Brookhaven National Laboratory (INEL/BNL) program was instituted to design, construct, and install a neutron beam filter in the BMRR. Aluminum oxide was selected as the filter material because it provided the desired neutron spectrum characteristics within the physical limitations of the BMRR filter irradiation port. The results of neutron spectral measurements performed on this filter are presented in this paper. Foil activation techniques using resonance, threshold, and pseudo-threshold response reactions were used in conjunction with adjustment and matrix analysis techniques to derive spectral data over the energy range of 0.5 eV to 7 MeV.

MEASUREMENT TECHNIQUE AND APPARATUS

Neutron energy spectrum measurements are obtained through the neutron activation of foil materials with differing energy responses. The induced gamma activity is then measured and information concerning the intensity and neutron energy distribution in the filtered neutron beam is deduced. Measurements in the epithermal energy region, 0.5 eV to 10 keV, were accomplished via a set of foil materials with single dominant resonances. The measurement apparatus configuration consisted of a stack of three 1.27-cm diameter activation foils of a given material. Foil thickness was the same for each foil position in the stack, but was different for each foil material. The specification of foil thickness depended on the resonance peak cross section. A side shield, composed of the same material, was used to minimize side leakage of neutrons into the foil set. The foil stack and shield were encapsulated in 0.102-cm thick cadmium (Figure 1). Epithermal pseudo-threshold measurements, ranging from approximately 0.6 eV through 20 keV, were performed with ^{63}Cu detectors, whose response was modified through spectrum tailoring with ^{6}Li-shielding (Figure 2). A third measurement in the epithermal and fast regions was accomplished by tailoring the neutron spectrum with a ^{10}B 1/V-shield and using the ^{235}U fission reaction as the detector. This arrangement exhibits a pseudo-threshold response at approximately 176 eV (Figure 3). Measurements in the fast-neutron energy range, 10 keV to 7 MeV, are determined from threshold reactions of ^{115}In (n,n') and ^{238}U fission chamber measurements beginning at 530 keV and 740 keV, respectively. This series of filtered-beam, spectral measurements did not include thermal-neutron intensities.

Epithermal Resonance Measurement Technique

Foil materials with a dominant resonance cross-section peak are useful for epithermal-neutron spectra measurements in that the majority of the detector response is due to the primary resonance. The foil materials utilized and the corresponding resonance peak energies are: (^{115}In - 1.46 eV), (^{197}Au - 5 eV), (^{186}W - 18 eV), (^{59}Co - 132 eV), (^{55}Mn - 340 eV), and (^{63}Cu - 580 eV). As expected, there was significant neutron self-absorption as a function of depth in the material about the resonance energy, which, in turn, complicated the determination of the incident flux in that particular energy group. If a correction for this self-absorption is not performed, the detector foil reaction rate, reactions per second per atom, will be artificially low and the calculated flux in the corresponding energy group will be underestimated. In addition, the response from neutron reactions in other energy regions must be corrected in order to isolate the dominant-resonance response and the neutron flux at that energy. The use of a foil stack provides a means to experimentally correct for self-shielding and secondary-resonance responses.

To account for self-shielding in the foil, the reaction rate formalism is treated as:

$$I_n = NV \sum_g {}_0\phi_g \, {}_nR_g \, {}_n\!<\!\bar{\sigma}\!>_g \tag{1}$$

where:

I_n	is the reaction rate in foil,
N	is the number density of target atoms,
V	is the volume of foil,
${}_0\phi_g$	is the flux in energy group g incident on the first foil,
${}_n\!<\!\bar{\sigma}\!>_g$	is the effective group cross section for foil n and energy group g (reactions per atom), and
${}_nR_g$	is the the ratio of the average group flux in foil n to the incident group flux.

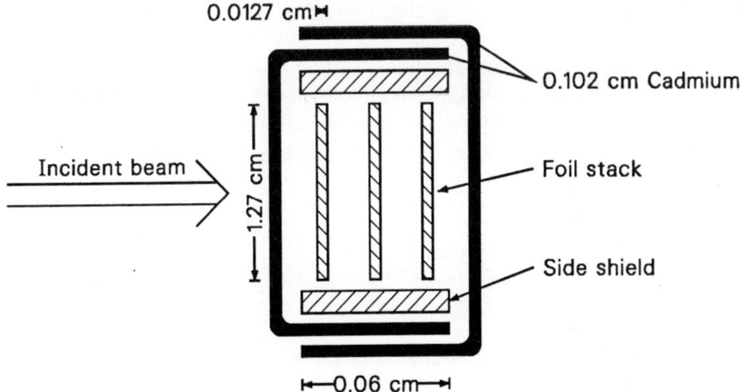

Figure 1. Resonance Foil Stack Assembly

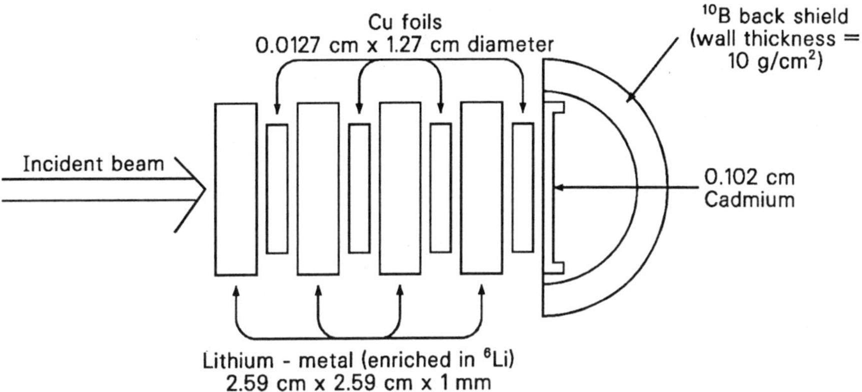

Figure 2. ^{6}Li Metal Stack Assembly.

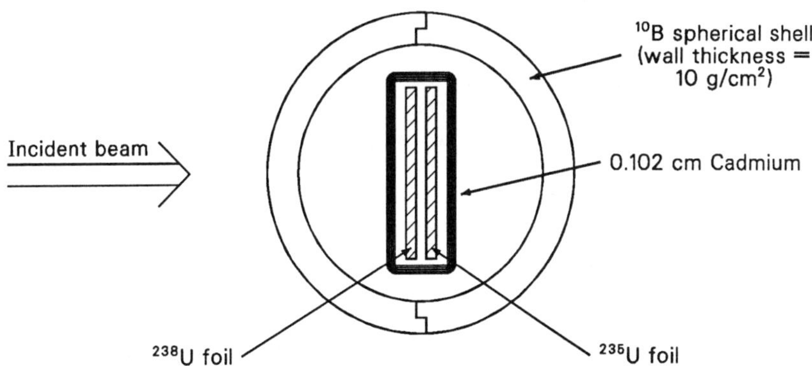

Figure 3. ^{10}B Sphere Assembly.

Equation (1) represents the reaction rate summed over all energy groups corrected for flux depression in a given foil. This expression is applicable to all foils in the stack. The values R_g and $_n\langle\bar\sigma\rangle_g$ were derived for each resonance material and each foil in the stack by the RAFFLE-V Monte Carlo transport code [1]. The Monte Carlo code uses a complete model of the incident beam geometry and the foil stack, and produces $_nR_g$ and $_n\langle\bar\sigma\rangle_g$ values for each energy group in each of the three foils. The product of these two terms is the useful and convenient term in all the analytical procedures involving absorption corrections. This term has been called the P-factor, $_nP_g$. This is the effective cross section of the foil and can be interpreted as the number of reactions per target atom in foil n per incident neutron flux with group g energy bounds. The units of this P-factor are barns per atom, the same as for microscopic cross sections.

As an example, assuming a single dominant resonance in the first foil with the calculated values of $_1P_g$, $_0\phi_g$, and the measured reaction rate, I_n, a first approximation flux for the primary resonance group can be obtained:

$$_0\phi_g = I_n/_1P_g. \tag{2}$$

Of course, additional corrections must be made for the secondary-resonance response in the first foil and any other contributions, such as response due to room return neutron flux. The second and third foils of the stack provide a means to make these corrections and will be addressed in the section on matrix data analysis.

Epithermal Pseudo-Threshold Neutron Measurement Technique

A pseudo-threshold response in the epithermal neutron energy range was accomplished by tailoring the neutron spectrum and folding that with a $1/V$ ^{63}Cu-detector response. The apparatus is an alternating shield-detector arrangement of 0.1-cm ^{6}Li and 0.0127-cm ^{63}Cu. The ^{6}Li-shield effectively creates a pseudo-threshold energy in ^{63}Cu response due to preferential attenuation of the low energy neutrons by the strong $1/V$ absorption cross section for ^{6}Li. Increasing the thickness of the ^{6}Li shield will increase the pseudo-threshold energy. This is the basis of the ^{6}Li/^{63}Cu stack assembly where the quantity of shielding in front of each Cu foil increases from the front to the back of the stack. The calculated 90% response ranges for each of the ^{63}Cu detectors are: (foil 1: 0.6 eV - 10.9 keV, (foil 2: 1.9 eV - 14.8 keV), (foil 3: 6.1 eV - 17.4 keV), and (foil 4: 19.6 eV - 20.1 keV). The measured response of the ^{63}Cu reactions can then be used in the spectrum adjustment procedure described below. The pseudo-threshold cross sections are derived using the same approach used to derive the P-factors in the resonance foil approach described in the previous section.

Fast-Neutron, Pseudo-Threshold Measurement Technique

The fast-neutron, pseudo-threshold theory is the same in principle as that of the epithermal pseudo-threshold approach described above. The differences are the materials used for the shield and detector. In this case, a 10-g/cm^2 thick ^{10}B-spherical shell is the shield material and ^{235}U fission and ^{238}U capture are the detector reactions. The 90% response ranges for these two reactions inside the boron sphere are: (^{235}U: 176 eV - 77.4 keV) and (^{238}U: 94 eV - 37 keV).

Fast-Threshold Measurement Technique

A fast-threshold measurement is obtained through the ^{115}In(n,n') and ^{238}U fission reactions. To suppress resonance energy reactions which would interfere with the ^{115}In(n,n') gamma measurement (in particular the ^{115}In(n,γ) reaction) the In foils were shielded in a ^{10}B-sphere. This, in effect, allowed only the fast component of the incident beam to interact with the foil material.

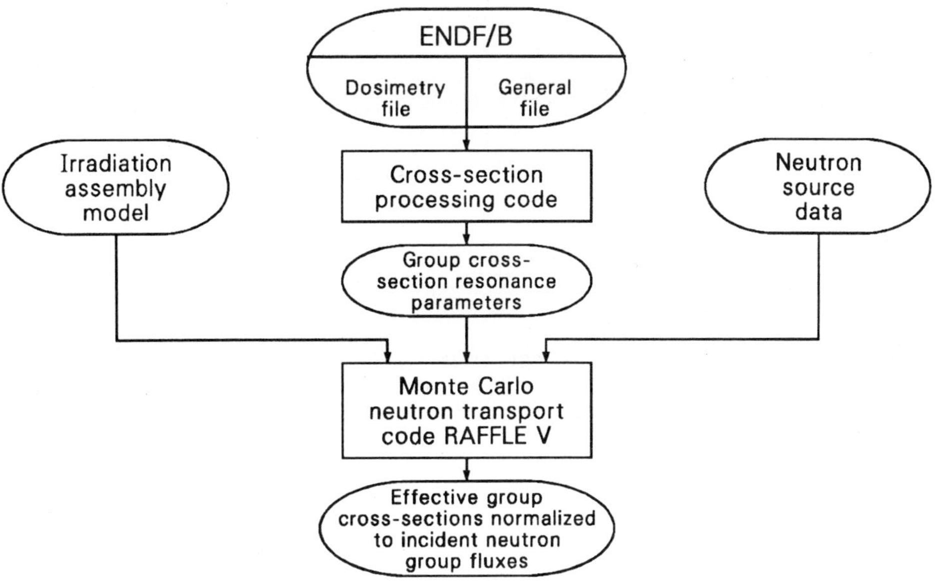

Figure 4. Monte Carlo Cross Section Generation.

DATA ANALYSIS

General Process

The data analysis process begins with measured gamma emissions of the activated foil sets. The measured gamma spectra are analyzed to determine the induced activity in the foil. From the foil activity, the reaction rate per atom is readily obtained. The reaction rates are used in two different analytical techniques to derive neutron flux data. One utilizes a matrix manipulation to determine the reaction rate contribution in the first foil due to the primary resonance. This procedure yields a flux value centered about the peak energy of the primary resonance. The second technique provides for an analysis of the neutron spectrum over the energy response range of all the dosimeters, not only the resonance foils. This is accomplished by adjusting the neutron spectrum within the constraints of the associated uncertainties until consistency is obtained between the measured and calculated reaction rates.

Monte Carlo Generated Activation Foil Cross Sections

The flow diagram for the generation of specialized cross sections, P-factors, is shown in Figure 4. The data source for this process is the ENDF/B-V Dosimetry and General Purpose Data Files. An intermediate step is required to produce group cross-section and resonance parameter data in the format required for the Monte Carlo neutron transport code, RAFFLE-V. In addition to the cross-section data, the RAFFLE-V input requires a descriptive model of the foil assembly and a specification of the incident neutron flux as a function of energy and incident angle. The output of the RAFFLE-V calculation (as far as cross-section generation is concerned) are P-factor values as previously defined.

FERRET Multigroup Spectrum Analysis

Multigroup characterization of the neutron spectrum through the energy range of 0.1 eV to 7 MeV was obtained by a least squares adjustment analysis using the FERRET code [2, 3]. In this analysis, adjustment of *à priori* estimates of the group fluxes was made to achieve consistency between the measured reaction rates and the reaction rates calculated using the equation:

$$_jR_i = \sum_g {}_{ij}C_g \; {}_i\sigma_g \; \phi_g \tag{3}$$

where: $_jR_i$ is the reaction rate for reaction i, foil j;

         $_{ij}C_g$ is the group correction factor to account for neutron self shielding for reaction i, foil j;

         $_i\sigma_g$ is the infinite dilute group cross section for reaction i; and

         ϕ_g is the group value for the neutron flux spectrum.

The analysis was made in a group structure employing 44 groups with group energy boundaries chosen to cover in some detail the major structure in the dosimeter cross sections and in the neutron flux spectrum. To provide a realistic input spectrum for the analysis, a DOT transport calculation was made for the BMRR filtered-beam facility. A spline interpolation scheme was subsequently used to convert the 47-group transport calculated spectrum into the 44-group values.

The infinite, dilute cross sections used as the base data set are from the ENDF/B-V Dosimetry (620 groups) or the General Purpose Data Files (47 groups). Using spectral weighting functions appropriate for this application, the fine group, infinite, dilute cross sections were collapsed into 44-group data sets required for the spectrum adjustment analysis.

To account for neutron-flux depression and resonance self-shielding effects in the dosimeter foil assemblies, Monte Carlo transport calculations were made with the RAFFLE-V code for the various dosimeter materials. These calculations yielded effective group cross sections, termed P-factors, normalized to the incident-neutron group fluxes for each dosimeter foil. For the FERRET analysis, group correction factors (C_g for each reaction and foil) were derived by dividing the group P-factor values by corresponding infinite, dilute group cross-section values.

Covariances for the input cross-section and flux data were prescribed. The PUFF code [4] was used to process ENDF/B-V covariance files for ^{197}Au(n,γ), ^{235}U(n,f), and ^{115}In(n,n') and to generate the appropriate 44-group covariance matrices. Because incomplete covariance prescriptions are given in ENDF/B for the remaining dosimeters, a parametric expression was used to generate the covariance matrix elements [2, 3].

<u>Matrix Resonance Foil Stack Analysis</u>

The matrix analysis approach utilizes the response of the second and third foils in the resonance foil stack to correct the response of the first foil in the energy group centered about the resonance peak. The induced activity in a foil of a given resonance material is due primarily to three contributions: (1) the activity produced by the primary resonance, (2) the activity produced by all secondary-resonance absorptions, and (3) the activity produced by room-return neutrons incident on the back of the foil assembly. The absorption constants for the primary-resonance energy group and the total number of absorptions over all groups per foil are obtained from the RAFFLE-V calculation results. The secondary-absorption response is derived by subtracting the primary-resonance response from the total response in the particular foil. The back-flux constant is the ratio of the total number of absorptions in a foil to the total absorptions in the third foil. The stack order of the back-flux constants is then inverted such that the ratios indicate a flux incident from the back of the foil assembly. A 3x3 matrix is then formed where the rows correspond to the three foils and the three columns correspond to the primary, secondary, and back-flux response constants, respectively. The matrix coefficients are normalized such that the values in the first row are unity. The measured reaction rate vector is equal to the product of the 3x3 response matrix and the vector whose elements are the contributions in the first foil from the primary-resonance, the secondary-resonance, and the room-return neutrons, respectively. This equation is then solved for the reaction rate contribution in the first foil due to the primary resonance. The incident flux for the primary-resonance group

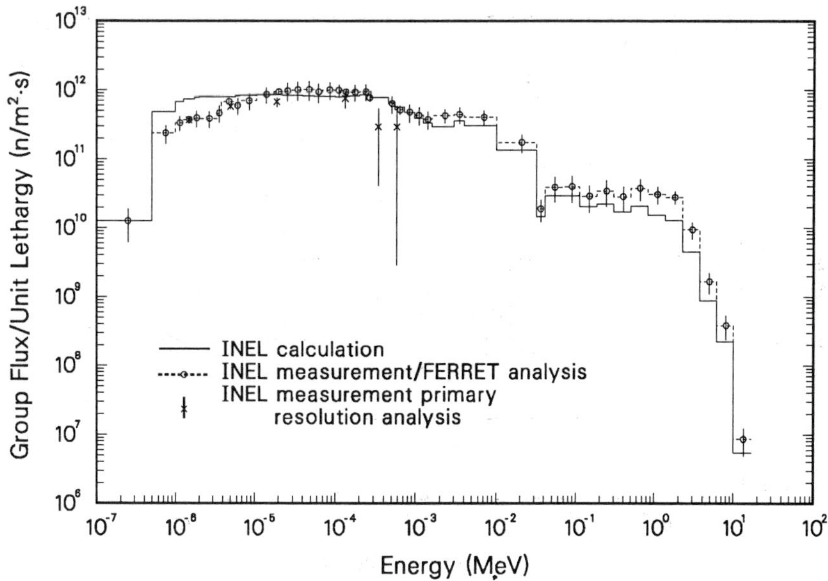

Figure 5. BMRR Neutron Spectrum.

is then obtained by dividing the primary-resonance reaction rate contribution in the first foil by the corresponding RAFFLE-V calculated P-factor, P_g. The group flux is then converted to flux per unit lethargy by dividing by the lethargy width of the primary-resonance group.

RESULTS

Pertinent results from the FERRET multigroup spectrum analysis are illustrated in Figure 5. This spectral plot shows a comparison of the DOT transport calculated spectrum and the spectrum that has been adjusted to achieve consistency with the measured reaction rates. The vertical lines through the adjusted group values indicate the uncertainties as obtained from the FERRET analysis. Also shown on the plot are the primary-resonance flux values and uncertainty bars as calculated by the matrix-analysis technique. Figure 6 shows the ratio of the adjusted group fluxes to the unadjusted or calculated group values, along with an explicit illustration of the fractional uncertainties in the input and adjusted group flux values. This figure demonstrates the influence of utilizing integral reaction rate data to effect an adjustment in the input group fluxes and to provide a significant reduction in the uncertainties in the group flux values. This analysis indicates a downward adjustment of 50% to 20% in group flux values between 0.5 eV and 6 eV, an upward adjustment of 20% to 50% in group flux values between 3 keV and 100 keV, and an upward adjustment of 50% to 100% in group flux values above 100 keV.

The relationship of the adjusted neutron spectrum based on measured parameters to the neutron tissue dose has been evaluated. Estimates of the neutron dose were calculated by folding both calculated and adjusted group spectra with a 44-group KERMA data set generated from the 98-group KERMA data file for tissue (as given in ICRU-26, Appendix A [5]). The results of these computations are shown in Figure 7, where the neutron dose per group per MW for the BMRR is plotted as a function of energy. The double-humped character of the neutron dose plot is due to the step function change in the flux spectrum at 25 keV. The uncertainties shown for the adjusted histogram correspond to a propagation of the adjusted spectrum uncertainties, derived from the FERRET analysis. The analysis clearly indicates that neutron dose predictions based on the DOT transport calculation underpredict the absorbed dose by as much as 40%.

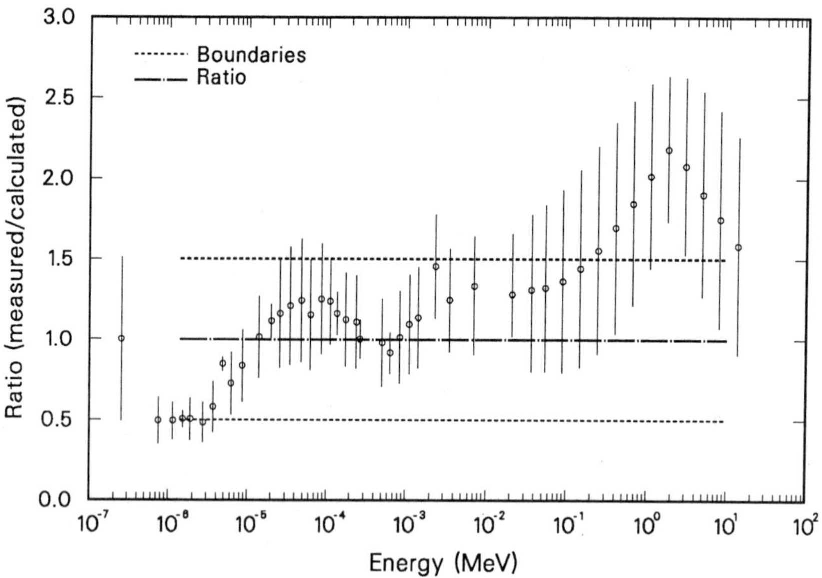

Figure 6. Ratio of Adjusted-to-Calculated Neutron Flux.

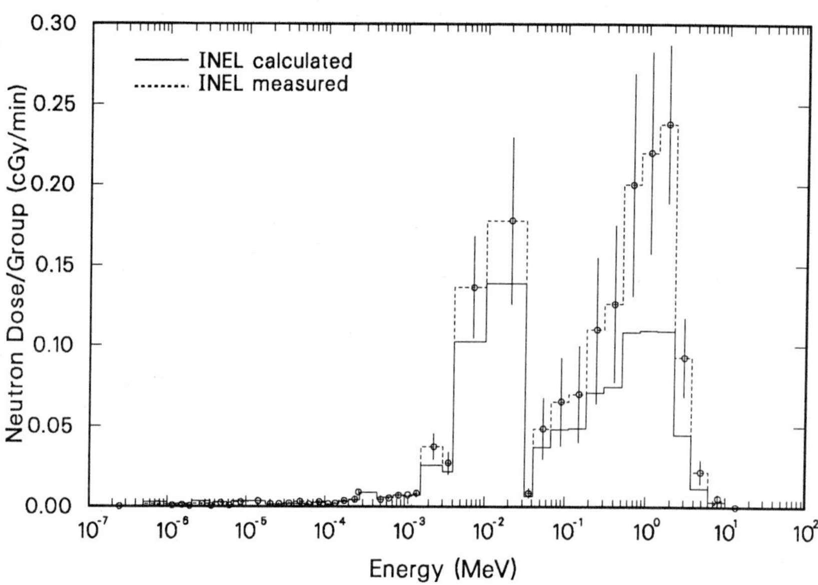

Figure 7. Fast-Neutron Dose Response.

The calculated and measured integral data results are tabulated in Tables One and Two. Table One contains the reaction-rate data from the measurements, the DOT transport calculation, and the FERRET analysis. These data indicate the degree of consistency achieved by the adjustment procedure. Table Two allows for a comparison of the INEL-measured, INEL-calculated, and BNL-measured integral data. Good agreement is evident between measurements and the calculation for the total and epithermal neutron-flux values. In the fast energy region, the BNL and INEL-measured data agree within the uncertainties

Table One

Reaction Rate Data

Reaction		Measured		Calculated		Calculated/	FERRET Fit		
ID	Foil	Rate	σ	Rate	σ	Measured	Rate	σ	Final/Meas.
^{115}In NG	1	0.488E+11	0.060	0.894E+11	0.45	1.832	0.486E+11	0.075	0.998
^{115}In NG	2	0.234E+11	0.067	0.396E+11	0.41	1.694	0.230E+11	0.058	0.986
^{115}In NG	3	0.178E+11	0.071	0.297E+11	0.39	1.673	0.177E+11	0.058	1.000
^{197}Au NG	1	0.369E+11	0.043	0.484E+11	0.45	1.313	0.403E+11	0.048	1.095
^{197}Au NG	2	0.182E+11	0.047	0.230E+11	0.40	1.267	0.187E+11	0.043	1.032
^{197}Au NG	3	0.141E+11	0.049	0.179E+11	0.39	1.272	0.143E+11	0.043	1.019
^{186}W NG	1	0.281E+11	0.086	0.273E+11	0.47	0.973	0.298E+11	0.090	1.062
^{186}W NG	2	0.156E+11	0.110	0.134E+11	0.44	0.861	0.142E+11	0.081	0.915
^{186}W NG	3	0.117E+11	0.120	0.106E+11	0.43	0.910	0.111E+11	0.079	0.955
^{59}Co NG	1	0.375E+10	0.060	0.367E+10	0.38	0.980	0.387E+10	0.071	1.032
^{59}Co NG	2	0.238E+10	0.070	0.231E+10	0.33	0.971	0.228E+10	0.058	0.959
^{59}Co NG	3	0.200E+10	0.075	0.200E+10	0.31	1.002	0.192E+10	0.057	0.964
^{55}Mn NG	1	0.731E+09	0.063	0.863E+09	0.32	1.182	0.783E+09	0.064	1.072
^{55}Mn NG	2	0.597E+09	0.066	0.644E+09	0.29	1.079	0.561E+09	0.049	0.940
^{55}Mn NG	3	0.526E+09	0.068	0.574E+09	0.29	1.092	0.491E+09	0.045	0.935
^{63}Cu NG	1	0.181E+09	0.085	0.246E+09	0.30	1.363	0.213E+09	0.053	1.179
^{63}Cu NG	2	0.152E+09	0.087	0.203E+09	0.28	1.338	0.172E+09	0.044	1.137
^{63}Cu NG	3	0.136E+09	0.091	0.184E+09	0.28	1.358	0.155E+09	0.042	1.143
^{235}U NF		0.204E+09	0.130	0.164E+09	0.36	0.804	0.201E+09	0.080	0.987
^{115}In NN		0.519E+06	0.087	0.228E+07	0.47	0.440	0.505E+06	0.018	0.973
^{238}U NG		0.483E+08	0.080	0.429E+08	0.35	0.889	0.516E+08	0.070	1.070
^{63}Cu NG	L11	0.169E+09	0.086	0.190E+09	0.33	1.128	0.179E+09	0.061	1.065
^{63}Cu NG	L12	0.108E+09	0.091	0.105E+09	0.35	0.978	0.108E+09	0.057	1.006
^{63}Cu NG	L13	0.912E+08	0.093	0.751E+08	0.37	0.824	0.804E+08	0.058	0.882
^{63}Cu NG	L14	0.792E+08	0.095	0.552E+08	0.38	0.697	0.608E+08	0.060	0.768
^{107}Au NG	Bare	0.116E+12	0.043	0.123E+12	0.47	1.066	0.104E+12	0.053	0.898

Table Two

Integral Data Comparisons

Data Classification	INEL-Meas.	INEL-Calc.	BNL-Meas.	INEL-Calc./ INEL-Meas.	BNL-Meas./ INEL-Meas.
Total Neutron Flux (10^{12} n/m^2-s)	6.65	6.75	6.6	1.015	0.99
Fast Neutron Flux (10^{12} n/m^2-s)	0.344	0.246	0.33	0.715	0.96
Epithermal Neutron Flux 0.5 eV – 10 keV (10^{12} n/m^2-s)	6.20	6.39	6.0[a]	1.031	0.97
Fast Neutron Dose (cGy/min)	1.67[b]	1.04[b]	1.75	0.623	1.05
Fast Neutron Dose / Epithermal Flux (10^{-15} cGy-m^2)	4.48	2.72	4.87	0.607	1.09

[a] Energy range covered by BNL measurement is 0.4 eV – 10 keV.

[b] Dose generated by folding neutron spectrum with tissue approximation KERMA as contained in ICRU No. 26, Appendix A.

in the measurements. The calculated fast-region neutron data are low compared with the BNL and INEL-measured data by a factor of 0.72 for the fast-neutron flux and a factor of 0.62 for the fast-neutron dose.

SUMMARY AND CONCLUSIONS

Foil activation techniques were used to obtain neutron spectral data in the energy range from 0.5 eV to 7 MeV for the BMRR aluminum oxide filter assembly. These data were obtained to support technology development associated with the PBF/BNCT program. Resonance reaction foils were used to obtain reaction-rate data with responses primarily in the epithermal energy region. Pseudo-threshold response reactions were used to obtain reaction-rate data with responses primarily in the high-epithermal and fast-neutron energy regions. Threshold reactions were used to obtain reaction-rate data in the fast-neutron energy region. The resulting spectral data have uncertainties which range from 20% to 40% in the fast-neutron region and from 5% to 30% in the epithermal region. The integral data derived from the measured spectral data are:

Epithermal Flux (0.5 - 10 keV)
6.20 E+12 (± 5.1%) n/m^2-s·MW

Fast Flux (10 keV - 7 MeV)
3.44 E+11 (± 21%) n/m^2-s·MW

Fast Neutron Dose
1.67 (± 16%) cGy/min·MW.

The integral results agree well with those measured by BNL [6]. The calculated integral results agree with the measured values for the epithermal flux, but predict a lower fast flux and, consequently, a lower fast-neutron dose. Future measurements will emphasize obtaining more data in the fast-neutron region.

ACKNOWLEDGMENT

This work was performed under the auspices of the U.S. Department of Energy, DOE Contract No. DE-AC07-76ID01570.

COPYRIGHT

REFERENCES

1. F. J. Wheeler et al., "RAFFLE-V General Purpose Monte Carlo Code for Neutron and Gamma Transport," EGG-PHVS-6003, Rev. 1 (Oct. 1983).

2. F. Schmittroth, "FERRET Data Analysis Code," Hanford Engineering Development Laboratory, US-DOE Report HEDL-TME 79-40 (Sept. 1979).

3. F. Schmittroth, "A Method for Data Evaluation with Lognormal Distributions," Nucl. Sci. Eng., 72:19 (1979).

4. "PUFF2 Determination of Multigroup Covariance Matrices from ENDF/B-V Uncertainty Files," RSIC Computer Code Collection, PSR-157 (1980).

5. "Neutron Dosimetry for Biology and Medicine," International Commission on Radiation Units and Measurements, Bethesda, MD, ICRU Report 26, Reprinted August 15, 1984.

6. S. K. Saraf, J. Kalef-Ezra, R. G. Fairchild, B. H. Laster, S. Fiarman, and E. Ramsey, "Epithermal Beam Development at the BMRR: Dosimetric Evaluation." (These Proceedings.)

ACCELERATOR-BASED NEUTRON BEAMS

THE POSSIBLE USE OF A SPALLATION NEUTRON SOURCE FOR

NEUTRON CAPTURE THERAPY WITH EPITHERMAL NEUTRONS

E. Grusell, H. Condé, B. Larsson, T. Rönnqvist, O. Sornsuntisook
Department of Radiation Sciences, Uppsala University
Uppsala, Sweden

J. Crawford and H. Reist
Paul Scherrer Institute, Villigen, Switzerland

B. Dahl and N. G. Sjöstrand
Department of Reactor Physics, Chalmers University of Technology
Gothenburg, Sweden

G. Russel
Los Alamos National Laboratory, Los Alamos, NM, USA

ABSTRACT

Spallation is induced in a heavy material by 72-MeV protons. The resulting neutrons can be characterized by an evaporation spectrum with a peak energy of less than 2 MeV. The neutrons are moderated in two steps: first in iron and then in carbon. Results from neutron fluence measurements in a perspex phantom placed close to the moderator are presented. Monte Carlo calculations of neutron fluence in a water phantom are also presented under some chosen configurations of spallation source and moderator. The calculations and measurements are in good agreement and show that, for proton currents of less than 0.5 mA, useful thermal-neutron fluences are attainable in the depth of the brain. However, the dose contribution from the unavoidable gamma background component has not been included in the present investigation.

INTRODUCTION

The eventual aim of the present joint Swedish-Swiss project is to construct an accelerator-based intermediate energy neutron source that would permit irradiation of neoplasms in the central nervous system by an intermediate energy neutron fluence rate of at least 10^9 n cm^{-2} s^{-1}. The accelerator should be of a moderate size to permit accommodation in a hospital environment. Therefore, the rather low proton energy of 72 MeV was chosen.

The clinical interests of this collaboration are primarily the treatment of vascular malformations in the central nervous system. In a longer perspective, the treatment of malignant brain tumors is given priority over other malignancies, such as melanomas and colorectal carcinomas.

Work on the project has thus far been devoted to studies of different moderator materials and configurations useful for combination with neutron production by 72 MeV-protons stopped in heavy materials [1]. The aim is to optimize the performance of a neutron source for neutron capture therapy (NCT). The required characteristics are first

Neutron Beam Design, Development, and Performance for Neutron Capture Therapy
Edited by O. K. Harling *et al.*
Plenum Press, New York, 1990

that the bulk of the neutrons should have an energy between 1 and 100 keV and second that the useful intensity of thermal neutrons should be at least 10^{12} n cm^{-2} h^{-1}. A third requirement, which has not been studied thus far, is that the dose from the gamma background should be kept within tolerable limits.

EXPERIMENTAL RESULTS

The iron and graphite moderator option was studied experimentally at the 72-MeV Injector I Cyclotron at the Paul Scherrer Institute (PSI) from August 29 to September 2, 1988. Neutrons were produced by stopping 72-MeV protons in a tungsten block. The moderator consisted of an iron block, approximately 50 cm by 60 cm by 60 cm. On one side, it was covered with 13 cm of graphite (see Figure 1).

The neutron field in two plastic phantoms (20 cm by 20 cm by 20 cm) was probed with different foil detectors: gold activation with and without Cd-shielding to measure the thermal-neutron flux, and by plastic recoil track detectors (Neutrak 144, Landauer Inc., [2]) to measure neutrons with energies above 144 keV.

The results from the gold activation foils are presented in Table One and compared with Monte Carlo calculated values. The Monte Carlo code MCNP was used [3]. It is evident that there is a fair agreement. The values are normalized to an integrated proton current of 1 mC to the target, which corresponds to the experimental conditions.

The results from the proton recoil track detector measurements are given in Table Two together with calculated values. As in Table One, the values are normalized to an integrated proton current of 1 mC. The upper limit for a meaningful readout of these detectors is $5.7 \cdot 10^5$ tracks cm^{-2}. Hence, some detectors received an overdose. The calculated neutron fluence values were converted to detector track density using the values for detector sensitivity as a function of neutron energy that are given in [2]. The agreement is within a factor of two. The deviation may be due to uncertainties in the calculation of the neutron spectrum, uncertainties in the detector sensitivity values, and to a directional dependence of the detectors.

In Table Three, the result from a measurement of neutrons of energy above 10 MeV is shown. It was made with an NE 213 liquid scintillator with pulse shape discrimination to eliminate gamma pulses. The detector was placed 200 cm from the iron moderator, perpendicular to the beam direction, on the side with no graphite. The result is given per coulomb of integrated proton current to the target.

Also shown are Monte Carlo calculated values at two different distances from the moderator. The agreement at 200 cm is nearly within a factor of two, which is reasonable considering the possible sources of error: uncertainties in the setting of the detector threshold, uncertainties in the source spectrum, and uncertainties in the calculation of the detector efficiency. The calculated flux of neutrons above 10 MeV at the surface of the moderator corresponds to a dose of 8 Gy/C at a depth of 5 cm in a plastic phantom. The calculated dose from all epithermal and fast neutrons is 210 Gy/C for the conditions that are expected to be present in the actual experiment. Compared to that figure, a correction of 8 Gy/C for neutrons with energies above 10 MeV is small. Thus, while the fast neutrons are less effectively stopped by the addition of more moderating material, this result indicates that, even in a more realistic design, the dominating background neutron dose will be produced by neutrons below 10 MeV.

RESULTS OF MONTE CARLO CALCULATIONS

For the Monte Carlo calculations presented here, the well known code MCNP, developed at Los Alamos, was used [3]. The neutron source was an evaporation source, with the neutron spectrum given by:

$$\Delta N/\Delta E = CE \exp(-E/E_0), \quad \text{where } E_0 = 1.29 \text{ MeV.}$$

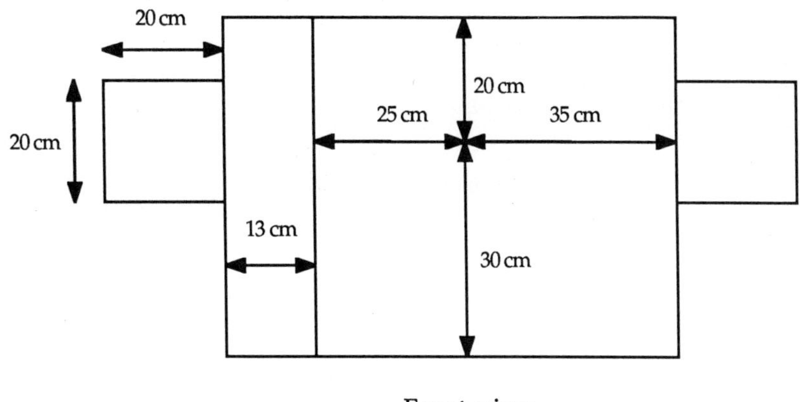

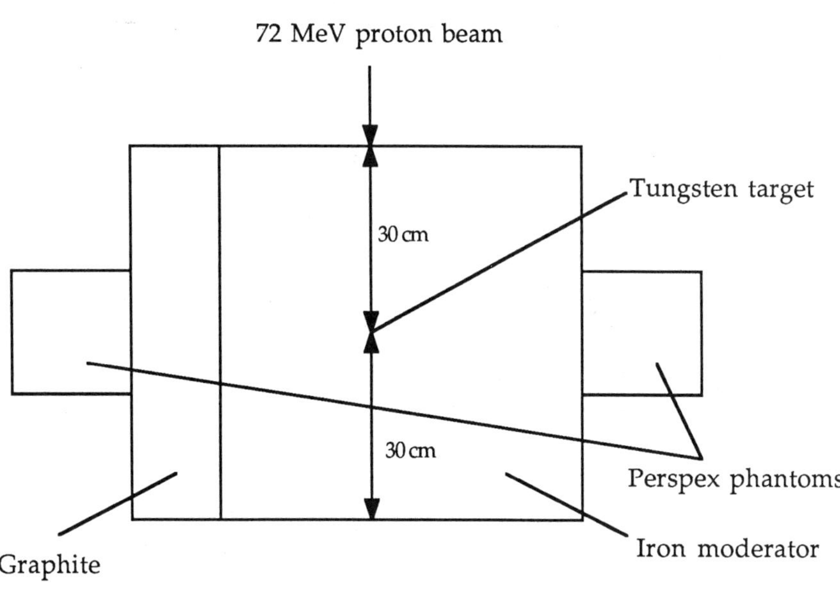

Figure 1. The Moderator and Phantom Arrangement Used for the Experiments at PSI in 1988.

Table One

Results from Gold Foil Measurements

Phantom I, 34-cm Iron

Depth in Phantom (cm)	Thermal-Neutron Fluence (10^{10} cm^{-2}/mC)	
	MCNP	Gold Foil Activation
0	3.36	–
2	16.6	12.4
5	17.2	13.1
10	6.8	5.13

Phantom II, 25-cm Iron and 13-cm Carbon (Graphite)

Depth in Phantom (cm)	Thermal-Neutron Fluence (10^{10} cm^{-2}/mC)	
	MCNP	Gold Foil Activation
0	10.1	–
2	15	13.2
5	9.6	8.8
10	2.3	3.2

Table Two

Results from Track Detector Measurements

Phantom I, 34-cm Iron

Depth in Phantom (cm)	Track Density (10^{5} cm^{-2}/mC)	
	MCNP	Neutrak 144
2	80	>5.7
5	20	>5.7
10	3.5	1.6
15	1.1	0.59

Phantom II, 25-cm Iron and 13-cm Carbon (Graphite)

Depth in Phantom (cm)	Track Density (10^{5} cm^{-2}/mC)	
	MCNP	Neutrak 144
2	13	>5.7
5	4.0	2.0
10	0.7	1.1
15	–	0.39

Table Three

Results from Liquid Scintillator Measurements of Neutrons with Energy above 10 MeV

Distance from Iron Moderator	Neutron Fluence (cm^{-2}/C)	
(cm)	Calculated	NE 213
0	$1.25 \cdot 10^{11}$	–
200	$4.40 \cdot 10^9$	$1.6 \cdot 10^9$

This was shown to be a valid approximation for neutron energies below 10 MeV by comparison with calculations of the source neutron spectrum using the code HETC (cf [2]), from which the neutron yield was also computed. Neutrons of energies above 10 MeV were not taken into account because of limitations in the cross-section libraries of MCNP. However, as discussed earlier, neutrons above 10 MeV are not expected to make a major contribution to the background dose.

The neutron transport calculations were made for spherical iron moderators of three diameters: 50 cm, 100 cm, and 150 cm. Each was covered with 15 cm of graphite. Both the iron and the carbon contained 1 percent boron-10 to suppress the thermal-neutron flux at the phantom surface. The arrangement of the moderators and the phantoms is shown in Figure 2. The spherical head phantoms were filled with water containing 1.84% nitrogen (see Table 4). The results are given in Figures 3 to 5.

It is seen that useful intensities of thermal neutrons can be obtained at depth in the phantom for proton currents of less than 0.5 mA and that the fast-neutron contribution is small if the iron moderator is thick enough. It is also clear that the useful depth increases with moderator thickness. However, the background dose is underestimated because the gamma dose component is not included. This component will be calculated as a part of the continued project. The RBE values used were 1.6 for fast neutrons and nitrogen capture and 2.3 for boron-10 capture. These values were chosen to facilitate comparisons with results of others and might not prove to be the best to use.

Table Four. Composition of Head Phantom

Density:	$1 \ g \ /cm^3$
Composition:	H_2O with 1.84 percent N.
Fractional Composition by Weight:	H 0.109; O 0.872; N 0.0184

CONCLUSION

A spallation neutron source may be a realistic option for the construction of an accelerator-based neutron source of reasonable size and cost. The next step in this work will be the construction of a full-scale prototype source where radiobiological as well as radiophysical studies can be made. Special attention must be paid to the cooling and maintenance of the target in which several kilowatts of heat will be produced.

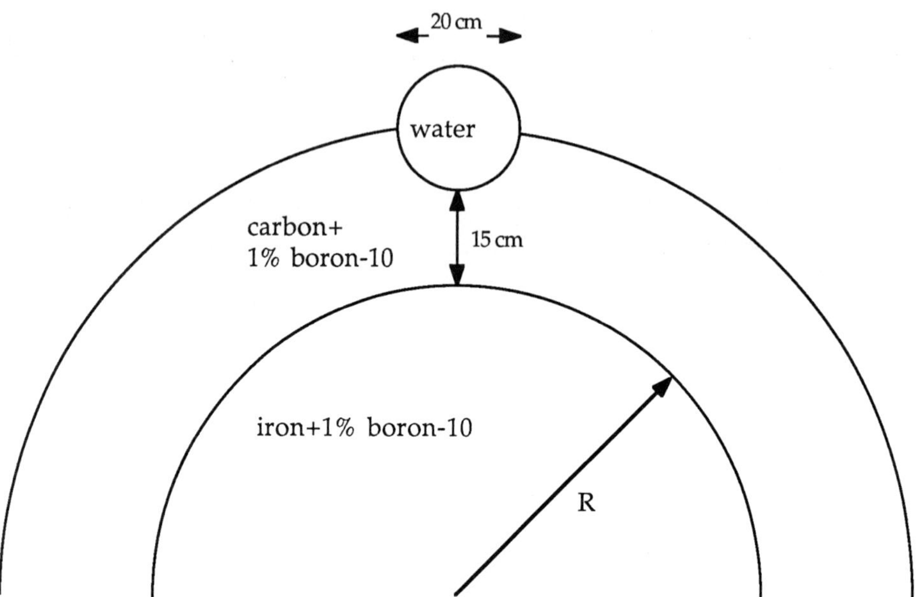

Figure 2.   Moderator and Head Phantom Arrangement Used in the Monte Carlo Calculations. R = 25, 50, or 75 cm.

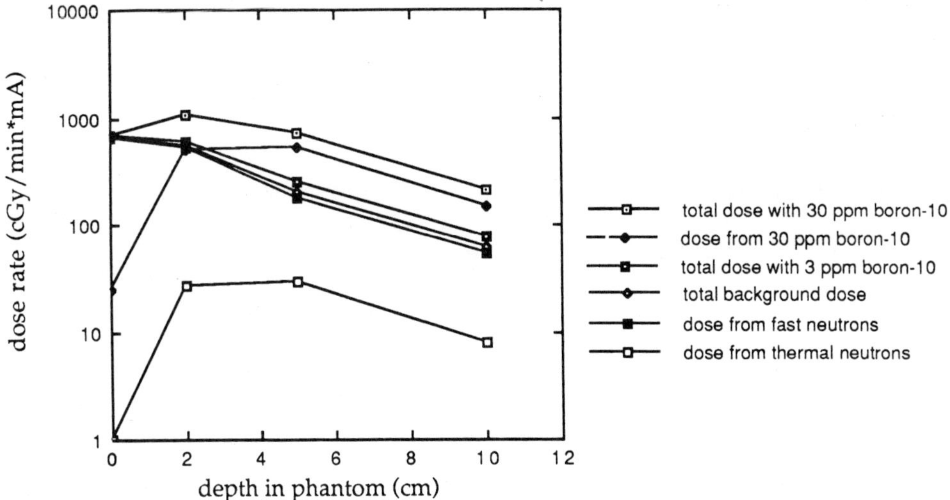

Figure 3a. Monte Carlo Calculated Depth-Dose Curves in a Head Phantom, Close to a 50-cm Diameter Iron Moderator with 15 cm of Graphite.

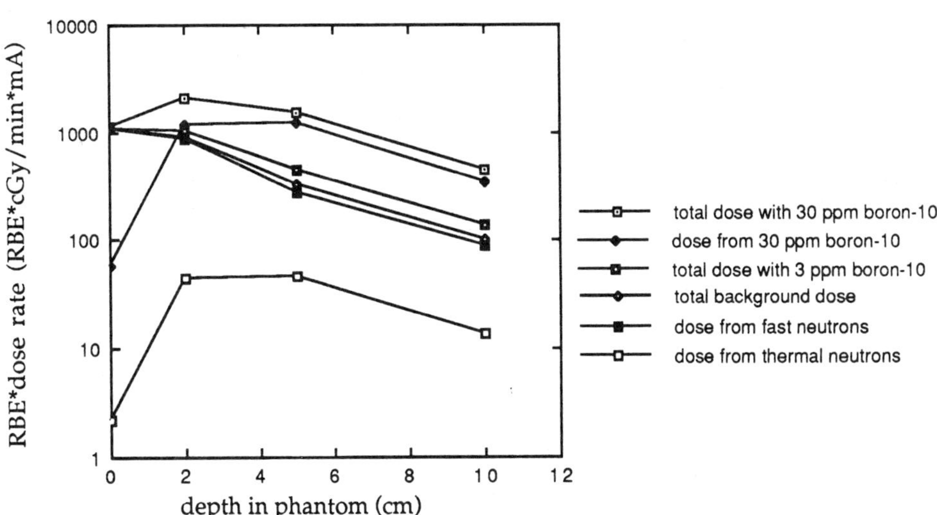

Figure 3b. RBE Times Depth Dose (RBE = 2.3 for Boron Capture, and 1.6 for Other Neutron Reactions). Moderator as in 3a.

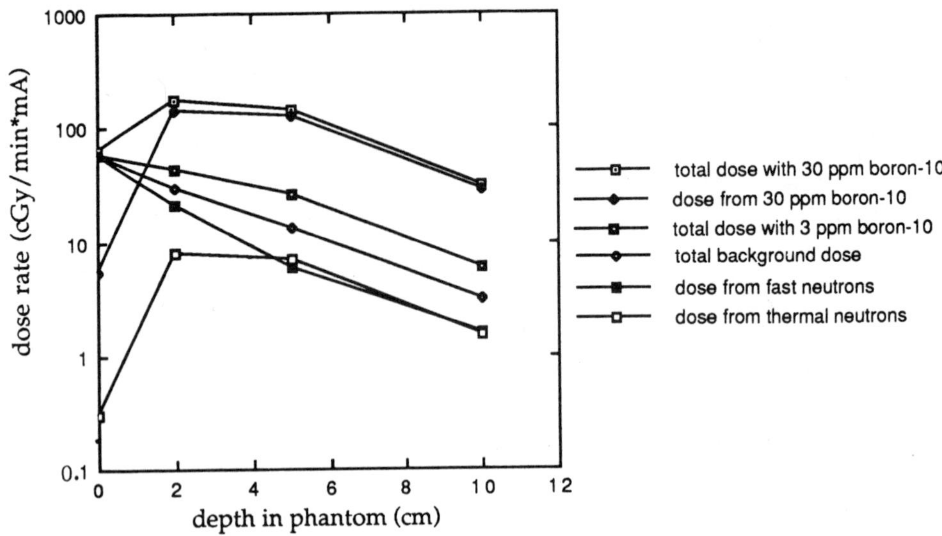

Figure 4a. Monte Carlo Calculated Depth-Dose Curves in a Head Phantom, Close to a 100-cm Diameter Iron Moderator with 15 cm of Graphite.

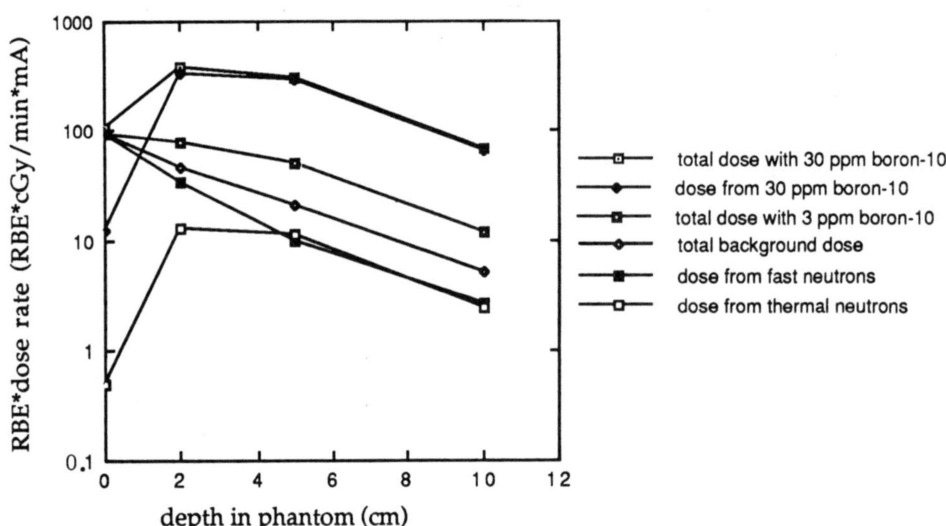

Figure 4b. RBE Times Depth Dose (RBE = 2.3 for Boron Capture, and 1.6 for Other Neutron Reactions). Moderator as in 4a.

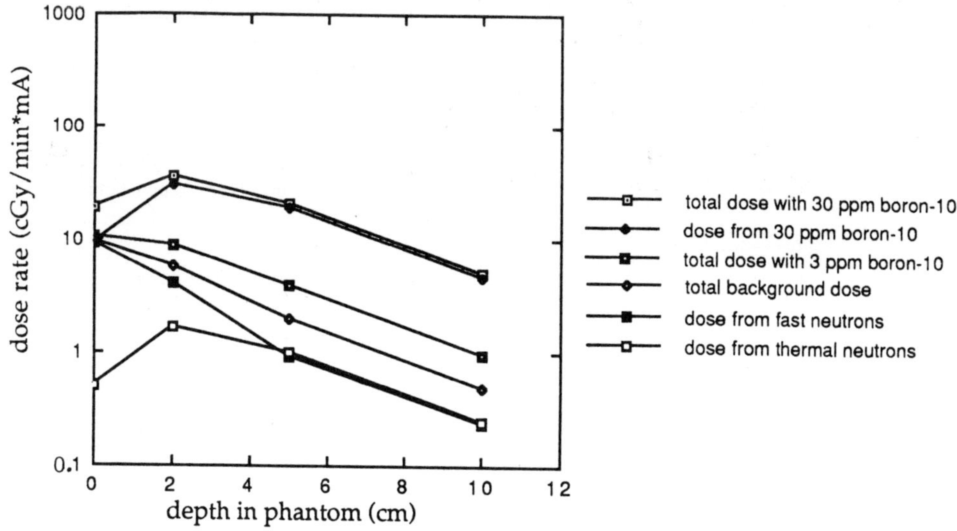

Figure 5a. Monte Carlo Calculated Depth-Dose Curves in a Head Phantom, Close to a 150-cm Diameter Iron Moderator with 15 cm of Graphite.

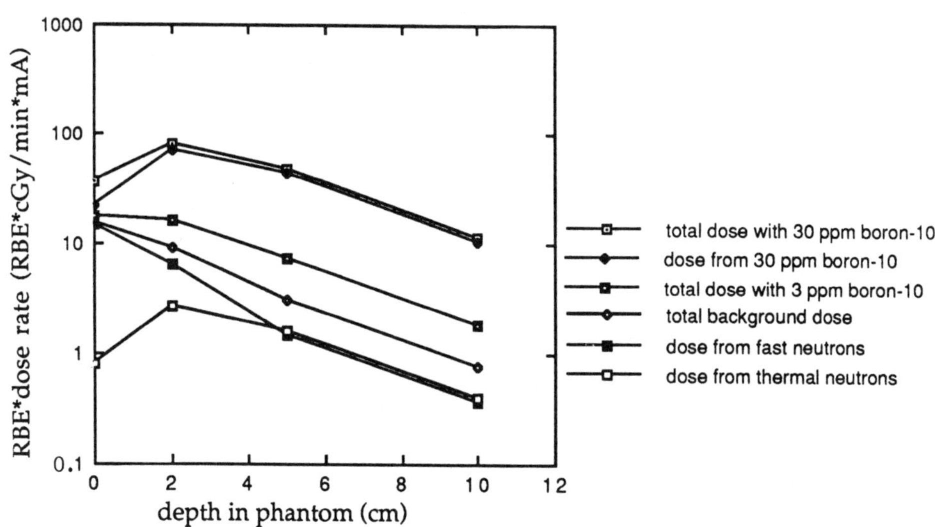

Figure 5b. RBE Times Depth Dose (RBE = 2.3 for Boron Capture, and 1.6 for Other Neutron Reactions). Moderator as in 5a.

REFERENCES

1. H. Condé, E. Grussel, B. Larsson, C. B. Pettersson, L. Thuresson, J. Crawford, H. Reist, B. Dahl, and N. G. Sjöstrand, "Time of Flight Measurements of the Energy Spectrum of Neutrons Emitted from a Spallation Source and Moderated in Water," <u>Nucl. Instruments and Methods in Phys. Res.</u>, A261:587 (1987).

2. E. V. Benton, R. A. Oswald, and A. L. Frank, "Proton-Recoil Neutron Dosimeter for Personnel Monitoring," <u>Health Physics</u>, 40:801 (1981).

3. J. F. Briesmeister, ed., "MCNP – A General Monte Carlo Code for Neutron and Photon Transport," Los Alamos National Laboratory, <u>LA-7396-M</u>, Rev. 2 (1986).

A VERSATILE, NEW ACCELERATOR DESIGN FOR BORON NEUTRON CAPTURE THERAPY: ACCELERATOR DESIGN AND NEUTRON ENERGY CONSIDERATIONS

R. E. Shefer,[1] R. E. Klinkowstein,[1] J. C. Yanch,[2] and G. L. Brownell[2]

[1] Science Research Laboratory
Somerville, MA

[2] Department of Nuclear Engineering
Massachusetts Institute of Technology
Cambridge, MA

INTRODUCTION

The development of a new, compact high-current proton accelerator by Science Research Laboratory (SRL) capable of producing large quantities of neutrons fits closely with the recent interest in the use of epithermal neutrons for Boron Neutron Capture Therapy (BNCT). Three important facets of this development are currently being investigated. These include 1) the production of neutrons by the $^7Li(p,n)$ reaction using a novel tandem cascade electrostatic accelerator, 2) the optimal neutron energy (or range of energies) for treatment of tumors at different depths in tissue, and 3) the extent of moderation and filtering required to reduce the energy of the accelerator neutrons to the energies deemed optimal in (2) above. The overall goal is to develop a versatile epithermal-neutron source which will be, initially, a valuable research tool in the development of BNCT, and ultimately, an essential component of a hospital-based patient irradiation facility.

The Tandem Cascade Accelerator (TCA) is a novel electrostatic accelerator which combines two existing technologies into a simple, compact proton accelerator suitable for operation in a hospital environment. Although the $^7Li(p,n)$ reaction has been well studied [1-3], there are many practical problems associated with its use as an epithermal-neutron source. The proton current requirements, heat dissipation in the target, and neutron energy spectra and angular distribution are all interrelated aspects of the overall design. Optimization of this design requires careful analysis. Similarly, the optimal neutron energy for BNCT has yet to be defined. The energy (or range of energies) considered most appropriate for NCT will determine the type and the extent of moderator and filter materials required to shift the spectrum of accelerator neutrons (where the average neutron energy is a few hundred keV) to a lower energy.

Calculations of neutron and gamma behavior in brain material for the purpose of determining optimal incident neutron energy have been carried out using a coupled neutron-photon transport code (MCNP) and preliminary results are given here. The information obtained through these calculations is currently being used to design the moderator/filter and collimator configurations which will become part of the entire neutron delivery system.

Neutron Beam Design, Development, and Performance for Neutron Capture Therapy
Edited by O. K. Harling *et al.*
Plenum Press, New York, 1990

ACCELERATOR NEUTRON SOURCE DESIGN

Isotopes and Reactions for Neutron Production with Accelerator Beams

Table One lists a number of possible charged particle reactions and their characteristics. Group I has medium atomic number nuclei as targets. These appear to be attractive because neutron energies obtained at 0° fall in a desirable range. Additionally, many of the elements are metals and can withstand intense proton beams with minimal degradation of the target provided proper cooling is maintained. The disadvantage of these isotopes, however, is their relatively low reaction cross sections. Group II consists of two light targets, ^{3}H and ^{7}Li. Although the neutron yields for both targets are comparable, the neutron energy values for the ^{3}H reaction are significantly higher than for ^{7}Li and thus more moderation will be required in order to render these neutrons suitable for patient irradiation. This is discussed under the section of this paper entitled, "Optimum Neutron Energies for Neutron Capture Therapy."

Considering the characteristics of the reactions listed in Table One, ^{7}Li appears to be the target of choice owing to the relatively low neutron energies and high neutron yields. This is directly attributable to the presence of a large cross-section resonance somewhat above the reaction threshold as shown in Figure 1. The proton threshold energy is 1.88 MeV, and a pronounced resonance exists at 2.25 MeV with a small resonance just above the threshold at 1.92 MeV; neutrons are preferentially emitted in the forward direction. The total cross section at 2.25 MeV is 580 mb. The low reaction threshold means that the proton beam power is low for a given proton current and that the energy which must be dissipated in the target, per proton, is relatively small.

By limiting the bombarding proton energy to 2.5 MeV, the neutrons are emitted with energies of less than 800 keV, yet full advantage is taken of the large resonance peak in the cross section. At this proton energy, a small amount of gamma radiation from proton reactions arises from inelastic proton scattering ^{7}Li(p,p')^{7}Li* and from the ^{7}Li(p,n)^{7}Be* reaction. Both ^{7}Li* and ^{7}Be* decay with the emission of a 478-keV photon with a lifetime of approximately 10^{-13} s. There is also a low probability of proton capture, i.e., ^{7}Li(p,γ)^{8}Be with emission of 16 and 19-MeV gamma radiations [3].

Several investigators [1,4,5,6] have examined the characteristics of the ^{7}Li(p,n)^{7}Be reaction as a source of epithermal neutrons for NCT. In the following paragraphs, analytic calculations based on published cross-section data [2] and semi-analytic computer simulations of neutron yields, spectra, and angular distributions from various lithium compound targets are presented.

For a Li target and a proton current of 1 mA, the absolute yield versus incident proton energy is shown in Figure 2. Yields are shown for 0° and 90° neutron emission angles, averaged over 5° intervals. Note the extremely rapid rise in neutron yield above threshold. From a proton energy of 2.0 to 2.5 MeV, the total yield for $\theta = 0°$ increases by more than an order of magnitude. The effect of target material on neutron production is shown in Figure 3. As expected, a pure metallic lithium target produces the highest relative yield. A possible target material may be LiH because of its high yield and high melting point (680 °C) as opposed to pure lithium (180 °C). LiF has the lowest yield due to its low lithium weight fraction (0.268) and high stopping power resulting from the relatively high Z component of fluorine. Figure 3 also shows the strong dependence of yield and average neutron energy on emission angle.

Figures 4 and 5 show neutron energy spectra and the dependence of maximum neutron energy on proton energy for 0° and 90° neutron emission angles respectively. These curves illustrate the relatively narrow band spectra obtainable by proton bombardment of lithium and the strong dependence of spectral width on bombarding energy and emission angle. Computed curves such as these aid in the determination of the optimum proton beam energy and emission angle for tissue irradiation and demonstrate the

Table One. Characteristics of charged particle reactions for neutron production.

Group I

Reaction	Threshold Energy (keV)	First Excited State (keV)	Energy Range of Monoenergetic Neutrons at 0° (keV)
^{45}Sc(p,n)^{45}Ti	2909 ± 2.5	37	5.6 - 52
^{51}V(p,n)^{51}Cr	1564 ± 1	749	2.36 - 786
^{57}Fe(p,n)^{57}Co	1648 ± 1	1378	2 - 1425

Group II

Neutron Energy Range (MeV) (for 2.5-MeV p)	Source Reaction	Target	0° Neutron Yield ($n \cdot sr^{-1} \cdot \mu A^{-1} \cdot s^{-1}$)
0.2 - 0.8	^{7}Li(p,n)^{7}Be	Solid Li	2×10^{7} - 2×10^{8}
0.7 - 3.0	^{3}H(p,n)^{3}He	Gaseous ^{3}H	3.3×10^{7} - 1.6×10^{8}
		Solid TiT*	5×10^{6} - 1.7×10^{7}

*TiT is Titanium Tritide (Ti^{3}H)

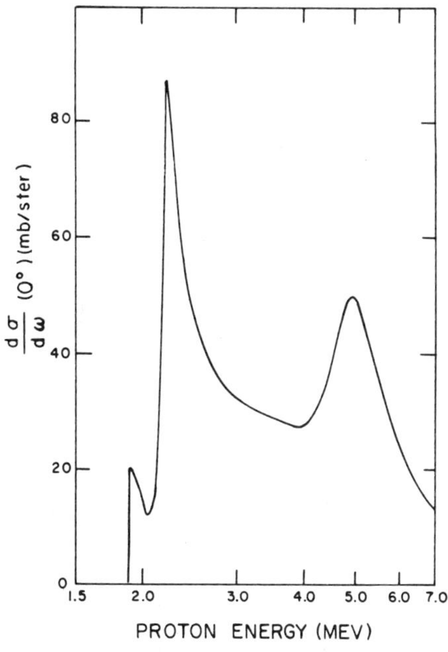

Figure 1. Experimental and recommended 0° center-of-mass cross sections for the reaction ^{7}Li(p,n)^{7}Be (Ref. 2).

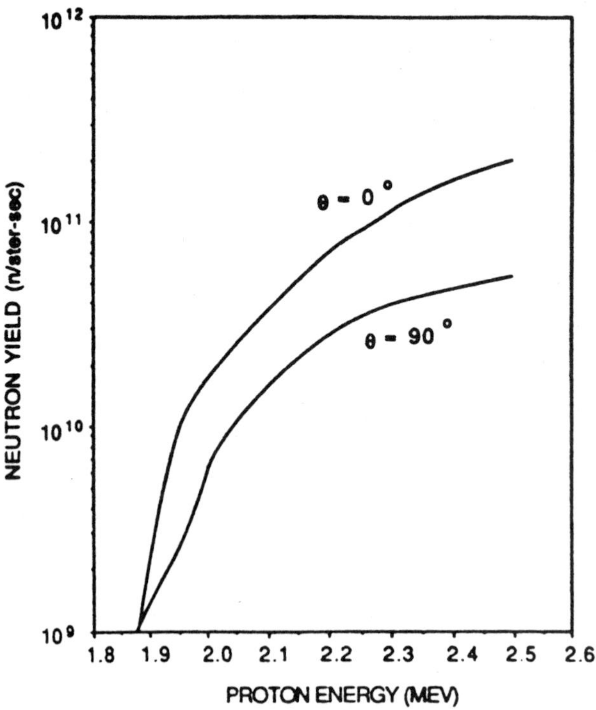

Figure 2.   Calculated neutron yield vs. proton energy at 0° and 90° for 1-mA proton current on Li target. The (p,n) reaction threshold is 1.88 MeV.

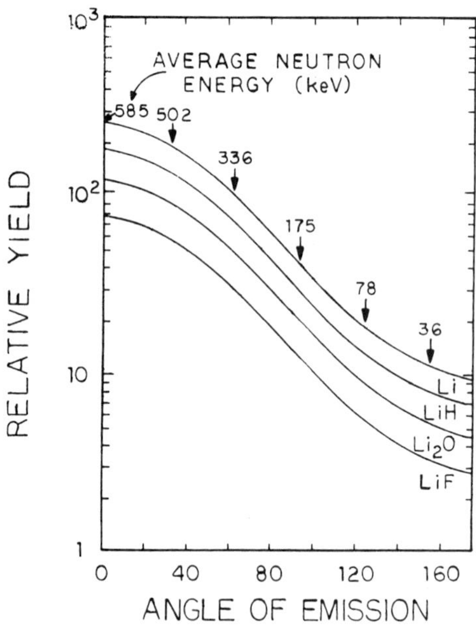

Figure 3.   Relative neutron yield curves for four lithium targets as a function of neutron emission angle. Curves were obtained by smoothing yields integrated over 5 intervals. Average neutron energies are noted for different angles for a proton energy of 2.5 MeV.

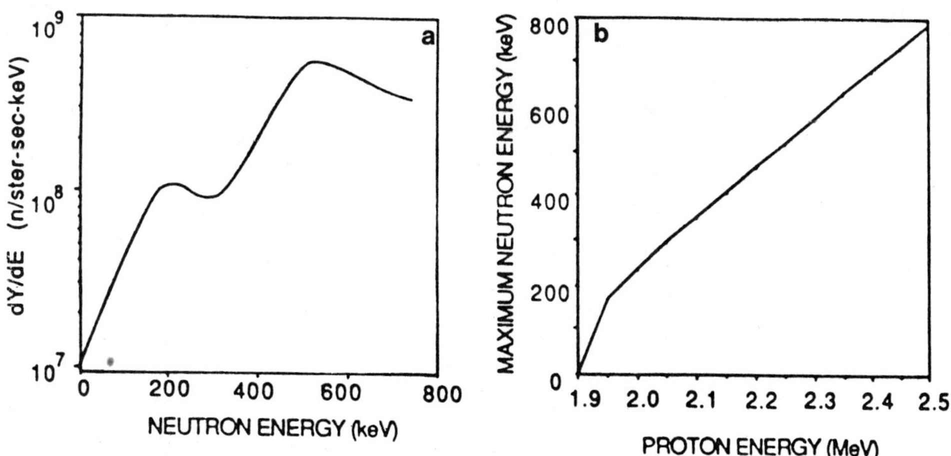

Figure 4. (a) Neutron energy spectrum at θ=0° for 1-mA, 2.5-MeV proton beam on Li.

(b) Maximum neutron energy vs. proton energy at θ=0°.

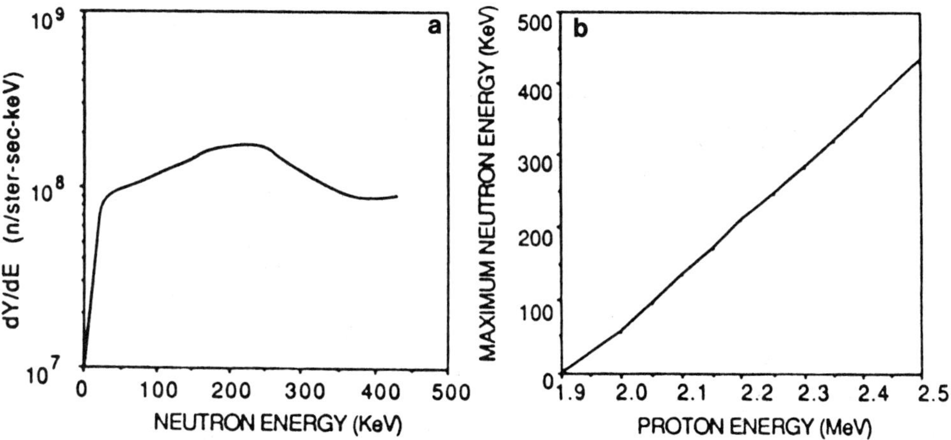

Figure 5. (a) Neutron energy spectrum at θ=90° for 1-mA, 2.5-MeV proton beam on Li.

(b) Maximum neutron energy vs. proton energy at θ=90°.

powerful potential of (p,n) reactions as variable neutron energy sources. For 2.0-2.5 MeV protons, maximum neutron energies are in the range of 200-800 keV.

As well as providing a variable neutron energy spectrum, an accelerator capable of delivering milliampere proton currents can potentially provide therapeutically significant neutron doses in reasonable irradiation times. For example, from Figures 2 and 3 it can be determined that a 2.5-MeV, 2-mA proton beam will produce 9×10^8 neutrons/cm^2-s with an average neutron energy of 590 keV at a distance of 30 cm from the source for a 0° emission angle. These intensities will be well suited for neutron capture therapy (where total neutron doses of order 10^{12}-10^{13} neutrons are desired) as long as the required neutron

Table Two. Accelerator Design Parameters.

Proton Beam Energy	2.0 - 2.5 MeV
Proton Current	≤ 4 mA
Terminal Voltage	1.0 - 1.25 MV
Overall Length	2.5 m
Height	1.5 m
Weight (excluding shield)	1000 lb
Power Consumption	25 kW

beam moderation and filtering does not result in excessive loss of yield. This will be discussed under the section of this paper entitled, "Optimum Neutron Energies for Neutron Capture Therapy."

<u>Production of Intense Proton Beams with a Tandem Cascade Accelerator</u>

Accelerator production of a sufficient neutron flux for the study of NCT requires a high current (>1 mA), low energy (2.0-2.5 MeV) proton beam. Two accelerator technologies are well suited for the production of high current ion beams in this energy range: radio frequency quadrupoles (RFQs) and high current electrostatic tandem accelerators. Cyclotrons are not capable of delivering the required beam current. The RFQ has recently received notoriety for its ability to accelerate low energy proton beams at peak currents of up to 100 mA and has been suggested as an accelerator source of neutrons for BNCT [5,6]. A disadvantage of the RFQ for the production of a carefully tailored neutron spectrum comes from the inability to easily vary the beam energy over a wide range. The quadrupole structure of the accelerator must be designed for a specific beam energy. Hence the neutron energy spectrum cannot be easily changed.

A high current tandem cascade electrostatic accelerator (TCA) is currently under development at SRL. This accelerator utilizes a recently developed high current negative ion source in conjunction with a high current solid state power supply to provide a compact, low cost, proton accelerator well suited for epithermal-neutron production. The inherent simplicity and flexibility of this accelerator provide several features which are desirable for laboratory and clinical applications requiring neutron generation. A negative hydrogen beam is continuously injected at low energy into the accelerating column of the TCA from a multicusp volume production negative ion source [7,8]. An integral high current symmetrical cascade rectifier power supply delivers a 1.0-1.25 MV accelerating potential at up to 10 mA to the high voltage terminal. An accelerating gradient of 1.57 MV/meter is maintained in the accelerating columns. Negative ions are stripped of two electrons in the terminal stopper assembly by a cryogenically pumped water vapor jet [14] and subsequently accelerated to ground potential where the proton beam attains a final energy equal to twice the terminal voltage.

Table Two summarizes the performance of the accelerator. A graph of the performance constraints of the TCA based on power supply design considerations and the demonstrated negative ion source performance is shown in Figure 6. A lithium or lithium compound target is being designed for use in the target assembly. Target material and geometry will be chosen for compatibility with the thermal loading due to proton beam bombardment. Target materials being investigated include Li, LiH, Li_2O, and LiF. The use of solid, liquid, and gas phase targets will be compared. Two specific target configurations have been identified for study. The first is a solid lithium compound target in which thermal loading of the target is reduced by a raster scan of a large target area. The second is a liquid lithium target heated to its boiling point by the thermal energy deposited by the proton beam itself. In the latter configuration, the target design must insure that the vaporized lithium condenses and flows back to a target reservoir without contaminating the accelerator column.

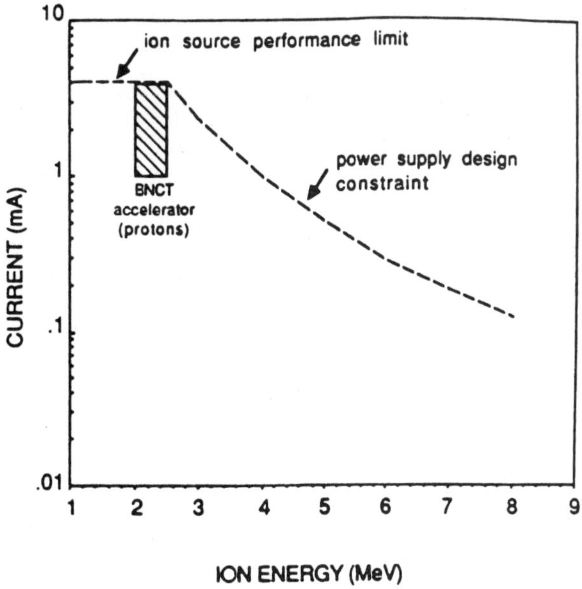

Figure 6. Operating regime of the TCA.

<u>Modification of Neutron Energies for Use in Neutron Capture Therapy</u>

Neutrons produced by the (p,n) reaction in the target must be subsequently moderated and/or filtered before becoming therapeutically useful (see below). The extent of moderation and filtration, however, will depend on the ultimate neutron energy desired for therapy. This energy, or range of energies, has been determined using Monte Carlo simulation studies the results of which are presented in the next section.

OPTIMUM NEUTRON ENERGIES FOR NEUTRON CAPTURE THERAPY

The benefits of utilizing neutrons with higher than thermal energy for BNCT are well known [9]. First, tissue penetration is greatly enhanced. Second the reduced ^{10}B capture cross section at higher energies results in a significant reduction in the occurrence of the capture reaction in surface tissues. Third, the high hydrogen content in tissue can be exploited to thermalize the beam as it penetrates the tissue. The combination of these factors results in skin sparing and a peak of thermal-neutron exposure at some depth beneath the surface. This implies that tumors need not be surgically exposed before neutron therapy is applied and opens up other therapeutic possibilities such as the use of fractionated doses. In addition, poor boron selectivity in tumor cells is partially mitigated by the use of epithermal-neutron beams if the neutron beam energy can be chosen so that energy is deposited preferentially at the tumor location.

If the energy of the epithermal neutrons is too high, however, the skin sparing due to the reduction in the capture cross section of ^{10}B (and ^{1}H and ^{14}N) will be offset by the unacceptably large surface dose caused by fast neutrons entering the tissue. A tradeoff must be realized, then, between maximizing the thermal-neutron flux at depth, and minimizing the dose to healthy tissues (particularly at the surface). The question of which energy (or range of energies) results in the optimum tradeoff for BNCT remains to be answered. One method of answering this question is to determine the energy that provides the largest depth at which healthy tissue will not receive a greater dose than the tumor to be treated, i.e., the energy which provides the greatest advantage depth (AD).

The best method of examining the dosimetry of monoenergetic uncontaminated neutron beams is by computer simulation. The Monte Carlo code, MCNP, is an excellent choice for such an application. MCNP (Monte Carlo for Neutron Photon transport) is a Monte Carlo code developed at the Los Alamos National Laboratory [10]. A copy is currently running on SUN 4 workstations in Whitaker College Biomedical Imaging Laboratory at the Massachusetts Institute of Technology (MIT). With the aid of its sophisticated built-in geometry package, this program permits complete three-dimensional freedom in the design of source and target geometries and neutron and/or photon fluence through any surface or into any volume can be calculated. Fluence-to-dose conversion is made possible by allowing the user to include specific KERMA factors for each radiation type encountered. Dose equivalence can then be determined by using appropriate RBE factors.

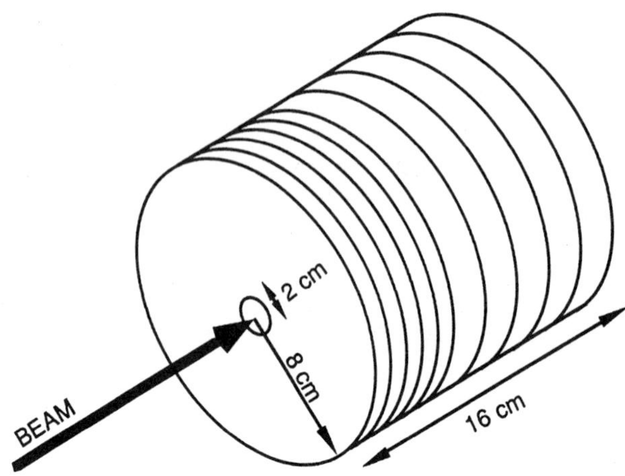

Figure 7. Illustration of cylindrical phantom and of tally values used
to assess neutron beams of varying sizes and energies.

To examine the dosimetric effects of neutron beams of varying energies on brain tissue (and hence to determine an optimal energy range for BNCT) a cylindrical phantom (16.0-cm diameter x 16.0-cm height) composed of brain-equivalent material was modeled. The brain formula was based on a 50%/50% by weight average of gray and white matter as suggested by the Guidance to Authors for this Workshop. Monoenergetic beams of neutrons were directed onto the surface of the cylinder as demonstrated in Figure 7. The disk-shaped source emitted neutrons only in the direction of the cylinder. The disk was centered on the axis of the cylinder and its diameter was varied to examine the effect of beam size on particle distribution in the phantom. Beam diameters considered were 20 cm, 12 cm, 6 cm, and a very thin pencil beam.

To estimate the contribution of individual dose components as a function of depth, the cylindrical phantom was divided into several 1-cm thick disks. In a 1-cm radial cylinder along the central axis the following flux components were assessed as a function of depth: thermal neutrons (cut-off energy: 0.36 eV), fast neutrons, and photons. The neutron and photon fluxes were modified by the KERMA-dose conversion factors of Caswell et al. [11], and Zamenhof et al. [12] respectively. Then to estimate the $^{10}B(n,\alpha)$ contribution to dose, the thermal-neutron flux was multiplied by either 3 (to represent 3 µg/g ^{10}B in healthy tissue) or 30 (to represent 30 µg/g ^{10}B in tumor) and then modified by ^{10}B-neutron fluence-to-KERMA conversion factors, also listed by Zamenhof et al. [12]. Individual dose components were variously combined to yield estimates of total dose (per neutron) to background (with no boron, or with 3 µg/g ^{10}B) and total tumor dose (per neutron). Maximum and minimum advantage depths and advantage ratios, as defined in

Table Three. RBE and non-RBE advantage depths (max/min) and advantage ratios calculated in the right-circular cylinder illustrated in Figure 7, for different neutron energies and different beam diameters.

Neutron Energy		Thin Beam		6 cm		12 cm		20 cm	
		No RBE	RBE	No RBE	RBE	No RBE	RBE	No RBE	RBE
0.025 eV	AD	2.5/2.0	3.2/2.2	3.9/3.1	4.8/3.4	4.6/3.7	5.7/4.3	4.9/4.0	6.0/4.5
	AR	7.8	13.6	4.9	8.8	3.4	6.7	3.0	6.0
0.5 eV	AD	4.7/3.8	5.4/4.2	5.5/4.8	6.5/5.3	6.2/5.4	7.4/6.1	6.4/5.6	7.6/6.4
	AR	6.3	11.4	4.4	7.9	3.5	6.2	3.3	5.7
1.0 eV	AD	5.1/4.3	6.0/4.6	5.9/5.1	7.0/5.6	6.4/3.7	7.6/6.4	6.6/5.9	7.9/6.6
	AR	6.0	10.7	4.4	7.8	3.5	6.0	3.3	5.8
10 eV	AD	6.0/5.1	7.0/5.6	6.7/5.9	7.8/6.5	7.2/6.5	8.4/7.2	7.3/6.6	8.4/7.4
	AR	5.1	9.2	4.2	7.5	3.4	6.1	3.3	5.8
100 eV	AD	6.9/6.0	8.0/6.6	7.3/6.6	8.5/7.3	7.6/7.0	8.9/7.8	7.7/7.0	9.0/7.8
	AR	4.6	8.2	4.0	7.0	3.4	5.9	3.3	5.8
1 keV	AD	7.0/6.5	7.7/7.1	7.9/7.2	9.1/7.9	8.4/7.7	9.7/8.4	8.3/7.6	9.7/8.5
	AR	3.7	6.3	3.7	6.6	3.4	5.9	3.2	5.5
2 keV	AD	5.9/5.6	6.6/6.2	8.0/7.3	9.2/8.0	8.2/7.5	9.5/8.3	8.4/7.8	9.8/8.6
	AR	3.3	5.2	3.6	6.3	3.3	5.8	3.2	5.5
10 keV	AD	1.1/1.0	1.4/1.3	6.9/6.6	7.7/7.4	8.6/7.9	9.6/8.9	8.7/8.0	9.9/8.9
	AR	1.3	1.5	3.1	4.9	3.1	5.2	3.0	5.2
24 keV	AD	0.7/0.7	0.8/0.7	4.6/4.4	5.6/5.4	7.5/7.2	8.3/8.0	7.9/7.7	8.6/8.3
	AR	1.1	1.1	2.3	3.4	2.8	4.5	2.8	4.5
35 keV	AD	0.6/0.6	0.7/0.6	2.9/2.7	4.4/4.2	6.6/6.4	7.6/7.3	7.0/6.8	7.9/7.6
	AR	1.1	1.1	1.8	2.6	2.6	4.1	2.7	4.3
100 keV	AD	0.5/0.5	0.6/0.6	1.0/0.9	1.3/1.2	3.3/3.1	4.9/4.7	4.1/3.9	5.6/5.4
	AR	1.0	1.0	1.2	1.2	1.7	2.5	1.9	2.8
800 keV	AD	0.5/0.5	0.5/0.5	0.6/0.6	0.7/0.7	1.1/1.1	1.6/1.6	1.1/1.0	1.7/1.6
	AR	1.0	1.0	1.0	1.0	1.1	1.1	1.1	1.1

the Guidance to Authors for this Workshop, were calculated and are given in Table Three for the four beam sizes. (Note: Refer to paper by Clement et al. in these proceedings for definitions as given in the guidance.) Table Three also presents the corresponding values when RBE doses are used. RBE values of 1.0, 1.6, and 2.3 were applied to gamma rays, neutrons, and the ^{10}B reaction products, respectively. Plots of neutron energy versus maximum and minimum RBE advantage depths for each beam size are shown in Figure 8.

The data in Table Three indicate that for a given beam size, advantage depth (AD) increases with increasing energy. A peak is reached at some energy between 1.0 and 10 to 20 keV at which point it decreases rapidly with further increases in energy. Also, as beam size increases, the energy at which the AD peaks (both maximum and minimum) also increases. The maximum AD obtained with a very thin beam is produced by an energy of 1.0 keV. However, the maximum occurs at 2.0 keV for a 6.0-cm beam and at 10 keV for a 12.0 or 20.0-cm beam. The 'optimal' beam energy is thus seen to be a function of beam size. Also note that the maximum AD itself actually increases with increasing beam size - a further 2.0-2.5 cm of advantage depth is realized in going from the thin beam to a beam diameter which is at least as large as the irradiated face of the phantom.

From Table Three and the plots of advantage depth versus energy in Figure 8, a prediction can be made concerning the optimal neutron energy for BNCT. If a wide irradiation beam is to be used, then maximum AD will be obtained with a 10-keV beam. However, the last plot of Figure 8 indicates that with this beam diameter, it will still be possible to reach the midline of the brain (assumed to be at a depth of 7.0 cm) with neutron energies from roughly 4 eV to 50 keV, even if a ^{10}B tumor/tissue ratio of 10 to 1 is assumed. Thus a fairly wide range of epithermal energies will be useful for neutron capture therapy.

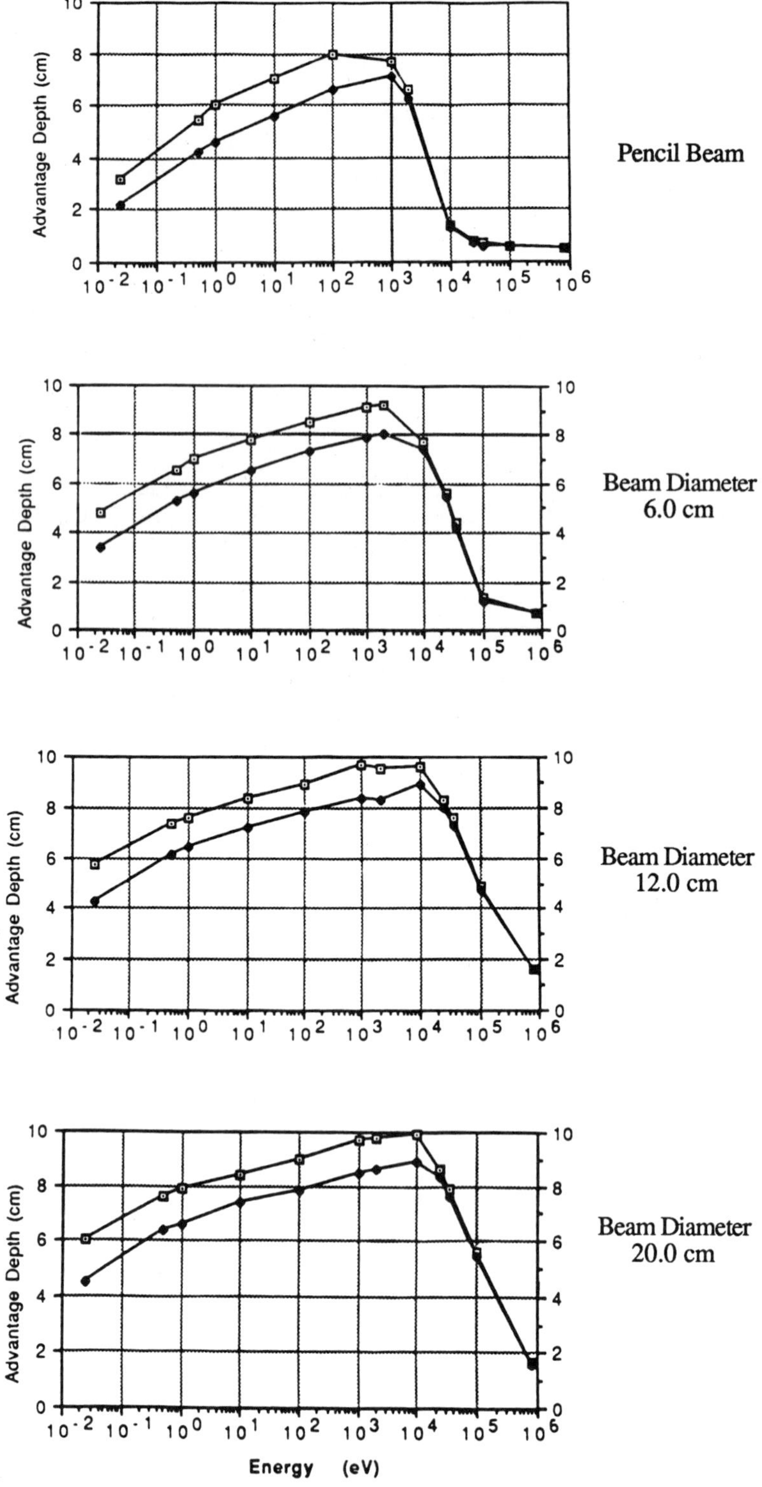

Figure 8.   Maximum (open boxes) and minimum (filled boxes) RBE advantage depths [as defined in the text] as a function of neutron energy for four different beam sizes.

Table Three also lists advantage ratios (AR) for all energies considered and for the four beam sizes. For any given beam size, the advantage ratio is a maximum at thermal energies and decreases as the energy is increased, reaching a value of 1.0 at high energies where the tumor dose becomes entirely swamped by the background dose component (primarily the fast-neutron dose). This decrease in AR illustrates the increasing effect that higher-energy neutrons have in healthy tissues, primarily at the surface. A smaller AR is an inevitable aspect of the improved penetration offered by neutrons with greater than thermal energies.

The relationship of AR and beam size is not as straightforward. For energies less than 2.0 keV, the advantage ratio decreases with increasing beam size, AR increases with increasing beam size if the beam energy is greater than 2.0 keV, and for E ~2.0 keV the AR is roughly constant regardless of beam diameter. This effect may be due to the drop in the interaction cross section of ^{14}N which occurs at roughly 1.0 keV [13]. Approximately 1.8% by weight ^{14}N is present in the brain phantom. The increase in AR with beam size occurring at 2.0 keV may be a reflection of a slight increase in pathlength of the higher energy neutrons. This improvement in advantage ratio above 2.0 keV suggests that even though a lower-energy cutoff of useful neutrons was found to be 4 eV on the basis of advantage depth comparisons, keeping the neutron energy above 2.0 keV may be desirable so that a larger fraction of the total dose is due to ^{10}B in the tumor.

CONCLUSIONS

Monte Carlo simulation studies examining the dosimetry of various monoenergetic neutron beams have predicted that 10-keV neutrons will provide the greatest advantage depth (with appropriate sparing of healthy tissues) for boron neutron capture therapy in the brain. This energy falls within a range of neutron energies, from 4 eV to 50 keV, which is capable of delivering therapeutic doses at least to the midline of the brain. Hence, a device (reactor or accelerator) producing neutrons in this range will be suitable for treating deep-seated brain tumors.

At MIT, studies are currently underway to design appropriate moderator and filter combinations which will reduce the energy of neutrons produced by the tandem cascade accelerator to energies in the range of 4 eV to 50 eV. These studies again make use of the precision and versatility of the Monte Carlo code, MCNP. Over 500 neutron interaction tables are included in this code for approximately 100 different isotopes or elements. It is anticipated that the simulation studies will lead to the design of an accelerator beam-line configuration which will offer a high yield of epithermal neutrons in the desired energy range.

ACKNOWLEDGMENTS

The authors, especially Jacquelyn Yanch, wish to thank Steve White of the Radiation Transport Group (X-6) at the Los Alamos National Laboratory and Steven Clement, of the MIT Nuclear Reactor Laboratory, for their assistance with the initial MCNP runs.

REFERENCES

1.   V. N. Kononov and E. D. Yurlov, "Absolute Yield and Spectrum of Neutrons from the $^{7}Li(p,n)^{7}Be$ Reaction," <u>Sov. At. Energy</u>, 43:947 (1977).

2.   H. Liskien and A. Paulsen, "Neutron Production Cross Sections and Energies for the Reactions $^{7}Li(p,n)^{7}Be$ and $^{7}Li(p,n)^{7}Be^{*}$," <u>Atomic Data and Nuclear Data Tables</u>, 15:57 (1975).

3.   J. H. Gibbons and H. W. Newson, "The $Li^7(p,n)Be^7$ Reaction," in <u>Fast Neutron Physics</u>, Vol. 1, J. B. Marron and J. L. Fowler, eds., Int. Science, New York, p. 133 (1960).

4.   G. L. Brownell, J. E. Kirsch, and J. Kehayias, "Accelerator Production of Epithermal Neutrons for Neutron Capture Therapy," in <u>Proc. Second Int. Symp. on Neutron Capture Therapy</u>, Tokyo, 1985, H. Hatanaka, ed., Nishimura Co., Ltd., Niigata, Japan, p. 127 (1986).

5.   C. K. Wang, T. E. Blue, and R. A. Gahbauer, "A Design Study of an Accelerator-Based Epithermal Neutron Source for Boron Neutron Capture Therapy," <u>Strahlenther. Onkol.</u>, 165(2/3):75 (1989).

6.   C. K. Wang, T. E. Blue, and R. Gahbauer, "A Neutronic Study of an Accelerator-Based Neutron Irradiation Facility for Boron Neutron Capture Therapy," <u>Nucl. Technol.</u>, 84:93 (1989).

7.   K. R. Kendall, M. McDonald, D. R. Mosscrop, P. W. Schmor, and D. Yuan, <u>Rev. Sci. Instrum.</u>, 57:7 (1986).

8.   R. K. York, R. R. Stevens, Jr., R. A. DeHaven, J. R. McConnell, E. P. Chamberlin, and R. Kandarian, "The Development of a High Current H-Injector for the Proton Storage Ring at LAMPF," <u>Nucl. Instruments and Methods in Phys. Res.</u>, B10/11:891 (1985).

9.   G. L. Brownell and W. H. Sweet, "Studies on Neutron Capture Therapy," in <u>Progress in Nuclear Energy Series VII, Vol. 2 - Medical Sciences</u>, Pergamon Press, London, p. 114 (1959).

10.   J. F. Briesmeister, ed., "MCNP – A General Monte Carlo Code for Neutron and Photon Transport, Version 3A," Los Alamos National Laboratory, <u>LA-7396-M</u>, Rev. 2 (1986).

11.   R. S. Caswell, J. J. Coyne, and M. L. Randolph, "KERMA Factors of Elements and Compounds for Neutron Energies Below 30 MeV," <u>Int. J. Appl. Radiat. Isot.</u>, 33:1227 (1982).

12.   R. G. Zamenhof, B. W. Murray, G. L. Brownell, G. R. Wellum, and E. I. Tolpin, "Boron Neutron Capture Therapy for the Treatment of Cerebral Gliomas: I. Theoretical Evaluation of the Efficacy of Various Neutron Beams," <u>Med. Phys.</u>, 2(2):47 (1975).

13.   D. I. Garber and R. R. Kinsey, <u>Neutron Cross-Sections, Vol. II, Curves</u>, Brookhaven National Laboratory, New York (1976).

AN EXPERIMENTAL STUDY OF THE MODERATOR ASSEMBLY FOR A LOW-ENERGY PROTON ACCELERATOR NEUTRON IRRADIATION FACILITY FOR BNCT

C. K. Wang,[†] T. E. Blue,[†] and J. W. Blue[*]

[†] Ohio State University, Columbus, OH
[*] Cleveland Clinic, Cleveland, OH

ABSTRACT

An accelerator-based neutron irradiation facility (ANIF), which has been proposed for BNCT, is based on a 2.5-MeV proton beam bombarding a thick lithium target. Neutrons which are emitted from the lithium target are too energetic for BNCT and must be moderated. A calculational study, which was done previously on the moderator assembly for an ANIF, shows that, with an optimized moderator assembly, an ANIF can produce a neutron flux which has quality and intensity sufficient for BNCT. In order to verify our previous calculational study, a lithium target and a non-optimized moderator assembly (a cylindrical tank of D_2O) have been constructed and tested at the Ohio State University Van de Graaff proton accelerator.

The neutron spectrum was measured for neutrons emerging from the moderator assembly. The measured neutron spectrum agrees reasonably well with that obtained from Monte Carlo calculations, except for neutrons with energies above 100 keV. For those neutrons, the measured spectrum is lower by a factor of two than the calculated one. In addition to the neutron spectrum measurement, the boron-10 absorbed dose was measured on the axis of the neutron field in a 20 cm x 20 cm x 20 cm water phantom, and the result agrees quite well with that obtained from calculation.

This experiment confirms that the calculated optimized moderator assembly, consisting of a 22.5-cm thick, 25-cm diameter cylinder of beryllia (BeO) surrounded by a 30-cm thick jacket of alumina (Al_2O_3), produces an epithermal neutron flux of 3.12×10^7 n/cm^2-s per mA of protons. For an accelerator delivering 30 mA of 2.5-MeV protons, the irradiation time for a single-session treatment can be as short as 50 minutes. The calculated ratio of absorbed neutron dose to fluence for the optimized moderator assembly is 4.9×10^{-11} cGy-cm^2/n, which is equal to that of a 5-keV neutron beam. Our experimental measurements indicate that the ratio of absorbed neutron dose to fluence may in fact be lower (better) than calculated.

INTRODUCTION

An accelerator neutron irradiation facility (ANIF), proposed at the Second International Symposium on Neutron Capture Therapy (NCT) [1], is based on a 2.5-MeV proton beam bombarding a lithium target. In boron neutron capture therapy (BNCT), neutrons with energies between 1 eV and 1 keV are most appropriate for treating deep-seated tumors. However, the neutrons which are emitted from the lithium target of the ANIF have energies between 100 and 800 keV, too energetic for BNCT. They must therefore be moderated. A moderator assembly serves this purpose. At the Third

International Symposium on NCT, a calculational study of a moderator assembly, based on Monte Carlo code results, was presented. The study suggests that a moderator assembly with a 22.5-cm BeO moderator provides an adequate neutron flux with acceptable beam quality for a 30-mA proton beam. More detailed discussions of the optimized (in terms of calculation) moderator assembly and its performance follow. Information is also given in a related paper [2].

BACKGROUND

Studies [3,4] show that epithermal-neutron beams with very little fast-neutron contamination can be extracted from a nuclear reactor by using a thick (~80 cm) moderator of alumina (Al_2O_3), or a mixture of aluminum (85% by volume) and D_2O (15% by volume). These moderator materials, however, will not work well for the proposed ANIF because the neutron flux at the irradiation point of the proposed ANIF would be too low to treat patients if such a thick moderator were used. This would occur because the neutron generation rate of the ANIF would be much lower than that of a nuclear reactor. The use of 10-20 cm of a low-Z moderator, such as H_2O, provides an intense neutron flux with an unacceptably large amount of fast-neutron contamination. For BNCT, a compromise between the neutron flux and the neutron energy spectrum has to be made in designing the ANIF moderator assembly. A previous calculational study [2], based on Monte Carlo code results, suggests that the optimized moderator assembly consists of a 22.5-cm thick, 25-cm diameter cylinder of BeO, surrounded by a 30-cm thick jacket of alumina (Al_2O_3). The detailed configuration of the optimized moderator assembly is illustrated in Figure 1. The additional D_2O and ^6LiF shown in Figure 1 serves as the neutron shield which reduces the patient's whole-body dose. With this optimized moderator assembly, the proposed ANIF provides an epithermal-neutron flux of 3.12×10^7 n/cm^2-s per mA of proton current. The calculated ratio of neutron absorbed dose to fluence is 4.9×10^{-11} cGy-cm^2/n, which is equal to the ratio for a 5-keV neutron beam.

If one considers RBE to be a function of neutron energy, then the quality of the accelerator-produced neutrons may be better than the quality of those from a 5-keV neutron beam because the majority of the accelerator-produced neutrons have energies less than 5-keV and should have RBE values smaller than that of 5-keV neutrons. Recent studies performed at Harwell show that 24-keV neutrons are too hard for BNCT and that 10-keV neutrons provide significant skin-sparing relative to 24-keV neutrons. Thus, the quality of the accelerator-produced neutrons, which have a fluence-averaged KERMA equal to that of a 5-keV neutron, may be adequate for BNCT.

Recent developments in radio frequency quadrupole (RFQ) accelerator technology at Los Alamos National Laboratory indicate that a 30-mA proton RFQ can be built for less than 3.5 million dollars [5]. At this proton current, and assuming that the boron-10 concentration is 30 µg/g in the tumor, the irradiation time for a single-session treatment with 20 Gy of total absorbed dose to tumor can be as short as 50 minutes. If the irradiation is divided into four equal fractions, then the irradiation time is reduced to a very reasonable length of 10-15 minutes.

In order to validate the calculational methods of our design study of the ANIF moderator assembly, an experiment was performed. The following sections provide descriptions of the experiment, the measurement techniques, and a comparison of measurements with calculations.

METHODOLOGY

Description of the Experiment

A lithium target and an experimental moderator assembly have been constructed and tested using the Ohio State University Van de Graaff proton beam. The moderator

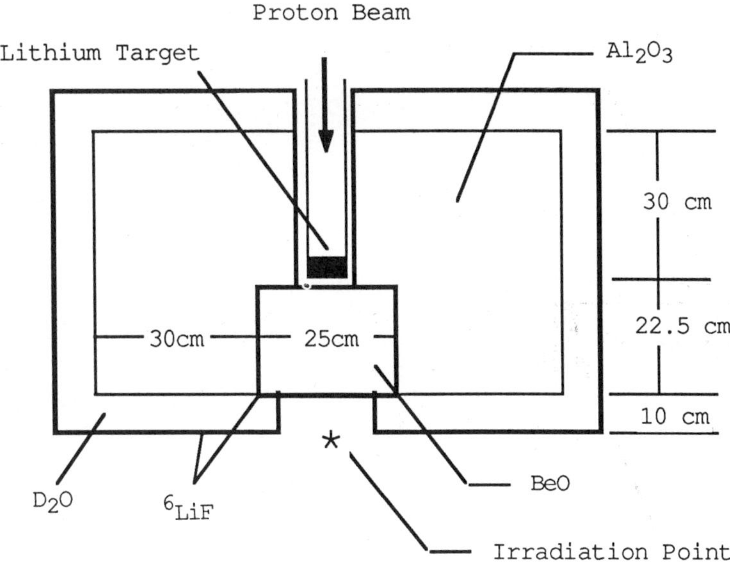

Side View

Figure 1. Configuration of the optimized moderator assembly for the low-energy proton accelerator neutron irradiation facility for BNCT.

assembly is a D_2O-filled, cylindrical tank that is made of aluminum. The tank is 36.8 cm in diameter and 20 cm in height. Its wall thickness is 0.318 cm. The neutron spectrum was measured for neutrons emerging from the moderator assembly. The experimental setup for this measurement is illustrated in Figure 2.

In addition to the spectrum measurement, the boron-10 absorbed dose was measured at various depths in a water phantom (20 cm x 20 cm x 20 cm) on the axis of the neutron field. The experimental setup is shown in Figure 3. The measurement of the boron-10 absorbed dose in the phantom indirectly verifies the neutron flux spectrum impinging upon the phantom. Also, it is a quantity of fundamental clinical importance.

Neutron Spectrum Measurement Techniques

Two measurement techniques were used to cover the full energy range of the neutron spectrum (1 eV - 800 keV). One technique was based upon a proton-recoil proportional detector and the other technique upon a boron-shell spectrometer, a new technique which was developed by us.

Proton-Recoil Proportional Detector. For neutrons above 10 keV, we used a proton-recoil proportional detector which is a sphere that is 5 cm in diameter and filled to 10 atmospheres with H_2. The detector is made by LND Inc. (detector type 27050). The energy calibration of the detector was done in the thermal column of the Ohio State University Research Reactor, using the small amount of ^{3}He which was incorporated into the detector during its manufacture. The ^{3}He(n,p)^{3}H reaction has a Q value of 765 keV. The energy resolution of the detector for the ^{3}He(n,p)^{3}H reaction is 10% at this energy, for a bias of 2500 Volts. The gas multiplication increases from one to about twenty as the detector bias is increased from 2000 to 4750 Volts. Sets of measurements were made with the detector operating at 2500, 4000, and 4750 Volts. The measurement, with the detector biased at 2500 Volts, covered the spectrum between 200 and 800 keV. The measurement, with the detector biased at 4000 Volts, covered the spectrum between 50 and 200 keV. The

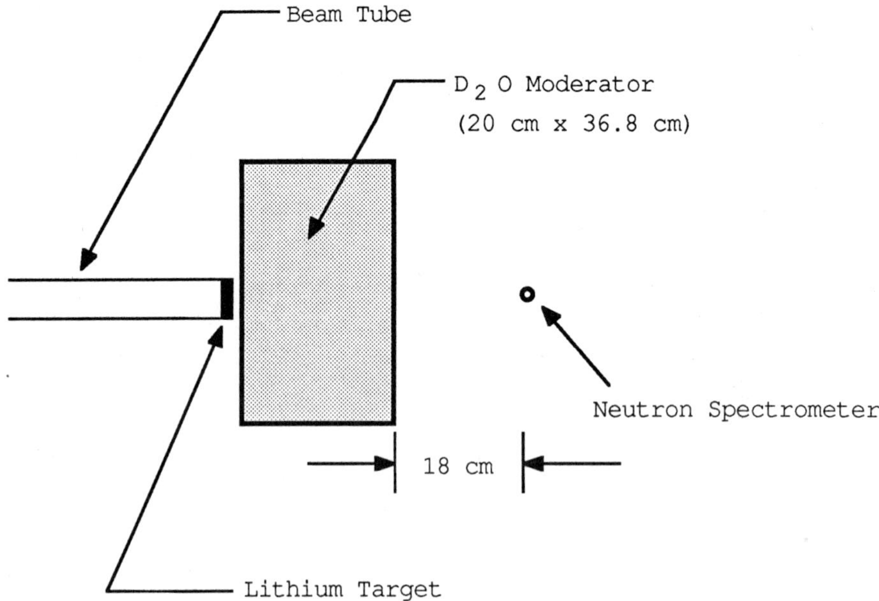

Figure 2. Experimental setup for the neutron spectrum measurement, for neutrons emerging from a tank of D_2O. The centerline of the D_2O tank and the neutron spectrometer are colinear with the centerline of the beam tube.

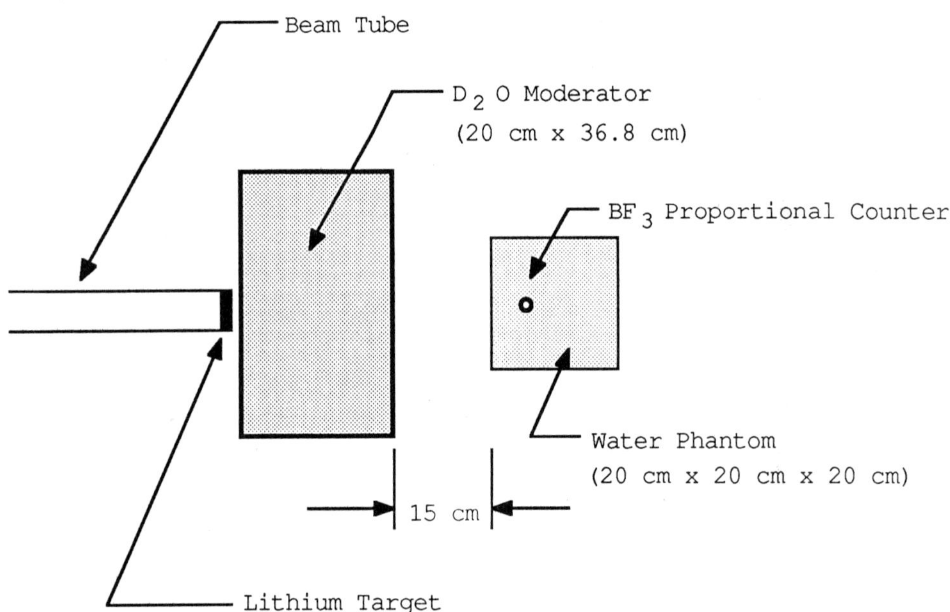

Figure 3. Experimental setup for the boron-10 absorbed dose measurement inside a 20 cm x 20 cm x 20 cm water phantom. The moderator, the phantom, and the BF_3 counter are colinear with centerline of the beam tube.

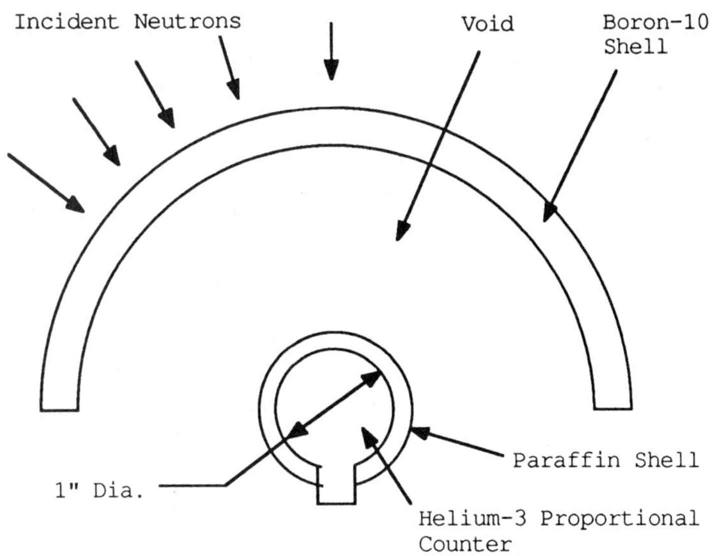

Figure 4. Geometric configuration of the boron-shell neutron spectrometer.

measurement, with the detector biased at 4750 Volts, covered the spectrum between 10 and 50 keV. The signals for neutrons with energies less than 10 keV are buried in the preamp noise and thus cannot be recovered.

As occurs in many other experiments, neutrons from our moderator assembly are accompanied by gamma rays. In this experiment, 1.78-MeV gamma rays arise from the ^{27}Al(n,γ)^{28}Al reactions which occur in the moderator assembly wall. Because the proton-recoil detector responds to both neutrons and gamma rays, a discrimination technique must be used to remove the gamma-ray signals from the recorded pulse height spectrum (PHS). The discrimination of gamma rays was accomplished by using an additional measurement made with a small GM counter (model GM-2 made by Far West Technology Inc.). The GM counter, which responds only to gamma rays, was placed beside the proton-recoil detector during the measurement. The gamma-ray counts recorded by the GM counter were converted to a PHS, which would have been recorded by the proportional counter in the absence of neutrons. The conversion of GM counts to a gamma-ray PHS was based on a calibration experiment in which both the GM counter and the proportional counter were exposed to a Na-22 source. This gamma-ray PHS was then subtracted from the PHS measured by the recoil detector to obtain the net neutron PHS. The net neutron PHS was used to unfold the neutron spectrum for neutrons with energies between 10 and 800 keV, using the unfolding code SPEC-4 [6].

Boron-Shell Neutron Spectrometer. For neutrons with energies less than 10 keV, we developed a new spectrometer to measure the spectrum [7]. The spectrometer is based on a set of interchangeable hemispherical shells which contain various amounts of ^{10}B and a small (one inch in diameter) spherical ^{3}He proportional counter (detector type 27066, made by LND Inc.) which is located at the focus of the hemispherical shells. Also, for the hemispherical shells which contain larger amounts of ^{10}B, the ^{3}He proportional counter is surrounded by a spherical paraffin shell. The thickness of the paraffin shell is larger for shells with larger amounts of ^{10}B. Figure 4 illustrates the geometric configuration of the boron-shell neutron spectrometer. Table One provides the material compositions of the spectrometer.

Arguments which support this choice of composition and geometry are provided

Table One

Material compositions of the boron-shell neutron spectrometer.

Detector Number	Boron-10 Loadings (g/cm^2)	Paraffin Thickness (cm)
1	0.367	0.0
2	0.075	0.0
3	0.15	0.0
4	0.39	0.64
5	0.55	0.64
6	0.79	0.95
7	0.94	0.95

below. Incoming neutrons are filtered by the ^{10}B shell before they reach the ^{3}He counter. Because of the spherical geometry of the spectrometer, the neutrons that strike the ^{3}He counter are restricted to those for which the angle of incidence is nearly perpendicular to the ^{10}B shell. All of these neutrons pass through nearly the same thickness of shell material in reaching the counter. Therefore, the filtering of spectra and hence the response function of the spectrometer, depends only on the neutron energy (E_n) and not on the neutron angular distribution. Because the neutron capture cross section of ^{10}B is proportional to $E_n^{-1/2}$, the ^{10}B shell filters out low-energy neutrons in preference to high-energy neutrons. However, because the cross section for the ^{3}He(n,p)^{3}H reaction is also proportional to $E_n^{-1/2}$, of the neutrons reaching the proportional counter, the proportional counter responds most strongly to the neutrons with the lowest energies. Therefore, by using a set of ^{10}B shells with various ^{10}B loadings and a ^{3}He proportional counter, one can design a neutron spectrometer that has a set of response functions which peak at various energies between 1 eV and 10 keV. Unfortunately, a simple spectrometer based on a set of ^{10}B shells and a ^{3}He proportional counter only responds very weakly to neutrons with energies above 100 eV. In order to enhance the response function for neutron energies above 100 eV, a shell of paraffin is placed around the ^{3}He counter. Then neutrons with energies above 100 eV are moderated somewhat by the paraffin, thus increasing the probability that they will be registered by the ^{3}He counter.

The spectrometer response function was calculated using the Monte Carlo radiation transport code MORSE-CG [8] in conjunction with the BUGLE-80 cross-section library [9]. In Figure 5, the calculated response functions are graphed for each of the seven combinations listed in Table One. Each of the seven response functions has a gentle peak. In addition, on a logarithmic energy scale, these peaks are evenly distributed in the energy range of 1 eV to 100 keV. The counts measured by the ^{3}He counter for the seven spectrometer combinations were used to unfold the spectrum by means of the calculated spectrometer response functions. The unfolding code used in this process is called SPUNIT, which is based on an iteration technique developed by Doroshenko et al. [10].

Boron-10 Absorbed Dose Measurement in Phantom. The boron-10 absorbed dose was measured using a small cylindrical BF$_3$ proportional counter (made by Reuter-Stokes) which is 0.635 cm in diameter and 2.5 cm in length. The BF$_3$ counter was first calibrated by a gold-foil activation in the thermal column of the Ohio State University Research Reactor to obtain the conversion factor between BF$_3$ counts and thermal-neutron fluence. Using this conversion factor, the counts measured by the BF$_3$ counter at various depths in the phantom were first converted to the thermal-neutron fluences and then multiplied by the corresponding fluence-to-KERMA factors to obtain the boron-10 absorbed doses.

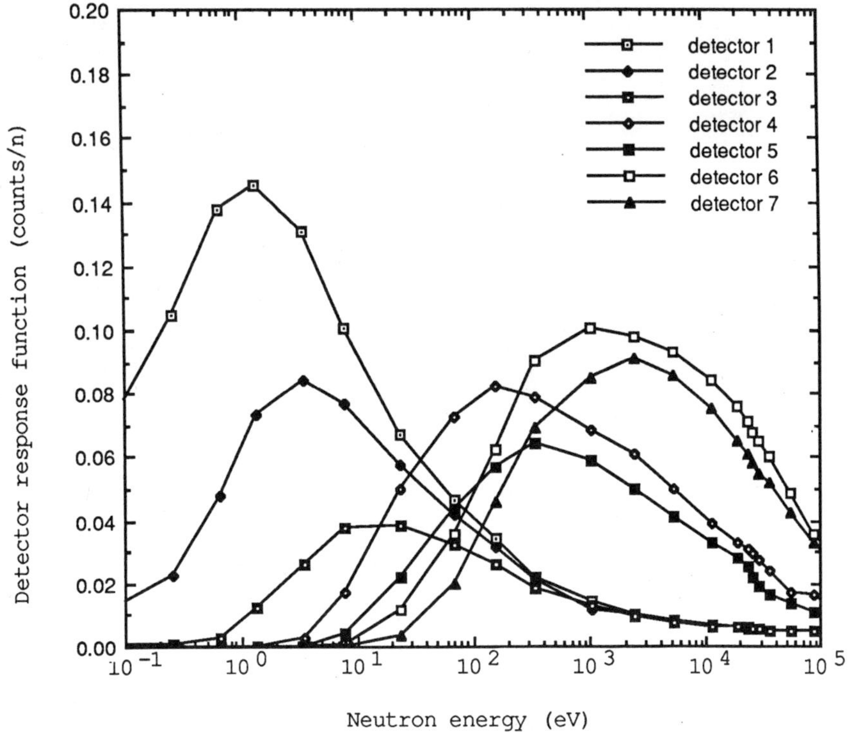

Figure 5. Calculated neutron response functions for the seven combinations of the boron-shell neutron spectrometer, which were listed in Table One.

RESULTS AND DISCUSSIONS

The measured spectrum for neutrons emerging from the experimental moderator assembly is presented in Figure 6. Figure 6 also shows the neutron spectrum obtained from a Monte Carlo calculation for the experimental moderator assembly. As it is shown, the measured spectrum agrees reasonably well with the calculation, except at energies above 100 keV, where the measured spectrum is lower by a factor of two than the calculated spectrum.

Figure 7 shows the measured and the calculated boron-10 absorbed dose-to-depth distributions. They appear to be in good agreement. The boron-10 absorbed dose inside the phantom indirectly confirms the neutron flux impinging upon the phantom.

So far, the cause of the factor of two discrepancy between the calculated and measured neutron spectra above 100 keV is not clear to us. However, because the measured and the calculated spectra do agree reasonably well for neutrons with energies less than 100 keV and for the boron-10 absorbed dose in the phantom, the discrepancy may be attributable to the uncertainties of the neutron response function of the proton-recoil detector.

CONCLUSIONS

The overall results of this experiment have validated the calculational methods of our previous design study on the moderator assembly for a low-energy proton accelerator neutron irradiation facility for BNCT. Our calculational study shows that the optimized moderator assembly consists of a 22.5-cm thick, 25-cm diameter cylinder of BeO

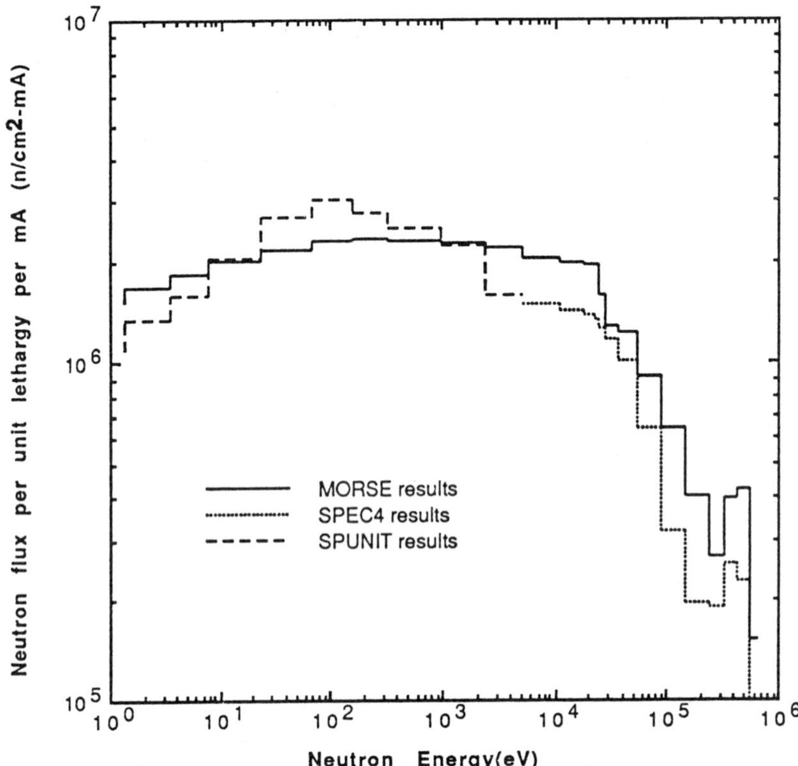

Figure 6. Measured and calculated neutron spectra for neutrons emerging from an experimental moderator assembly which consists of a 20-cm high and 36.8-cm diameter cylindrical tank of D_2O.

surrounded by a 30-cm thick jacket of alumina (Al_2O_3). With this optimized moderator assembly, the accelerator facility provides an epithermal-neutron flux of $3.12 \times 10^7 n/cm^2$-s per mA of proton current. This is confirmed by the results of our boron-shell spectrometer measurements. For an RFQ accelerator delivering 30 mA of 2.5-MeV protons, the irradiation time for a single-session treatment can be as short as fifty minutes. This irradiation time is confirmed within 30% by our BF_3-detector measurements. Finally, for our optimized moderator assembly, we have calculated a ratio of absorbed neutron dose to fluence of 4.9×10^{-11} cGy-cm^2/n, which is equal to the ratio of a 5-keV neutron beam. Our measurements with a proton-recoil detector indicate that the absorbed dose to neutron-fluence ratio is at least as good as calculated and is perhaps better. The factor of two discrepancy between the calculated and measured neutron spectra above 100 keV is yet to be resolved.

ACKNOWLEDGMENT

This work was supported in part by the National Cancer Institute under Grant No. 1-R01-CA47298-01.

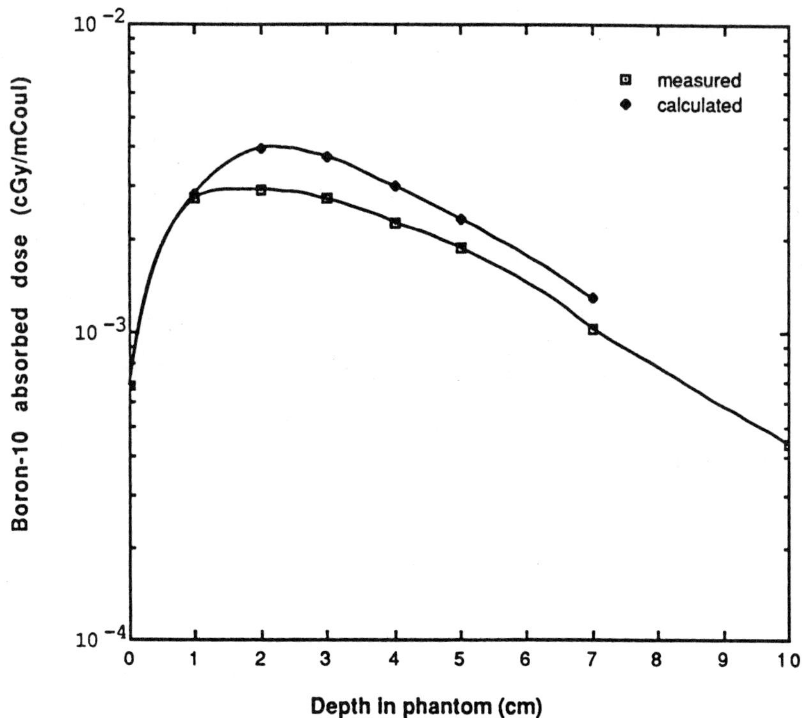

Figure 7. Measured and calculated boron-10 absorbed doses in a 20 cm x 20 cm x 20 cm water phantom on the axis of the neutron field.

REFERENCES

1.   J. W. Blue, W. K. Roberts, T. E. Blue, R. A. Gahbauer, and J. S. Vincent, "A Study of Low Energy Proton Accelerators for Neutron Capture Therapy," in <u>Proc. Second Int. Symp. on Neutron Capture Therapy</u>, Tokyo, 1985, H. Hatanaka, ed., Nishimura Co., Ltd., Niigata, Japan, p. 147 (1986).

2.   Y. Oka, I. Yanagisawa, and S. An, "A Design Study of the Neutron Irradiation Facility for Boron Neutron Capture Therapy," <u>Nucl. Technol.</u>, 553:642 (1981).

3.   R. M. Brugger, T. J. Less, and G. G. Passmore, "An Intermediate-Energy Neutron Beam for NCT at MURR," <u>U.S. Dept. of Energy 1986 Workshop on Neutron Capture Therapy</u>, R. G. Fairchild and V. P. Bond, eds., Brookhaven National Laboratory, <u>BNL-51994</u>, p. 83 (1987).

4.   C. K. Wang, T. E. Blue, and R. Gahbauer, "A Neutronic Study of an Accelerator-Based Neutron Irradiation Facility for Boron Neutron Capture Therapy," <u>Nucl. Technol.</u>, 84:93 (1989).

5.   T. P. Wangler J. E. Stovall, T. S. Bhatia, C. K. Wang, T. E. Blue, and R. A. Gahbauer, "Conceptual Design of an RFQ Accelerator-Based Neutron Source for Boron Neutron Capture Therapy," in <u>Proc. 1989 IEEE Particle Accelerator Conf.</u>, Chicago, IL, Vol. I, p. 678 (1989).

6. P. W. Benjamin C. D. Kemshall, and A. Brickstock, "SPEC-4: A Neutron Spectrum Unfolding Code for a Spherical Proportional Counter Containing Hydrogen," Radiation Shielding Information Center, Oak Ridge National Laboratory, PSR-99 (1976).

7. C. K. Wang and T. E. Blue, "A Neutron Spectrometer for Neutron Energies Between 1 eV and 10 keV," Trans. Am. Nucl. Soc., 57:79 (1988).

8. M. B. Emmett, "The MORSE Monte Carlo Radiation Transport Code System," Oak Ridge National Laboratory, ORNL-4972 (Feb. 1975).

9. R. W. Roussin, "BUGLE-80: Coupled 47-Neutron, 20-Gamma-Ray Group, P_3, Cross-Section Library for LWR Shielding Calculations," Radiation Shielding Information Center, Oak Ridge National Laboratory, DLC-75 (1980).

10. J. J. Doroshenko S. N. Kraitor, T. V. Kuznetsova, K. K. Kushnereva, and E. S. Leonov, "New Methods for Measuring Neutron Spectra with Energy from 0.4 eV to 10 MeV by Track and Activation Detectors," Nucl. Technol., 33:296 (1977).

DOSIMETRY AND TREATMENT PLANNING

MONTE CARLO BASED DOSIMETRY AND TREATMENT PLANNING FOR NEUTRON CAPTURE THERAPY OF BRAIN TUMORS

R. G. Zamenhof,* S. D. Clement,† O. K. Harling,† J. F. Brenner,* D. E. Wazer,* H. Madoc-Jones,* and J. C. Yanch†

* Department of Radiation Oncology
Tufts – New England Medical Center
Boston, MA

† Nuclear Reactor Laboratory
Massachusetts Institute of Technology
Cambridge, MA

ABSTRACT

Monte Carlo based dosimetry and computer-aided treatment planning for neutron capture therapy have been developed to provide the necessary link between physical dosimetric measurements performed on the MITR-II epithermal-neutron beams and the need of the radiation oncologist to synthesize large amounts of dosimetric data into a clinically meaningful treatment plan for each individual patient. Monte Carlo simulation has been employed to characterize the spatial dose distributions within a skull/brain model irradiated by an epithermal-neutron beam designed for neutron capture therapy applications. The geometry and elemental composition employed for the mathematical skull/brain model and the neutron and photon fluence-to-dose conversion formalism are presented. A treatment planning program, NCTPLAN, developed specifically for neutron capture therapy, is described. Examples are presented illustrating both one and two-dimensional dose distributions obtainable within the brain with an experimental epithermal-neutron beam, together with beam quality and treatment plan efficacy criteria which have been formulated for neutron capture therapy. The incorporation of three-dimensional computed tomographic image data into the treatment planning procedure is illustrated. The experimental epithermal-neutron beam has a maximum usable circular diameter of 20 cm, and with 30 ppm of B-10 in tumor and 3 ppm of B-10 in blood, it produces (with RBE weighting) a beam-axis advantage depth of 7.4 cm, a beam-axis advantage ratio of 1.83, a global advantage ratio of 1.70, and an advantage depth RBE-dose rate to tumor of 20.6 RBE-cGy/min (cJ/kg-min). These characteristics make this beam well suited for clinical applications, enabling an RBE-dose of 2,000 RBE-cGy/min (cJ/kg-min) to be delivered to tumor at brain midline in six fractions with a treatment time of approximately 16 minutes per fraction. With parallel-opposed lateral irradiation, the planar advantage depth contour for this beam (with the B-10 distribution defined above) encompasses nearly the whole brain. Experimental calibration techniques for the conversion of normalized to absolute treatment plans are described.

INTRODUCTION

An important factor in the advance of cancer therapy is an improvement in the local control of tumors that do not frequently metastasize. These comprise one sixth of all cancers [1] and are a specific area where neutron capture therapy (NCT) might make its

Neutron Beam Design, Development, and Performance for Neutron Capture Therapy
Edited by O. K. Harling *et al.*
Plenum Press, New York, 1990

most valuable impact. One example of this category of cancers is glioblastoma multiforme, an intracranial cancer to which NCT has been most frequently applied. The basic principles and a historical review of NCT have been presented elsewhere [1], and a comprehensive report of the state of the art of the various multidisciplinary aspects of NCT may be found in [2].

Since its original conception by Gordon Locher in 1936 [3] up to the first successful clinical trials of NCT initiated in Japan in 1968 [4], successful implementation of NCT was principally hindered by the lack of adequate means for boron-10 delivery to tumor. Following the introduction of second-generation B-10 labeled compounds in the late 1950s, such as sodium mercaptoundecahydrododecaborate (BSH) [5] (the B-10 compound which has been used exclusively in the Japanese clinical trial of NCT), it was generally recognized that, in parallel with further development work on B-10 compounds, there was also an urgent need for the development of improved neutron beams [6,7]. In particular, it was felt by many investigators that the focus of such beam development should be in the area of epithermal-neutron beams. In principle, the use of epithermal-neutron beams should permit NCT of brain tumors to be accomplished through the intact scalp and skull rather than intraoperatively, as was the technique employed in the early unsuccessful U.S. trials and is the current NCT technique employed in Japan.

Specifically, the mainly thermal-neutron beams that are being used for the clinical trials in Japan lack adequate penetration to permit the treatment of deeply-seated intracranial tumors through the intact scalp and skull. Patients participating in the Japanese NCT trial are treated intraoperatively in an attempt to offset to some extent the poor penetrability of thermal neutrons, but those patients in whom the deepest tumor margins exceed 6 cm in depth from the cortical surface are classed as "not favorable cases" [4]. Development work on epithermal-neutron beam design and construction has been continuing for the past three years at the Massachusetts Institute of Technology Research Reactor, MITR-II, as part of a broad interinstitutional NCT research program between New England Medical Center (the primary teaching hospital of the Tufts University School of Medicine) and the Massachusetts Institute of Technology. One part of the research program has focused on the development of Monte Carlo simulation-based dosimetry and clinical treatment planning for the NCT treatment of intracranial tumors, and is the principal topic of this paper.

The sequence of steps, starting with the development of "raw" dosimetric data by Monte Carlo simulation and ending with the calibration of Monte Carlo derived treatment plans by experimental mixed-field dosimetry, is outlined below:

(1) The distribution within the patient's brain of all significant radiation components resulting from neutron beam irradiation is computed using Monte Carlo simulation. The simulation model includes all features of the irradiation which are expected to have a significant influence on the resulting dose distributions.

(2) A four-dimensional dose matrix is calculated from the radiation fluxes generated by Monte Carlo simulation; three of the dimensions are spatial position, while the fourth dimension comprises the six separate dose-rate values computed at each spatial position within the brain and skull for each radiation dose component present.

(3) A dedicated treatment planning algorithm named NCTPLAN, developed specifically for NCT, allows total or differential dose distributions to be examined in either one or two-dimensions, along any arbitrary axis, or within any arbitrarily oriented plane. Each dose component can be arbitrarily weighted to allow the incorporation of, for example, fractionation modifying factors or RBE values. Defined figures of merit, such as advantage depths and advantage ratios (to be defined shortly), are computed in one, two, or three dimensions and, specifically in the case of advantage depth, may be visually displayed in one or two dimensions. Because the treatment planning algorithm interpolates the Monte Carlo dose input data into a fine 1 mm x 1 mm x 1 mm matrix, it is possible to provide pseudo-dose files to the treatment planning program which actually contain three-dimensional diagnostic image data, such as CT or MRI. By merging such pseudo-dose files with real-dose

files, NCTPLAN can superimpose quantitative dose data with corresponding plane CT or MRI image data.

(4) After a radiation oncologist and other supporting physicians have considered the various treatment options available to them through interaction with the treatment planning program NCTPLAN, they will decide what macroscopic total dose, RBE dose, and/or fractionation-modified dose should be delivered to a specific percent isodose contour on a normalized treatment plan. In addition, they will examine off-axis two-dimensional treatment plans to ensure that specific tissues at risk such as, for example, retina and pituitary gland, are adequately protected in the chosen treatment plan (these tissues would be expected to contain similar B-10 concentrations to those in blood).

(5) The final step in the treatment planning process is to convert the normalized treatment plans to rationalized treatment plans, i.e., to assign absolute dose-rate values to the relative isodose contours. This final step is achieved by performing a calibration measurement with a neutron beam configuration and a skull/brain phantom designed to match the configuration of the original Monte Carlo computational model. Gold foils and paired ionization chambers for mixed-field dosimetry are placed in the skull/brain phantom at locations which can be matched, respectively, to relative gold iso-activation rate, and epithermal plus fast-neutron dose-rate contours obtainable from the results of the Monte Carlo simulation. These data, together with the knowledge of B-10 concentrations in blood, which would be measured immediately prior to and following irradiation by the prompt-gamma neutron activation analysis facility at MITR-II, will be used to determine the final patient irradiation time.

This paper describes our approach to implementing the above sequence of tasks in preparation for our anticipated commencement of a pilot NCT treatment trial of high-grade astrocytoma brain tumors at the New England Medical Center and the Massachusetts Institute of Technology in the fall of 1990.

MATERIALS AND METHODS

<u>Definition of NCT Dosimetric Terminology</u>

Specific dosimetric criteria and figures of merit have been developed for NCT to facilitate neutron beam quality evaluation and clinical treatment planning [7–13]. These include: "advantage depth" (AD), "advantage ratio" (AR), "global advantage ratio" (GAR), and "advantage depth dose-rate" (ADDR). Each is briefly defined below.

The AD is useful for evaluating the ability of a neutron beam to treat deeply-seated tumors or tumor extensions. AD is defined as the depth in tissue at which the dose to tumor equals the maximum dose to non-tumor tissue [7]. The AD is influenced by many factors, such as the pharmacologic distribution of B-10 at the time of treatment [7,14], the anatomical configuration, elemental composition and bulk B-10 content of the tissues being irradiated [7,10], the size and configuration of the neutron beam [15], and the radiation characteristics of the neutron beam and resulting radiation fields within the irradiated tissues [6,7,8]. The AD may be thought of as describing the distal boundary of a dose "window" in which normal brain tolerance and tumoricidal doses coexist.

The AR has been developed to measure a particular treatment beam's ability to minimize integral dose to normal brain when a tumoricidal dose is delivered to brain tumor. Two ARs are used in the present paper. The one-dimensional AR is defined as the integral dose that would be delivered to tumor tissue if it were uniformly distributed within the brain, divided by the integral dose that would be delivered to normal brain, along a particular one-dimensional axis through the brain. Normally, the axis of greatest interest corresponds to the axis of the treatment beam.

The GAR is fundamentally defined as above, but is computed in three-dimensions for the entire brain volume, thereby representing a more clinically relevant quantity from the viewpoint of normal brain tolerance considerations. For example, the GAR is very sensitive to the size of the neutron beam employed, whereas the AR (measured along the neutron beam's axis) is much less affected by it. The AR and GAR are intended to quantitate the integral dose (to normal tissue) of a particular treatment configuration, normalized to a predefined integral dose to tumor assuming that tumor is uniformly distributed throughout the brain volume.

The ADDR is the RBE dose rate to tumor defined at the advantage depth. From the previous definition of AD, the ADDR is also the maximum RBE dose rate to normal tissue. The ADDR was developed primarily as a clinically meaningful neutron beam intensity criterion in our epithermal-neutron beam design studies [12,13].

<u>Models for Monte Carlo Based Dose Computations</u>

In NCT, the large degree of dependence of dose distributions within the brain on factors such as beam characteristics, beam dimensions, beam orientation, and brain/head geometrical configuration, requires that a calculational dosimetric technique capable of accurately accounting for all these degrees of freedom be employed. Monte Carlo simulation is the only available technique that is capable of meeting these objectives.

MCNP ("Monte-Carlo, Neutron-Photon") [16] is an exceptionally well designed and experimentally validated Monte Carlo code which we decided was the most appropriate computational tool for applications to NCT, both in the areas of patient dosimetry and neutron beam design [8,10,12,13]. MCNP is a three-dimensional, point-wise continuous energy cross-section code, which is rigorous in the physics principles it employs for the computation of neutron and photon transport. Other features of MCNP of particular importance to its application to NCT include: coupled treatment of neutrons and photons, the very flexible geometry that it can accommodate, its ability to compute reaction integrals, its continuous cross-section treatment, its ability to utilize the full accuracy of the best and most up-to-date nuclear cross-section data available, its flexible capability for utilizing a variety of user-specified variance reduction techniques, and the tremendous amount of scientific user support that is provided for this code at the Los Alamos National Laboratory. Further discussions of our reasons for selecting MCNP are given in [13].

The mathematical model that we have employed to represent the human brain and its surrounding structures is based on the geometry of a Monte Carlo brain model originally described by Snyder [17]. Because Snyder's model was developed for purposes of internal radioisotope dosimetry computations, we have made certain necessary modifications to improve the model's suitability for NCT dosimetric computations [10,18,19].

The neutron-photon brain-equivalent [NPBE] Monte Carlo model consists of two concentric three-dimensional ellipsoids. Using a Cartesian coordinate system, the Z-axis is defined as the patient's superior-inferior axis, the X-axis as the patient's lateral axis, and the Y-axis as the patient's antero-posterior axis. The inner ellipsoid, which defines the volume of the brain, is represented by the equation:

$$(X/6)^2 + (Y/9)^2 + (Z/6.5)^2 - 1 = 0$$

The outer ellipsoid, which encloses between its own surface and that of the inner ellipsoid a volume of bone representing the skull, is represented by the equation:

$$(X/6.8)^2 + (Y/9.8)^2 + ([Z+1]/8.3)^2 - 1 = 0$$

The dimensions for X,Y, and Z are in cm and the denominators of the X,Y, and Z terms in the equations above are the half-axes of the respective ellipsoids. The outer ellipsoid is displaced inferiorly along the Z-axis by 1 cm relative to the inner ellipsoid to represent more realistically the actual thickness of skull at different positions around the head. Thus, the skull thickness is 0.8 cm at the vertex, while the posterior fossa and base of skull are represented by a bone thickness of up to almost 3 cm. Brain cavity dimensions are 12 cm

Table One

Elemental Compositions and Densities Employed for Brain and Skull Components of the NPBE Skull/Brain Model.

	Weight Percent by Element									Density
Tissue	H	C	N	O	Ca	P	Cl	K	Na	(g/cc)
BRAIN	10.6	14.0	1.84	72.6	---	0.39	0.14	0.39	0.14	1.047
SKULL	5.0	14.0	4.0	45.0	21.0	11.0	---	---	---	1.5

(lateral), 18 cm (antero-posterior), and 13 cm (superior-inferior). The brain volume enclosed by the inner ellipse is 1470 cc, while the total bone volume is 847 cc. These conform to the brain and bone volumes of the average human skull [17]. Figure 1 shows the geometrical design of the NPBE phantom in transverse, sagittal, and coronal planes with each plane passing through the origin of the smaller ellipsoid. Figure 2 shows approximately midbrain transverse, coronal, and sagittal views of a typical patient's brain (obtained from reformatted CT scans) with the two ellipses of the NPBE model superimposed for comparison. Numerous measurements of CT scans of patients' heads have indicated that the NPBE model is indeed closely representative in dimensions to the average human adult's brain and skull and was determined, therefore, to be an adequate model for these initial investigations. In addition, a 12-cm diameter, 12-cm high cylindrical "neck" was included as part of the NPBE model. The neck section is contiguous to the inferior outer ellipsoidal surface of the skull/brain model, is oriented with its axis coincident with the Z-axis of the skull/brain model, and is composed of soft tissue-equivalent material. The neck section of the NPBE model is not shown in Figure 1 or Figure 2.

The elemental composition and physical density of the brain-equivalent material were obtained by averaging the elemental compositions and densities of white and gray brain matter as published by Brooks et al. [20], and are listed in Table One. The elemental composition of cranial bone is the same as that published by Weissberger et al. [21] for human cortical bone, and is also listed in Table One. An average cranial bone density of 1.5 g/cc was used based on an examination of a large number of CT scans of human skulls. This value reflects an average of both the compact and trabecular bone present in the human skull. A careful examination of neutron cross-section data was made to ensure that no element actually present in human brain or skull which would have more than a 1% influence on either the total neutron absorption or total neutron scattering cross-sections was omitted from the tissue-equivalent formulations.

Figure 1 also shows a 1 cm x 1 cm x 1 cm three-dimensional grid of "cells," which share the Cartesian coordinate system of the skull ellipsoids. These cells are the elementary volumes used by the Monte Carlo simulation to record the fluxes resulting from the transport through the model of each individual neutron or photon history. It is well recognized that within the neutron energy range under consideration in this study, i.e., 0.0001 eV – 15 MeV, the most dominant factors influencing neutron transport in tissue are its geometrical configuration and its hydrogen content. Because the hydrogen content per cc differs by only about 28% between the brain and bone equivalent materials, the resultant neutron transport will not be too dissimilar over distances of 1 cm – characteristic of cell dimensions. Thus, the neutron fluxes recorded in cells which encompass both brain and bone are assumed to represent the true neutron fluxes existing at their centers, whether brain or bone actually exist at those points, to within the statistical accuracy of the Monte Carlo simulation. However, cells which extend beyond the outer boundaries of the skull are excluded. In total, 1971 useful 1 cm^3 cells are used to define the spatial distribution of each dose component within the NPBE model.

The choice of 1 cm^3 for the cell volume was based on a compromise between a desire to maximize the spatial resolution of the Monte Carlo computations and the computer time required to obtain acceptable statistical accuracy. Although, by conventional radiation therapy treatment planning standards, a 1-cm spatial resolution for dose estimation would be very coarse, such a resolution in the present case required approximately 100 hours of CPU time on a DEC Micro-VAX computer for each Monte Carlo simulated configuration.

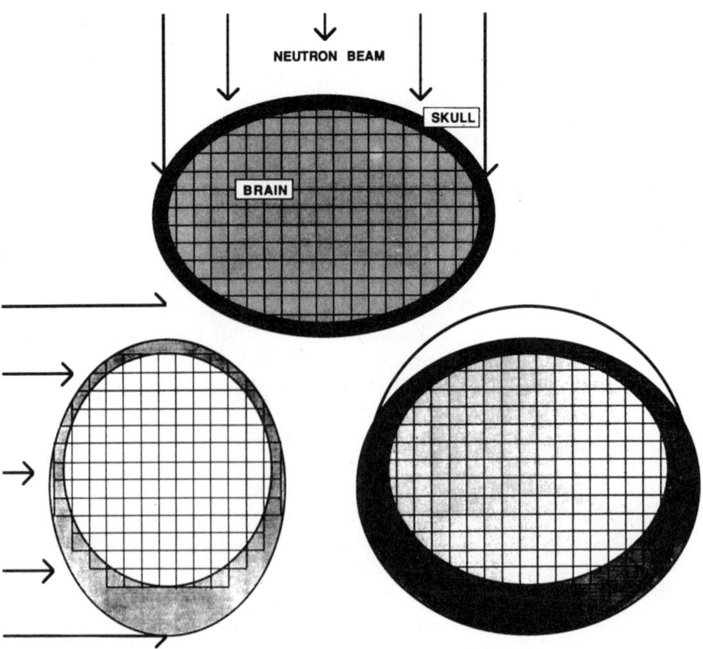

Figure 1. Geometric design of NPBE skull/brain model for Monte Carlo simulation; transverse plane (top), coronal plane (left), sagittal plane (right). 1 cm x 1 cm x 1 cm squares in each plane represent the dose tally cells. Arrows and circles represent orientation of 18.4-cm diameter neutron beam.

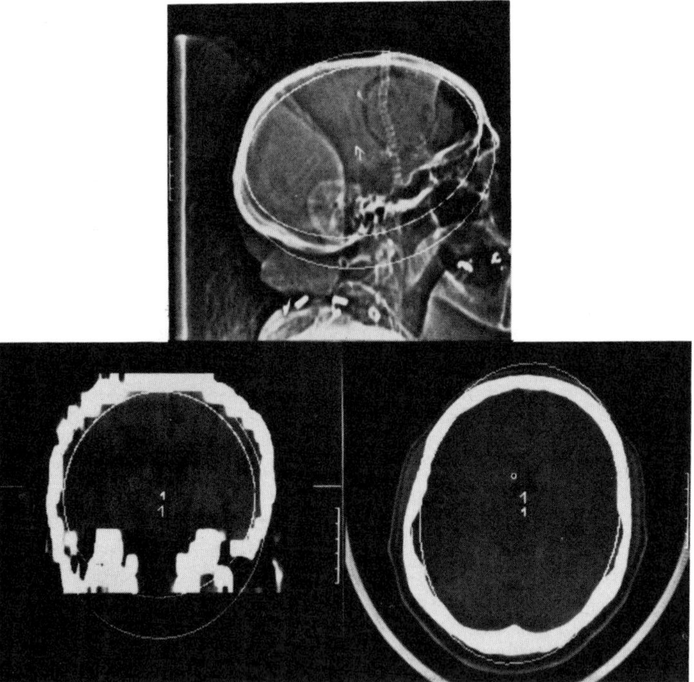

Figure 2. Superposition of the skull and brain ellipsoids of the NPBE model on corresponding anatomical planes reformatted from CT scans of a typical adult patient.

To fully understand the performance of a particular neutron beam for use in NCT, it is necessary to analyze the dose not only in terms of its spatial distribution within the brain but also to disassemble it into its constituent components. This is also necessary if dose modifiers, such as RBE factors or LET-specific fractionation factors, are to be included.

Consider first an "idealized" neutron beam, consisting of a current of monoenergetic epithermal neutrons incident on the NPBE model and comprising no undesirable contaminating radiation. The dose components resulting from such a beam which need to be examined separately include:

1. Thermal-neutron dose, principally resulting from $^{14}N(n,p)^{14}C$ thermal-neutron capture reactions,

2. Epithermal-neutron dose, principally resulting from $^{1}H(n,n')^{1}H$ epithermal-neutron scattering reactions,

3. Tissue induced gamma dose, principally resulting from $^{1}H(n,\gamma)^{2}H$ thermal-neutron capture reactions, and

4. Boron-10 dose, resulting from $^{10}B(n,\alpha)^{7}Li$ thermal-neutron capture reactions.

Consider next an actual epithermal-neutron beam from a nuclear reactor, such as one of those that are currently under evaluation at the Massachusetts Institute of Technology Research Reactor, MITR-II [9,12,13]. In addition to the four dose components listed above, such realistic neutron beams would also include three additional dose components:

5. Fast-neutron dose, principally resulting from $^{1}H(n,n')^{1}H$ fast-neutron scattering reactions, but possibly including significant amounts of recoil reactions on other tissue elements, such as carbon and oxygen,

6. Reactor core gamma dose, principally from gamma rays originating in the uranium fission process in the reactor's core (core gammas), and

7. Neutron-induced gamma dose, resulting from gamma rays produced by the capture and inelastic scattering of neutrons by the structural materials surrounding the beam line (structure-induced gammas).

To separate the seven dose components listed above it was necessary to perform three separate Monte Carlo computations for each case of interest, as follows:

<u>Simulation Type</u>	<u>Dose Component Present</u>						
Neutrons Only	1	2	3	4	5	*	7
Core Gammas Only	*	*	*	*	*	6	*
Structure Gammas Only	*	*	*	*	*	*	7

While dose components 1, 2, 3, 5, 6, and 7 were obtained explicitly (i.e., derived from actual neutron and gamma tallies), dose component 4 (the boron-10 dose) was obtained implicitly, as will be described shortly.

The Monte Carlo simulations were performed on a VAX Station-II/GPX computer. One source neutrons-only and two source gammas-only simulations were run each with 300,000 starting histories to generate a complete case of interest. Each composite set of three Monte Carlo runs required approximately 100 hours of CPU time. Under these conditions, typical Poisson statistical errors, estimated for both shallow and deep regions, were as follows:

| Dose Component | Typical Statistical Errors (%) | |
	Shallow (1-2 cm)	Deep (11-13 cm)
1	4 - 5	7 - 10
2	20 - 30	40 - 50
3	9 - 12	10 - 15
4	4 - 5	7 - 10
5	20 - 30	40 - 50
6	4 - 5	7 - 10
7	7 - 10	9 - 12

In reality, because individual dose components were highly spatially correlated with corresponding dose values in neighboring cells (i.e., the dose values for each dose component were known to be described physically by a smooth depth-dose curve), the final estimate of error in each dose component is somewhat lower than the values tabulated above would suggest.

The B-10 doses were computed implicitly, i.e., boron-10 was not actually included as part of the brain material composition. Instead, a feature of MCNP which permits nuclear reaction rates to be computed was employed to obtain the total neutron spectrally-integrated KERMA rates for the B-10 reaction (described more fully in the next subsection). The rationale for this approach was based on the fact that, for a given patient, the actual spatial distribution of B-10 in the brain would not be known, either pre or post-irradiation, and thus could not be explicitly simulated by Monte Carlo. However, one may argue on the basis of the facts presented below that it is unnecessary in most practical situations to explicitly simulate B-10 in the brain.

Given that less than 5% of the volume of the brain is occupied by vasculature [22], the impact of B-10 present in the blood on neutron transport can be modeled by less than 5% of the blood B-10 concentration being uniformly distributed throughout the brain. Thus, because maximum practical blood concentrations of B-10 would not be expected to exceed approximately 50 ppm (parts-per-million by weight), the equivalent uniform B-10 brain concentration would therefore be less than 2.5 ppm. On the other hand, although tumor tissue would also be expected to contain B-10 at concentrations not exceeding approximately 50 ppm, the equivalent uniform B-10 concentration in the region of gross tumor could, in principle, be as high as approximately 50 ppm. However, following tumor debulking surgery, the total volume of such residual gross tumor would be small relative to the volume of the brain.

To examine the above arguments quantitatively, Monte Carlo simulations were performed on the NPBE model with gradually increasing concentrations of B-10 explicitly present in the brain composition. B-10 concentrations of 0, 5, 10, 20, 50, and 100 ppm were uniformly included in the brain composition and the resulting B-10 doses (dose component 4) were examined as a function of depth in the NPBE model. A monoenergetic 35 eV idealized epithermal-neutron beam was employed for this computational experiment. Figure 3 shows the percent B-10 dose depression per 1 ppm of B-10 as a function of depth in brain. It can be seen that a uniform B-10 concentration of 2.5 ppm produces negligibly small B-10 dose depression, even at large depths. Given that, as argued above, only comparatively small volumes of tumor tissue would normally be expected to approach B-10 concentrations of 50 ppm, one would not expect to observe B-10 dose depressions of more

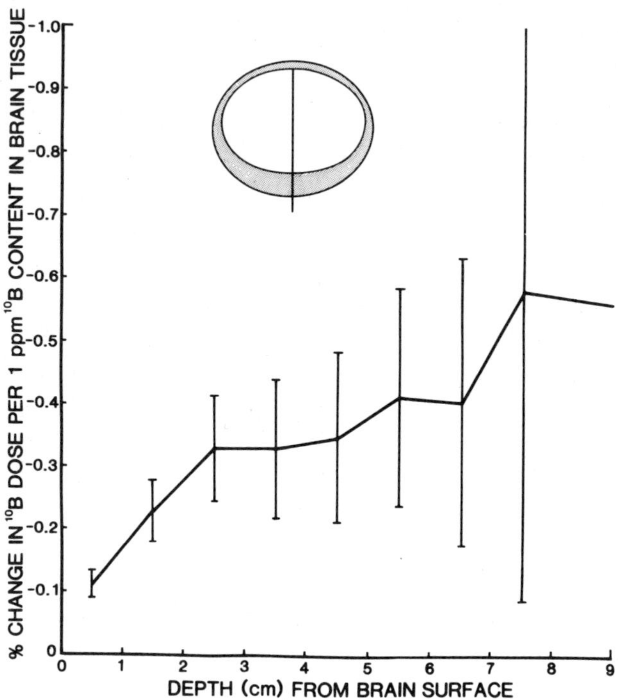

Figure 3. Degree of B-10 dose depression obtained from 0, 5, 10, 20, 50, & 100 ppm of B-10 uniformly distributed within the brain; computed by Monte Carlo simulation at different depths in the NPBE model, and expressed as percent change in B-10 dose per 1 ppm of B-10. Icon shows orientation of 6.5 cm x 9.5 cm 35-eV epithermal-neutron beam relative to sagittal view of NPBE model.

than a few percent. Thus, the error in excluding B-10 from the NPBE model would be similar to, or less than, the estimated statistical errors of the Monte Carlo simulations themselves. However, if in the future non-invasive methods are developed to measure macroscopic B-10 concentrations in gross tumor shortly prior to irradiation, it would be a simple matter to represent such additional detail in the NPBE model.

<u>Flux-to-Dose-Rate Conversion and Choice of RBE Factors</u>

To convert the photon and neutron fluxes computed for each cell of the NPBE model to dose rates, a feature of MCNP was employed which allows spectrally integrated atomic and nuclear reaction rates to be computed. This feature enables the use of flux-to-dose-rate conversion factors (sometimes referred to as KERMA-rate factors) to be included as pseudo cross-section tables where the photon or neutron flux at each energy in each cell is multiplied by the interpolated value of the flux-to-dose-rate conversion factor to yield dose rate in cGy/min. B-10 KERMA-rate factors were calculated from cross-section data and branching ratios given in BNL-325 [23] and are listed in Table Two. Photon KERMA-rate factors were calculated from the mass energy-absorption coefficients of Hubbel [24] and are listed in Table Three. Neutron KERMA-rate factors were calculated from the data of Caswell, Coyne, and Randolph [25] and are listed in Table Four. MCNP employs a log-log interpolation to compute KERMA-rate factors at energies not explicitly specified in the KERMA-rate tables.

RBE factors chosen followed the recommendations presented as part of the guidance to authors submitting manuscripts for these proceedings. The RBE factors were agreed upon through discussions with the principal NCT research groups in the U.S. and abroad, and are as follows:

Radiation Type	Recommended RBE
Gamma Rays	1.0
Neutrons (no B-10)	1.6
B-10 Reaction	2.3

However, in this paper, we have employed an additional dose-modifying factor of 0.5 applied to gamma dose, to represent the reduced biological effect of gamma radiation on the brain (with respect to late radiation injury) when a fractionation schedule of six daily fractions is employed, as we propose to implement in clinical trials. Justification for this dose-modifying factor is based on experimental human brain tolerance data presented in [27].

<u>Monte Carlo Dosimetry of an Experimental Epithermal-Neutron Beam</u>

As part of our group's research on the development of epithermal-neutron beams for NCT, the dosimetry of over 100 epithermal-neutron beam designs has been studied by the Monte Carlo simulation technique [13]. As well as examining the performance of "practical" epithermal beam designs (i.e., those which we believe can be physically implemented at MITR-II under our funding constraints), we have also examined the performance of "idealized" neutron beams (i.e., beams which according to physical principles should exhibit favorable characteristics as measured by the criteria of AD, AR, and GAR). The philosophy behind this was to define the goals that practical epithermal beam design should approach, and thereby to better assess the compromises between beam quality, beam intensity, and beam development cost. These ideas are described in greater detail in [12,13]. In this paper, only the dosimetry of an existing experimental epithermal beam has been reported in detail, although pertinent results for two idealized epithermal beams (35 eV and 2 KeV) are tabulated for purposes of comparison. A Monte Carlo simulation analysis of idealized neutrons beams for NCT applications is presented in [13].

A clinically usable experimental epithermal-neutron beam has been designed and constructed at the MIT Nuclear Research Reactor, MITR-II, for clinical NCT applications. The design of this beam was carried out by a combination of experimental measurements and Monte Carlo simulations. This beam, hereafter referred to as the "experimental epithermal beam," was designed to enable the treatment of deeply-seated tumors or tumor remnants by single-field or parallel-opposed irradiations through the intact scalp and skull.

Monte Carlo simulations were performed to dosimetrically characterize the clinical performance of the experimental beam and also the 35 eV and 2 keV idealized epithermal beams. The beams were oriented in a lateral aspect with respect to the NPBE model, as illustrated in Figure 1, and were circular in cross section with a diameter of 18.4 cm. The three-dimensional dose matrices thereby obtained were subsequently interpolated into a fine 1 mm x 1 mm x 1 mm matrix by the treatment planning program NCTPLAN. The dose interpolation involved a linear three-dimensional interpolation between neighboring cell values. With the dose matrices thus interpolated into 1 mm^3 voxels, two important features of NCTPLAN are facilitated: one-dimensional and two-dimensional (planar) dose plots in arbitrary orientations can easily be displayed and pseudo-dose matrices can be created using three-dimensional MRI or CT image data, allowing planar isodose displays to be superimposed on their corresponding image planes.

<u>NCTPLAN, A Neutron Capture Therapy Dosimetry and Treatment Planning Program</u>

Because of the complexity of the radiation fields that need to be considered in the clinical implementation of NCT, a simplified and more intuitive approach to the evaluation of patient dosimetry needs to be developed. There is also the need to examine the influence on dose distribution of the uncertainties in the various dose-modifying factors, such as varying tumor and blood B-10 concentrations, endothelial cell and tumor cell RBEs, the effect of fractionation on the reduction of the biological effectiveness of the low LET dose

Table Two

Photon Fluence-to-Dose Conversion Factors
(KERMA-Factors) for Brain-Equivalent Material.

Photon Energy (KeV)	KERMA (cGy-cm^2)
1.00E+00	5.99E-08
2.00E+00	1.80E-08
5.00E+00	3.24E-09
1.00E+01	7.75E-10
2.00E+01	1.75E-10
5.00E+01	3.42E-11
1.00E+02	4.04E-11
2.00E+02	9.46E-11
5.00E+02	2.63E-10
1.00E+03	4.94E-10
2.00E+03	8.29E-10
5.00E+03	1.52E-09
1.00E+04	2.48E-09
2.00E+04	4.38E-09

Table Three

Neutron Fluence-to-Dose Conversion Factors
(KERMA-Factors) for Brain-Equivalent Material.

Neutron Energy (KeV)	KERMA (cGy-cm^2)
2.50E-05	1.50E-11
2.00E-03	1.73E-12
2.00E-02	7.51E-13
2.00E-01	2.29E-12
2.00E+00	2.10E-11
2.00E+01	1.89E-10
1.05E+02	6.95E-10
1.05E+03	2.56E-09
1.05E+04	5.98E-09
1.55E+04	7.03E-09

Table Four

Neutron Fluence-to-Dose Conversion Factors (KERMA-Factors)
for Boron-10 Reaction, for 100 ppm by Weight B-10.

Neutron Energy (KeV)	KERMA (cGy-cm^2/100 ppm)
1.00E-05	1.38E-09
2.50E-05	8.72E-10
3.60E-05	7.27E-10
2.51E-04	2.92E-10
6.84E-04	1.78E-10
1.86E-03	1.09E-10
5.00E-03	6.56E-11
1.37E-02	4.00E-11
3.73E-02	2.33E-11
1.01E-01	1.47E-11
2.75E-01	8.99E-12
7.49E-01	5.48E-12
2.03E+00	3.27E-12

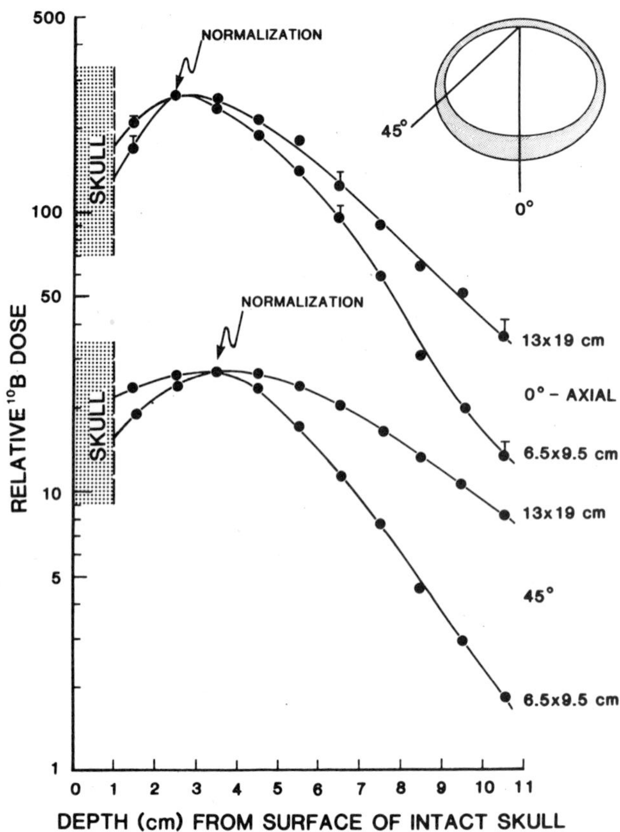

Figure 4. Monte Carlo computed B-10 dose-depth curves at 0° and 45° from the beam axis for two 35-eV idealized epithermal-neutron beams of different size.

components, etc. Given the large degree of uncertainty in many of these factors, a radiation oncologist must be assisted in visualizing the estimated uncertainties in the dose plans and thereby to intuit their effects from the clinical perspective. Finally, in those instances where NCT treatment proves unsuccessful, it is of critical importance to perform a detailed and complete retrospective correlation of pathological findings with accurate dose distributions so that maximum opportunity for learning from such failures is afforded. NCTPLAN was developed in consideration of the above requirements. NCTPLAN is written in FORTRAN and currently operates on a VAX-78O computer with VT241 and DeAnza ID 5424 image display terminals.

RESULTS AND DISCUSSION

Illustration of the Influence of Neutron Beam Size on B-10 Dose-Depth Distribution

Figure 4 shows the results of Monte Carlo simulations performed for the 35-eV idealized beam employing two rectangular neutron beam sizes of 6.5 x 9.5 cm and 13 x 19 cm. B-10 dose-depth profiles are displayed along the beam axis and at an angle of 45° to the beam axis. Along each of these two axes the curves for the two different beam sizes are normalized at their build-up peak to facilitate a visual comparison of their shapes. It can be seen that the peak B-10 dose occurs at depths of 1.5–2.5 cm below the surface of the brain (2.3–3.3 cm below the surface of the skull), thereby providing a sparing effect on the scalp and skull. Such a thermal-neutron flux build-up is a familiar feature of epithermal-neutron beams and is the principal reason for their superior penetration relative to thermal-

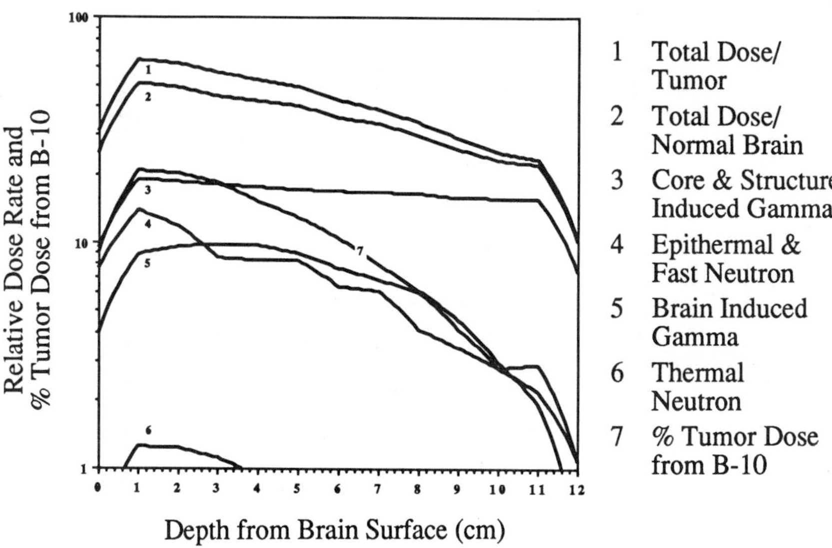

Depth from Brain Surface (cm)

Figure 5. Relative dose rate (cGy/min) along beam axis vs. depth in brain for MITR-II experimental epithermal-neutron beam showing: total dose rate to tumor (with 30 ppm B-10); total dose rate to normal brain (from blood containing 10 ppm B-10); induced gamma dose in brain; epithermal and fast-neutron dose to brain; combined core and structure-induced gamma dose to brain; and thermal-neutron dose to brain. Also shown is a curve indicating the percent of tumor dose due to B-10.

neutron beams [6,7,12,13]. At a depth of 6 cm below the brain surface the curves show a 25–55% increase in B-10 dose for the large beam relative to the small beam (the increase being dependent on angle). Since Figure 4 shows that the larger beam size provides a more penetrating B-10 dose distribution, and hence a more desirable beam configuration for the treatment of deep tumors, all subsequent examples presented in this paper employ an 18.4-cm diameter circular neutron beam which essentially irradiates the entire skull and brain, as illustrated in Figure 1. A beam diameter of approximately 20 cm is currently available for the experimental epithermal beam at MITR-II at the exit of the medical therapy beam portal. However, a larger diameter beam could be obtained with little difficulty if needed.

<u>Dosimetric Characteristics of an Experimental Epithermal-Neutron Beam</u>

Figure 5 shows the constituent dose components for the experimental beam in the NPBE model computed by Monte Carlo simulation. The surface tissue-sparing characteristics of this beam can be appreciated by observing that the total dose to normal brain peaks at approximately 1.8 cm below the skull surface. Although the 1-cm cell-size employed for these Monte Carlo simulations was too coarse to permit a reliable quantitative estimate of the scalp-sparing effect, calculations using a finer cell structure and measurements suggest that the total scalp surface dose is approximately 80% of the peak brain tissue dose, thus indicating a desirable scalp-sparing property for this beam [12,13]. At the AD, 25-30% of the total tumor dose (with no RBE weighting) is due to B-10. Table Five summarizes the AD, AR, GAR, and ADDR characteristics of the experimental beam.

The RBE-ADDR for the experimental beam is 25.9 RBE-cGy/min (cJ/kg-min) (for 30 to 10 ratio of tumor-to-blood B-10 concentrations and single-field irradiation), as shown in Table Five. From this, one may estimate the irradiation time required to treat a patient using such a beam at MITR-II in the following way.

Table Five

Tabulation of AD, AR, GAR, and ADDR (with RBE weighting) for Experimental Epithermal Neutron Beam in Service at MITR-II.

T = B-10 Concentration in Tumor (ppm by Weight)
B = B-10 Concentration in Blood (ppm by Weight)
AD's are measured from the Skull Surface

T	B	T/B	AD	AR	GAR	RBE-ADDR (cJ/kg-min)
30	10	3*	6.1	1.51	1.44	25.9
30	3	10	7.4	1.83	1.70	20.6
30	0	inf	8.1	2.02	1.85	18.3
60	20	3*	6.9	1.81	1.72	83.5
60	6	10	8.7	2.52	2.31	22.9
60	0	inf	9.6	3.03	2.70	18.3
Parallel-Opposed**						
30	10	3*	11.6	1.51	1.44	16.6
30	3	10	11.8	1.83	1.70	13.7
30	0	inf	12.0	2.02	1.85	12.5

* These tumor-to-blood values are intended to represent the most pessimistic estimate for the BSH B-10 compound; i.e., equal physical B-10 concentrations in tumor and blood, with an assumption of endothelial cell absorbed fraction of 1/3.

** The AD is symmetrical about the brain midline; the RBE-ADDR assumes that there is no delay between the sequential delivery of the two fields.

Assume that a conservative whole-brain tolerance dose (based on late radiation injury) for conventional low-LET radiation delivered in six daily fractions is 2000 cGy [27]. If the NCT irradiation were to be delivered in six daily fractions to an equivalent total macroscopic blood RBE-dose of 2000 RBE-cGy (cJ/kg), an irradiation time of 13 minutes per fraction would be required. The macroscopic RBE-dose of 2000 RBE-cGy (cJ/kg) resulting from 10 ppm of B-10 in blood probably represents an actual endothelial RBE-dose of approximately 667 RBE-cGy (cJ/kg), assuming that the actual endothelial B-10 dose would be approximately 1/3 the macroscopic B-10 dose due to B-10 in blood. There are theoretical arguments based both on absorbed dose fraction and nuclear hit probability distributions [1,28], and also on experimental data from the irradiation of non-tumor bearing beagle dogs [30] which support this approximate factor of 1/3. It should be noted that although the ADDR, as discussed above, is obtained from the results of Monte Carlo simulation, the individual peak dose rates for each of the dose components of the experimental beam were initially normalized to in-phantom experimental measurements [12,13].

A difficult aspect of assessing the performance of neutron beams for NCT is that different pharmacologic agents labeled with B-10 have been shown to produce widely differing tumor-to-blood ratios. The agent employed in the Japanese clinical trials (BSH) was estimated on the basis of almost 100 clinical cases to have an average B-10 tumor-to-blood ratio at the time of neutron irradiation of 2.35±0.36 to 1 [4]. Although the blood B-10 concentrations estimated for these patients are expected to be reasonably accurate, the tumor B-10 concentrations were based on gross tumor samples containing unknown ratios of tumor to non-tumor tissue. For example, if a "tumor" sample only contained 20% actual tumor, then the true tumor B-10 concentrations could be up to five times higher than actually measured for the sample. Table Five shows that even considering the most pessimistic situation of <u>equal</u> physical concentrations of B-10 in tumor and blood, corresponding to an <u>effective</u> concentration ratio of 30 to 10 (assuming the 1/3 factor for the endothelium, as described above), clinically useful AD, AR, and GAR values are maintained with the experimental beam. In fact, it will be demonstrated in the following section that such a situation would be quite adequate for NCT, despite this worst-case assumption.

Table Six

**Tabulation of AD, AR, and GAR (with RBE Weighting) for
35 eV Idealized Epithermal Beam.**

**T = B-10 Concentration in Tumor (ppm by Weight)
B = B-10 Concentration in Blood (ppm by Weight)**

T	B	T/B	AD	AR	GAR
30	10	3*	7.2	2.33	2.33
30	3	10	9.2	4.36	4.36
30	0	inf	10.4	6.97	6.95
60	20	3*	7.4	2.60	2.60
60	6	10	9.7	5.90	5.89
60	0	inf	11.8	12.9	12.9

* These tumor-to-blood values are intended to represent the most pessimistic estimate for the BSH B-10 compound; i.e., equal physical B-10 concentrations in tumor and blood, with an assumption of endothelial cell absorbed fraction of 1/3.

In order to appreciate the quality of the design of the experimental beam, it is useful to compare the figures of merit for this beam with those for the two idealized epithermal-neutron beams that were investigated. Tables Six and Seven show the AD, AR, and GAR values for the 35-eV and 2-keV beams. For the 30 to 10 tumor-to-blood B-10 distribution, the idealized beams manifest ADs that are 1.1–2.0 cm greater than that of the experimental beam under similar conditions (i.e., 7.2 and 8.1 cm, respectively, for the 35-eV and 2-keV beams, vs. 6.1 cm for the experimental beam). These differences, however, become smaller when an improved 30 to 3 tumor-to-blood B-10 distribution is considered.

Table Seven

**Tabulation of AD, AR, and GAR (with RBE Weighting) for
2 KeV Idealized Epithermal Beam.**

**T = B-10 Concentration in Tumor (ppm by Weight)
B = B-10 Concentration in Blood (ppm by Weight)**

T	B	T/B	AD	AR	GAR
30	10	3*	8.1	2.32	2.31
30	3	10	9.9	4.31	4.27
30	0	inf	11.3	6.81	6.72
60	20	3*	8.3	2.59	2.58
60	6	10	10.7	5.84	5.80
60	0	inf	12.1	12.6	12.4

* These tumor-to-blood values are intended to represent the most pessimistic estimate for the BSH B-10 compound; i.e., equal physical B-10 concentrations in tumor and blood, with an assumption of endothelial cell absorbed fraction of 1/3.

The principal factors responsible for the lower AD manifested by the experimental beam are the fast-neutron and core gamma-ray dose components present in the latter. These dose components could be reduced by further modifications in the design of this beam, but at the expense of some reduction in the ADDR. Beam design considerations at MITR-II are discussed in more detail in [12,13]. The most striking difference between the figures of merit for the idealized and experimental beams is the markedly inferior AR of the latter (i.e., 2.3 for the idealized beams vs. 1.5 for the experimental beam). This is almost entirely due to the penetrating core gamma-ray dose component in the latter which contributes strongly to the integral dose to normal brain. The optimal compromises between AD, AR, GAR, and ADDR are difficult to define at this early stage in the development of NCT, but the experimental beam described here is presently suitable for clinical use.

<u>Treatment Planning Strategies for NCT Using an Experimental Epithermal-Neutron Beam</u>

We have presented some criteria which have been developed for the evaluation of neutron beams for NCT. Although such one-dimensional evaluations are quite acceptable for the purposes of neutron beam design and optimization, a more comprehensive and clinically relevant approach requires the extension of the concepts of AD, AR, and GAR to two and three dimensions and the examination of isodose distributions in two-dimensional planes. We will present some examples of how clinically intuitive treatment planning data presentation and display can be obtained using NCTPLAN from the raw dosimetric data generated by Monte Carlo simulation.

Figure 6 shows one of a series of CT scans of a patient (treated at Tufts - New England Medical Center) with a glioblastoma multiforme arising from the thalamus. This is an example of a case which was considered inoperative and would, therefore, not be amenable to intraoperative NCT treatment with a thermal-neutron beam such as is employed in Japan. We will use this clinical case to illustrate two different approaches to NCT treatment planning using the experimental epithermal-neutron beam. Because of the deep tumor location, this case may be considered to represent the most challenging from the viewpoint of tumor access by epithermal-neutron beams. Also, due to the low 30 to 10 B-10 tumor-to-blood ratio assumed (corresponding effectively to equal physical B-10 concentrations in blood and tumor) and the comparatively low 30 ppm absolute B-10 concentration in tumor, this case may be considered to represent close to a worst-case scenario regarding B-10 distribution.

In the first treatment planning approach, we will assume that the enhanced tumor has actual margins which are 2 cm beyond the enhanced tumor boundary visible on CT. This is in fact an assumption often employed in conventional external beam radiation therapy of brain tumors. We will first examine the possibility of treating this tumor with a single lateral 18.4-cm diameter circular beam. Figure 7 shows a normalized isodose contour map (such that maximum in-plane dose is equal to 100%) within the NPBE model representing the relative total RBE-dose to normal brain based on blood containing 10 ppm B-10. The neutron beam is incident from the left. The RBE factors employed are, as described earlier, but with a "fractionation" factor of 0.5 applied to the gamma-dose components. The plane displayed is an anatomically transverse plane that is level with the lateral axis of the neutron beam (see Figure 1). The reformatted CT scan corresponding to this plane is shown superimposed on the isodose contour map. It is interesting to observe how relatively flat the isodose contours are, which is in part due to the wide neutron beam diameter used. The small triangle at the left side of the brain, slightly anterior to the midline, shows the location of the 100% (maximum) dose.

Figure 8 shows the same anatomical plane as before, but the contour seen is a planar AD contour referenced to a global normal tissue maximum RBE-dose; i.e., the total maximum RBE-dose to normal tissue containing 10 ppm B-10 anywhere within the NPBE model is used for computing the AD contour. The interpretation of this contour is that, anywhere within it, tumor will have higher dose than normal brain, while, anywhere outside it, tumor will have lower dose than normal brain. It can be seen from Figure 8 that with a 30 to 10 tumor-to-blood B-10 ratio, it is not possible to reach the point in brain where the deepest extent of tumor is assumed to exist (i.e., 2 cm to the right of the right-most enhanced tumor boundary). In this particular case study, one may thus conclude that single-field treatment under the specified conditions of B-10 distribution would not allow sufficient tumor margin coverage no matter from which side the beam was oriented.

Figure 9 shows planar AD contours for the same situation as in Figure 8 except that the irradiation is performed using two parallel-opposed beams incident in a coplanar

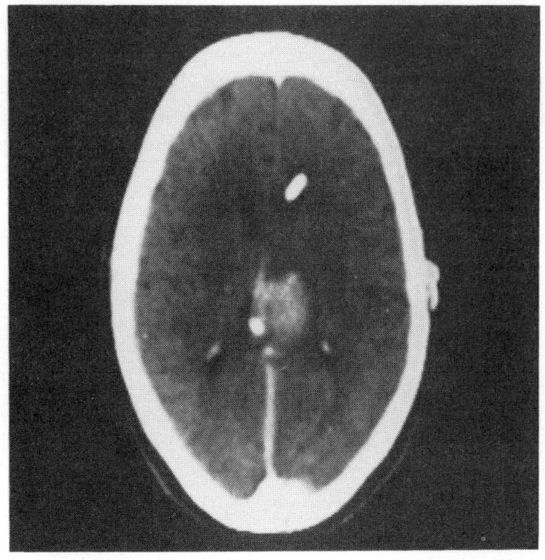 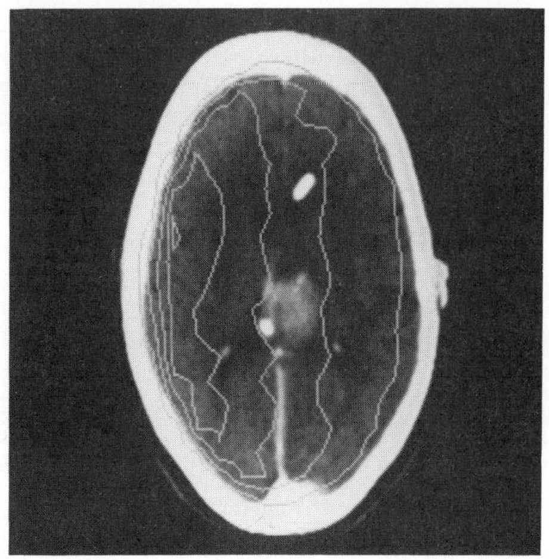

Figure 6. Figure 7.

Figure 6.   Contrast-enhanced transverse CT scan of patient with thalamic glioblastoma multiforme brain tumor visible at midline. The scalp surface is slightly visible in some areas as a discontinuous line overlying the skull.

Figure 7.   Monte Carlo computed iso-RBE-dose contours for normal brain with 10 ppm B-10. Single-field irradiation from the left side with experimental epithermal-neutron beam of 18.4-cm diameter. Iso-RBE-dose contours are (from right-to-left): 20%, 40%, 60%, 80%, 100% (small triangle on left side). Computed using NCTPLAN.

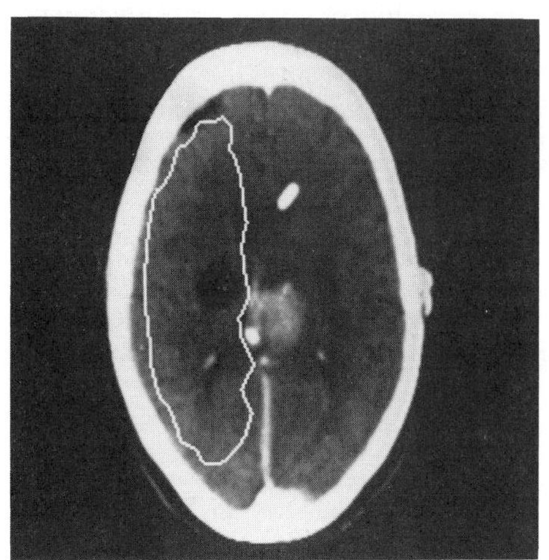 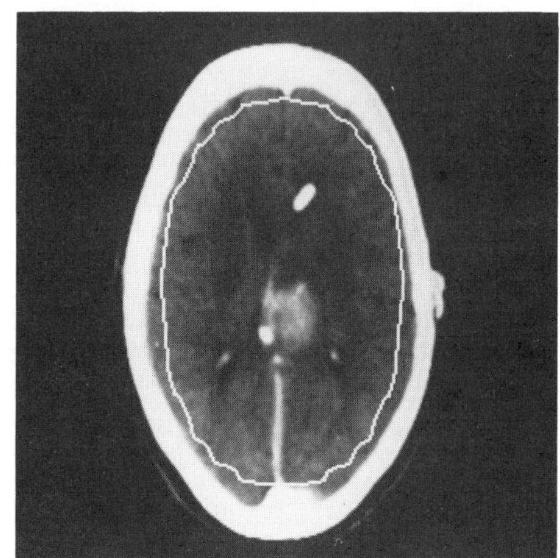

Figure 8. Figure 9.

Figure 8.   Same anatomical plane, beam size and configuration, and B-10 distribution as in Figure 7. Shown is a planar advantage depth contour, referenced to a global normal brain maximum RBE-dose. Computed using NCTPLAN.

Figure 9.   Same as Figure 8, except that neutron irradiation is parallel-opposed. Computed using NCTPLAN.

fashion from the left and right sides of the patient's head. In this case, the AD contour is seen to almost encompass the entire brain, thus indicating that with parallel-opposed irradiation the characteristics of the experimental beam are favorable for treating tumor tissue almost anywhere within the brain. Moreover, in this particular case, the off-axis planar AD distributions encompass the brain more completely than those shown in Figure 9.

It is important to observe from Table Five that neither the AR nor GAR values for a single-field irradiation change if parallel-opposed irradiation is employed. This might seem counter-intuitive if one interprets the AR and GAR as being related to the volume of normal tissue irradiated. However, with parallel-opposed irradiation one would aim to maintain minimum tumor dose equal to what it would be with a single irradiation, whether or not the latter would be a viable plan in this particular case. This would reduce normal tissue integral dose to the same integral dose as would be obtained with a single-field irradiation. Thus, in this case, the use of parallel-opposed irradiation would not increase the integral dose to normal brain and, therefore, would not necessarily lower its tolerance. Moreover, for the same reason the total irradiation time for parallel-opposed treatment would be only slightly longer than for a single-field treatment to the same tumor dose (see the corresponding ADDRs in Table Five).

The GAR is a parameter which could be useful to the radiation oncologist in evaluating the influence of different beam sizes and B-10 distributions on the prescription of brain tolerance dose after additional experience with NCT has been gained. For example, for the same epithermal-neutron beam, increasing beam size will decrease the GAR. A decrease in GAR means that for a given dose to tumor, the integral dose to normal brain would be proportionately higher, and hence the tolerance dose would be lower. Thus, the GAR might in the future provide a useful additional cue to the radiation oncologist for determining dose prescription in NCT.

One of the concerns that has been raised with regard to the use of epithermal-neutron beams for treating patients through the intact scalp and skull is that the scalp may exhibit low tolerance to the potentially high doses delivered to it [29]. This could result from fast-neutron or gamma contamination of the beams used, from the presence of undesirably high B-10 concentrations in the scalp, or from both causes. Although our experience with beagle dog irradiations through the intact scalp and skull using the MITR-II thermal-neutron beam suggests that with the use of the BSH boron compound radiation injury to the scalp should not be an issue at 2000 RBE-cGy (cJ/kg) [30], favorable surface-sparing characteristics in an epithermal-neutron beam would nevertheless be a desirable feature to further ensure that scalp tolerance considerations would not impose a limit on tumor dose. It may be seen from Figure 5 that the experimental epithermal beam provides a useful degree of surface-sparing. Moreover, because approximately 95% of the blood-born BSH compound has been shown to be bound to plasma albumins, it is unlikely that the blood B-10 present in the scalp would diffuse out of the vasculature and thereby increase the risk to cutaneous tissue other than to the blood vessels themselves. If the B-10 present in the scalp were assumed to remain confined to the blood, the RBE-dose to the scalp for the experimental epithermal beam would be approximately 80% that of the maximum RBE-dose to normal brain [12,13].

The second approach to NCT treatment planning involves the separate though concurrent evaluation of isodose distributions for normal tissue and tumor. Figure 10 shows normalized RBE-isodose contours for the case under study for tumor containing 30 ppm and blood containing 10 ppm of B-10, assuming a single-field irradiation. The bright white contour represents the planar AD contour, also defined here as the 100% contour; the thin gray contours (outside the AD contour) are the RBE-isodose contours for normal brain, while the thin gray contours (inside the AD contour) are the RBE-isodose contours for tumor. Figure 11 shows the equivalent situation for parallel-opposed irradiation. As before, the AD contour is seen to encompass almost the entire brain within this plane, while the 120% tumor RBE-isodose contour (see caption) encloses a sufficiently large region of brain to include approximately 2-cm minimum margins beyond the enhanced tumor boundary, i.e., more than fulfilling the previously defined treatment planning criteria. Within the 120% contour, tumor will receive at least 20% more RBE-dose than the

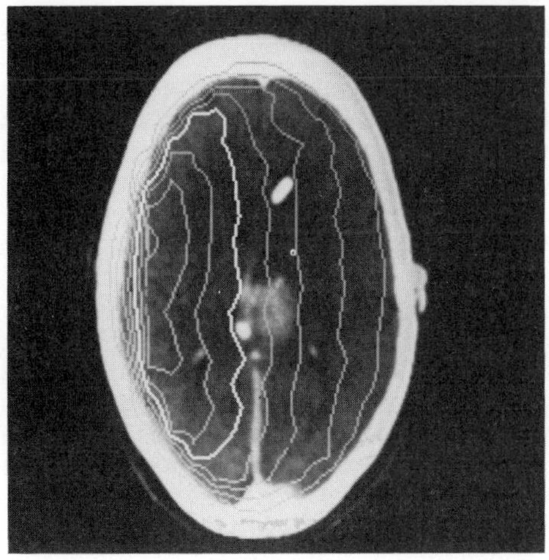 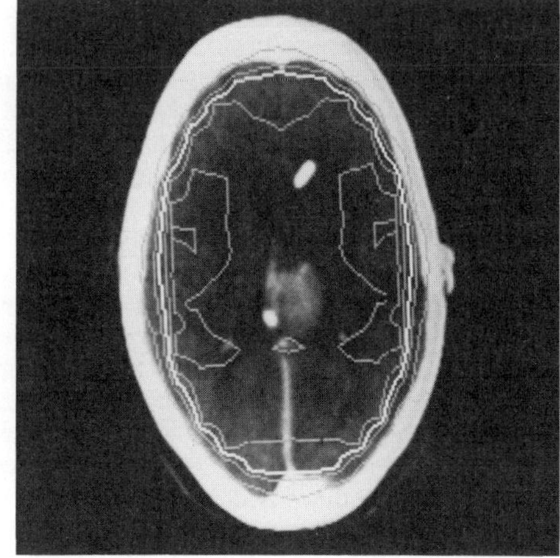

Figure 10.   Figure 11.

Figure 10. Combination display showing planar advantage-depth contour (thick white contour), normal brain RBE-isodose contours (thin gray contours outside AD contour), and tumor RBE-isodose contours (thin gray contours inside AD contour), for single-field irradiation with experimental beam incident from the left. AD contour is normalized to 100%. Normal-brain RBE-isodose contours are (right-to-left) 20%, 40%, 60%, 80%, 100% (AD contour); tumor RBE-isodose contours are (right-to-left) 100% (AD contour), 120%, 140%, and 160%. Computed using NCTPLAN.

Figure 11. Same as Figure 10 except that a parallel-opposed irradiation is assumed. Normal-brain RBE-isodose contours are highly bunched together, but from bottom-to-top at the posterior end of the skull/brain they are ordered as 40%, 80%, & 100% (AD contour); tumor RBE-isodose contours (inside AD contour) are (from brain midline laterally outwards) 140%, 160%, and 120%. Approximately 80% of the brain area is bounded by 120% tumor RBE-isodose contour, while larger left and right "lobes" and central "coin" region are the 140% tumor RBE-isodose contour. Computed using NCTPLAN.

maximum RBE-dose that would be received by any single location in normal brain. This tumor to normal tissue dose ratio is impressive when compared to conventional external beam radiation therapy, where such ratios are generally less than 1.

The two approaches to NCT treatment planning, as illustrated above, are fundamentally equivalent. However, the second approach has the advantage of allowing the radiation oncologist to examine the RBE-dose distribution within the tumor region and to also evaluate the RBE-doses to various normal tissue structures which may be at risk, such as the retina or the pituitary gland. These particular structures do not possess a blood-brain barrier and hence may contain B-10 levels substantially higher than normal brain (although most likely not exceeding the B-10 concentration in blood). We believe that, by using a combination of the treatment planning approaches illustrated above, the most complete characterization of an NCT treatment can be obtained.

An important fundamental consideration in NCT treatment planning strategy is whether treatment plans should be designed to encompass evident (CT or MRI-demonstrated) tumor boundaries with appropriate margins, or whether the decision should be made to treat the entire brain volume on the assumption that occult tumor cells may be present anywhere within the brain. Our clinical experience with high-grade astrocytomas and data on isolated tumor cells from the Mayo Clinic suggest that the latter approach might be the more judicious one, supporting our contention that an AD contour encompassing essentially the entire brain volume should be one of the treatment planning criteria in the NCT of glioblastoma multiforme. In addition to examining the dose distributions within the principal anatomical plane of the tumor, NCTPLAN allows the user to select any other plane within the brain for examination. Thus, after acceptable tumor coverage has been shown to exist within the principal plane, other anatomical planes should be examined to ensure that acceptable coverage also exists there and that sensitive structures are adequately protected.

Experimental Validation of Normalized NCT Treatment Plans

In addition to the experimental normalization of the Monte Carlo calculations as referred to earlier, it is also necessary to experimentally validate the dosimetric results obtained using NCTPLAN. Further, when an acceptable normalized treatment plan has been designed and approved by the radiation oncologist, it is necessary to convert the relative percent-isodose contours on the normalized plan to absolute RBE-dose-rate values. For example, consider the second treatment planning approach illustrated above. If the set of tumor and normal brain RBE-isodose contours shown in Figure 10 are considered acceptable, a decision must be made as to what absolute RBE-dose-rate the 100% contour is to represent. In principle, the dose-rate matrix computed by Monte Carlo simulation can be normalized to absolute neutron and gamma production rates in the reactor core, and thus to reactor thermal-power operating level. Thermal power is a parameter that is routinely and accurately monitored. Hence, one means of implementing NCT plans would be to correlate the Monte Carlo simulation results against the thermal power. However, many radiation oncologists would be uncomfortable with such an approach. Hence, even though a one-time successful experimental validation can be demonstrated, we believe that to ensure patient safety and to be certain of optimally accurate dose delivery, experimental calibration of the Monte Carlo results is desirable prior to each individual patient treatment by NCT.

At present, we are experimentally validating the Monte Carlo calculations carried out using the NPBE model. A solid neutron and gamma-equivalent brain-substitute material has been designed by our group and prototype physical skull/brain phantoms have been constructed. These physical phantoms precisely conform in geometry and elemental composition to the mathematical skull/brain model employed for the Monte Carlo simulations. A number of locations within the physical phantoms have been engineered to permit the measurement of gold activation rate using bare and cadmium-covered gold foils, and also the measurement of epithermal and fast-neutron dose rates using tissue-equivalent, graphite, and magnesium ionization chambers [12]. Using NCTPLAN, the normalized percent level of the gold activation-rate contour and epithermal and fast-neutron dose-rate contour passing through the location of the detectors are computed, thereby providing a comparison between Monte Carlo and experimental measurements on a relative basis. These comparative dosimetric data will be published elsewhere.

We are also investigating whether the relatively simple ellipsoidal phantoms employed thus far are adequate representations of actual human cranial anatomy. We have fabricated a set of three tissue-equivalent ellipsoidal skull/brain phantoms of varying sizes, and also filled an actual human skull with solid brain-equivalent epoxy resin. The ellipsoidal phantoms are intended to represent a realistic range of human skull/brain dimensions. We are performing paired-ionization chamber and gold foil measurements within the brain region of the "realistic" phantom and comparing them with similar measurements made within the most closely corresponding physical ellipsoidal phantom (based on CT scans of the former). These data will aid us in determining whether the simple ellipsoidal phantoms provide sufficiently accurate results for Monte Carlo based treatment planning and for its experimental calibration.

The calibration procedure that we propose to implement prior to the treatment of each patient is to select (on the basis of CT or MRI scans) the physical ellipsoidal phantom which most closely matches in size the patient's cranial anatomy. Similarly, the dimensions of the NPBE Monte Carlo model will be adjusted to match the specific patient's cranial dimensions. Experimental measurements of gold activation rate as well as neutron and gamma-ray dose rate will be made at representative locations within the physical phantom. These measurements will be made on a per unit response of a neutron beam monitor ionization chamber (such as a small moderated BF_3 chamber) placed within the neutron beam but away from the phantom itself. Thus, the relative iso-dose-rate contours on the patient's Monte Carlo based treatment plan will be calibrated (rationalized) to experimentally derived absolute dose rates.

SUMMARY AND CONCLUSIONS

We have developed computational techniques based on Monte Carlo simulation for performing dosimetry and treatment planning for neutron capture therapy. A detailed description of the elemental compositions and the geometrical design of the skull/brain model employed has been presented. We have discussed criteria for use in epithermal-neutron beam design and treatment planning, including: advantage depth (one- dimensional and planar), advantage ratio (one-dimensional and global), surface dose sparing characteristics, and beam intensity. Examples have been given to illustrate the influence of beam size and neutron energy on these criteria, and the results of calculations have been presented which show that the explicit inclusion of boron-10 as part of the tissue composition is unnecessary in most practical circumstances. The dosimetric characteristics of two idealized epithermal-neutron beams have been analyzed to develop realistic design goals for practical epithermal-neutron beam development. An experimental epithermal-neutron beam, which has been constructed and is undergoing experimental evaluation at the MITR-II Research Reactor Medical Therapy Facility, has been compared against the performance of idealized beams. A sophisticated multidimensional neutron capture therapy treatment planning program (NCTPLAN) has been developed which functions in conjunction with the Monte Carlo simulation program. NCTPLAN also has the capability of incorporating three-dimensional CT or MRI image data. Using this treatment planning capability, illustrative treatment plans for an actual patient with a midline thalamic glioblastoma multiforme have been developed. The experimental epithermal-neutron beam at the MITR-II has a maximum circular diameter of 20 cm and has dosimetric characteristics well suited to the treatment of brain tumors through the intact scalp and skull, even under poor conditions of tumor to blood boron distribution (30 ppm B-10 in tumor, 10 ppm B-10 in blood). Using parallel-opposed lateral brain irradiation to deliver a macroscopic RBE-dose of 2,000 RBE-cGy (cJ/kg) to blood containing 10 ppm of B-10, a treatment time of 10 minutes per field for each of six fractions is achieved, with the planar advantage depth contour enclosing almost the entire brain. Experimental validation and calibration techniques for comparing Monte Carlo simulation results with experimental mixed-field dosimetry are described.

ACKNOWLEDGMENTS

We are grateful to the graduate students who have contributed to this work. This research was supported by Grant No. DE-FG02-87ER-6060 from the U.S. Department of Energy, Office of Health and Human Assessments.

REFERENCES

1.   G. L. Brownell, R. G. Zamenhof, B. W. Murray, and G. R. Wellum, "Boron Neutron Capture Therapy," in <u>Therapy in Nuclear Medicine</u>, R. P. Spencer, ed., Grune and Stratton, Inc., New York (1978).

2.   Proc. Third Int. Symp. on Neutron Capture Therapy, <u>Strahlenther. Onkol.</u>, D. Gabel, ed., 165(2/3):5-257 (1989).

3.   G. L. Locher, "Biologic Effects and Therapeutic Possibilities of Neutrons," <u>Am. J. Roentgenol.</u>, 36:1 (1936).

4.   H. Hatanaka, "Clinical Experience of Boron-Neutron Capture Therapy for Gliomas – A Comparison with Conventional Chemo-Immuno-Radiotherapy," in <u>Boron-Neutron Capture Therapy for Tumors</u>, H. Hatanaka, ed., Nishimura Co., Ltd., Niigata, Japan, p. 349 (1986).

5.   A. H. Soloway, "Chemical Aspects of Neutron Capture Therapy," in <u>Radionuclide Applications in Neurology and Neurosurgery</u>, Y. Wang and P. Paoletti, eds., Charles Thomas, Springfield, IL (1970).

6.   R. G. Fairchild, "Development and Dosimetry of an 'Epithermal' Neutron Beam for Possible Use in Neutron Capture Therapy," <u>Phys. Med. Biol.</u>, 10(4):491 (1965).

7.   R. G. Zamenhof, B. W. Murray, G. L. Brownell, G. R. Wellum, and E. I. Tolpin, "Boron Neutron Capture Therapy for the Treatment of Cerebral Gliomas: I. Theoretical Evaluation of the Efficacy of Various Neutron Beams," <u>Med. Phys.</u>, 2(2):47 (1975).

8.   R. G. Fairchild and V. P. Bond, "Current Status of B-10 Neutron Capture Therapy: Enhancement of Tumor Dose via Beam Filtration and Dose Rate, and the Effects of these Parameters on Minimum Boron Content: A Theoretical Evaluation," <u>Int. J. Radiat. Oncol. Biol. Phys.</u>, 11(4):831 (1985).

9.   O. K. Harling, S. D. Clement, J. R. Choi, J. A. Bernard, and R. G. Zamenhof, "Neutron Beams for Neutron Capture Therapy at the MIT Research Reactor," <u>Strahlenther. Onkol.</u>, 165(2/3):90 (1989).

10.   R. G. Zamenhof, S. D. Clement, K. Lin, C. Lui, D. Ziegelmiller, and O. K. Harling, "Monte Carlo Treatment Planning and High-Resolution Alpha-Track Autoradiography for Neutron Capture Therapy," <u>Strahlenther. Onkol.</u>, 165(2/3): 188 (1989).

11.   R. G. A. Zamenhof, H. Madoc-Jones, O. K. Harling, and J. A. Bernard, Jr., "Clinical Considerations in the Use of Thermal and Epithermal Neutron Beams for Neutron Capture Therapy," in <u>Proc. 1988 Workshop on Clinical Aspects of Neutron Capture Therapy</u>, R.G. Fairchild, V. P. Bond, and A. D. Woodhead, eds., Basic Life Sciences Series, Vol. 50, Plenum Press, New York, p. 121 (1989).

12.   J. R. Choi, S. D. Clement, O. K. Harling, and R. G. Zamenhof, "Neutron Capture Therapy Beams at the MIT Research Reactor." (These Proceedings.)

13.   S. D. Clement, J. R. Choi, R. G. Zamenhof, and O. K. Harling, "Monte Carlo Methods of Neutron Beam Design for Neutron Capture Therapy at the MIT Research Reactor (MITR-II)." (These Proceedings.)

14.   G. R. Wellum, R. G. Zamenhof, and E. I. Tolpin, "Boron Neutron Capture Radiation Therapy of Cerebral Gliomas: An Analysis of the Possible Use of Boron-Loaded Tumor-Specific Antibodies for the Selective Concentration of Boron in Gliomas," <u>Int. J. Radiat. Oncol. Biol. Phys.</u>, 8(8):1339 (1983).

15.   T. Matsumoto and 0. Aizawa, "Depth-Dose Evaluations and Optimization of the Irradiation Facility for Boron Neutron Capture Therapy of Brain Tumors," <u>Phys. Med. Biol.</u>, 30(9):897 (1985).

16.   J. F. Briesmeister, ed., "MCNP – A General Monte Carlo Code for Neutron and Photon Transport, Version 3A," Los Alamos National Laboratory, <u>LA-7396-M</u>, Rev. 2 (1986).

17.   W. S. Snyder, M. R. Ford, G. G. Warner, and H. L. Fisher, Jr., "Estimates for Absorbed Fractions for Monoenergetic Photon Sources Uniformly Distributed in Various Organs of a Heterogeneous Phantom," MIRD, <u>J. Nucl. Med.</u>, Suppl. No. 3, Pamphlet 5, p. 47 (1969).

18.   B. W. Murray, O. L. Deutsch, R. G. Zamenhof, and G. L. Brownell, "New Approaches to the Dosimetry of Boron Neutron Capture Therapy at MIT-MGH," in <u>Biomedical Dosimetry</u>, IAEA, Vienna (1975).

19.	O. L. Deutsch, and B. W. Murray, "Monte Carlo Dosimetry Calculation for Boron Neutron Capture Therapy in the Treatment of Brain Tumors," <u>Nucl. Technol.</u>, 26:320 (1975).

20.	R. A. Brooks, G. DiChiro, and M. R. Keller, "Explanation of Cerebral White-Gray Contrast in Computed Tomography," <u>J. Comp. Assist. Tomog.</u>, 4(4):489 (1980).

21.	M. A. Weissberger, R. G. Zamenhof, S. Aronow, and R. M. Neer, "Computed Tomography Scanning for the Measurement of Bone Mineral in the Human Spine," <u>J. Comp. Assist. Tomog.</u>, 2:253 (1978).

22.	E. Betz, "Cerebral Blood: Its Measurement and Regulation," <u>Physiological Reviews</u>, 3:595 (1972).

23.	BNL-325, Suppl. No. 2, 6th Ed. (1988).

24.	J. H. Hubbel, "Photon Mass Attenuation and Energy Absorption Coefficients from 1 KeV to 20 MeV," <u>Int. J. Appl. Radiat. Isot.</u>, 33:1269 (1982).

25.	R. S. Caswell, J. J. Coyne, and M. L. Randolph, "KERMA Factors of Elements and Compounds for Neutron Energies Below 30 MeV," <u>Int. J. Appl. Radiat. Isot.</u>, 33:1227 (1982).

26.	A. K. Asbury, R. Ojemann, and S. L. Nielsen, "Neuropathologic Study of Fourteen Cases of Malignant Brain Tumors Treated by Boron-10 Slow Neutron Capture Therapy," <u>J. Neuropathol. Exp. Neurol.</u>, 31:278 (1972).

27.	L. E. Kun, "The Brain and Spinal Cord," in <u>Radiation Oncology: Rationale, Techniques, Results</u>, W. T. Moss and J. D. Cox, eds., 6th Ed., C. V. Mosby Co., St. Louis, MO, p. 597 (1989).

28.	K. Kitao, "Vascular Wall Dose from Boron Neutron Capture Reaction," in <u>Boron-Neutron Capture Therapy for Tumors</u>, H. Hatanaka, ed., Nishimura Co., Ltd., Niigata, Japan, p. 191 (1986).

29.	Rapporteurs' Report. (These Proceedings.)

30.	R. G. Zamenhof, W. C. Schoene, G. L. Brownell, G. R. Wellum, H. Hatanaka, A. Takeuchi, and M. Shalev, "An Investigation of the Tolerance of Canine Brain to Thermal Neutron Capture Therapy." (In Preparation.)

EPITHERMAL BEAM DEVELOPMENT AT THE BMRR:
DOSIMETRIC EVALUATION

S. K. Saraf,[1] J. Kalef-Ezra,[2] R. G. Fairchild,[1]
B. H. Laster,[1] S. Fiarman,[1] and E. Ramsey[3]

[1] Medical Department, Brookhaven National Laboratory, Upton, NY
[2] University of Ioannina, Ioannina, Greece
[3] Health Sciences Center, State University of New York, Stony Brook, NY

INTRODUCTION

The utilization of an epithermal-neutron beam for neutron capture therapy (NCT) is desirable because of the increased tissue penetration relative to a thermal-neutron beam. Over the past few years, modifications have been and continue to be made at the Brookhaven Medical Research Reactor (BMRR) to produce an optimal epithermal beam by changing filter components. An optimal incident epithermal beam should contain the minimum possible fast-neutron component and no thermal neutrons. Recently, a new moderator for the epithermal beam was installed at the epithermal port of the BMRR. With the installation of this moderator, an optimal beam has been realized [1]. This new moderator is a combination of alumina (Al_2O_3) bricks and aluminum (Al) plates. A 0.51-mm thick cadmium (Cd) sheet has reduced the thermal-neutron intensity drastically. Furthermore, an 11.5-cm thick bismuth (Bi) plate installed at the port surface has reduced the gamma-dose component to negligible levels. In order to compare various filter configurations for best optimization [2], the following parameters have been measured on the beam axis, directly in front of the epithermal port:

1) Thermal-neutron fluence rate free in air,
2) Epithermal-neutron fluence rate free in air,
3) Fast-neutron fluence rate free in air,
4) Thermal-neutron fluence rate in a polyethylene cylindrical head phantom as a function of distance along the axis of the phantom,
5) Fast-neutron dose rate in soft tissue, free in air, and
6) Gamma-dose rate in soft tissue, free in air.

Bare and cadmium-covered gold foils have been used to determine the thermal as well as the epithermal-neutron fluence. The fast-neutron fluence has been determined by activation of indium foils. Fast neutron and gamma dose in soft tissue, free in air, is being determined by the paired-ionization chamber technique, using tissue-equivalent (TE) and graphite chambers. Thermoluminescent dosimeters (TLD-700) have also been used to determine the gamma dose independently.

This paper describes the methods involved in the measurement of the above mentioned parameters. Formulations have been developed and the various corrections involved have been detailed.

FOIL ACTIVATION ANALYSIS

Quite frequently transmutations induced by neutrons produce radioactive species as

Neutron Beam Design, Development, and Performance for Neutron Capture Therapy
Edited by O. K. Harling *et al.*
Plenum Press, New York, 1990

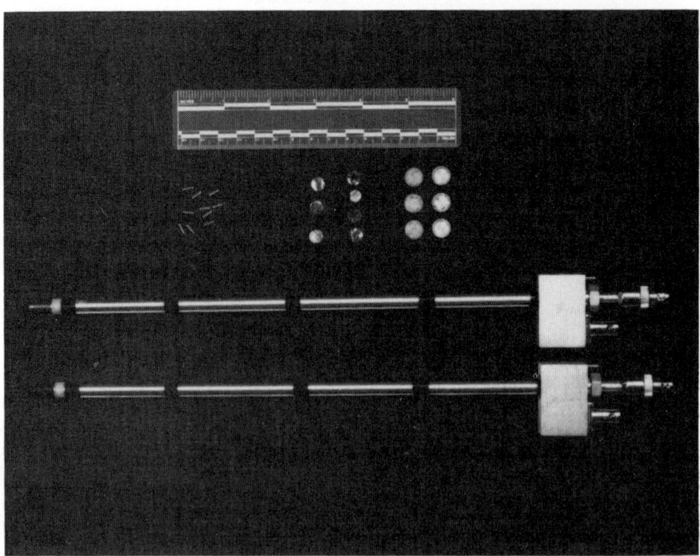

Figure 1. Dosimeters used in mixed-field radiation. Thermoluminescent dosimeters (TLD-700, 1 mm x 1 mm x 6 mm rods), gold and copper foils of 0.0127-mm thickness, and a cadmium cover of 0.76-mm thickness are shown. A Far West Technology chamber (0.1 cc) used for fast-neutron and gamma measurements is also shown.

product nuclides. The activity of the product radioisotope can often be used to measure the amount and energy distribution of the neutron flux which induced the radioactivity. A detector employing this principle is referred to as a neutron-activation detector.

The thermal, epithermal, and fast-neutron fluence rates at the BMRR epithermal port facility were determined using thin gold, copper, and indium foils. Here, the term 'thermal neutrons' (energy range from 0 to 0.4 eV) refers not only to the Maxwellian component of the spectrum, but also to the much weaker slowing-down component which ideally is proportional to l/E. Epithermal neutrons have been defined [3] as having neutron energies between 0.4 eV to 10 keV and fast neutrons from 10 keV to the end-point of the fission spectrum.

The thermal-neutron fluence rate, $\emptyset_{th}$, has been determined via measurements of the induced gamma activity in gold and copper foils. The gold foils were of thickness 0.0127 mm with an average mass of 9 mg and an average diameter of 6.8 mm as shown in Figure 1. The copper foils had an average mass of 5 mg with the same dimensions as the gold foils. Gold has a neutron absorption resonance at an energy of 4.9 eV. The presence of this resonance complicates the calculation of the thermal-neutron fluence rate ($\emptyset_{th}$). Fortunately, this difficulty can be eliminated by the cadmium difference method. A cadmium cover of thickness 0.76 mm was used to encase the gold foil, which was then irradiated at the reactor's epithermal port surface. The induced activity of the gold foil was measured with a NaI (Tl) well-type detector that had an absolute efficiency of 25%. The counting system was calibrated by absolute β and γ counting as well as with the BNL's sigma-pile standard neutron source. The thermal-neutron fluence rate ($\emptyset_{th}$) can be determined using standard relations from a knowledge of the effective microscopic neutron absorption cross section, the foil number density, and the saturated activity. The latter is given by:

$$A_{th} = A_{bare} - F_{Cd}\,A_{Cd} \qquad\qquad (1)$$

where: A_{th} is the saturated activity from the thermal flux density,
       A_{bare} is the activity measured with a bare gold foil,
       F_{Cd} is the correction factor that takes into account the absorption of resonance neutrons by cadmium, and
       A_{Cd} is the activity measured with a Cd-covered gold foil.

The correction factor, F_{Cd}, depends on a number of parameters including the thickness of the Cd cover, the thickness of the gold foil, and the angular distribution of the resonance neutrons. For the 0.76-mm Cd covers that were used with the 9-mg gold foils, the correction factor was 1.02 [3,4].

The ratio of the activity of a bare gold foil to that of the Cd-covered gold foil is defined as the cadmium ratio (CdR). This parameter is a measure of the degree of thermalization in the beam. The activities of bare and Cd-covered foils are measured at the same point on the port face. Ideally, the gold foil detector should be so thin that the flux would not be perturbed. However, practical limitations in foil thickness almost always result in some self-shielding. The method used to correct the flux depression caused by the self-shielding is:

$$CdR(0) = CdR(x) \cdot Q(x) + F_{Cd} (1 - Q(x)) \qquad (2)$$

where the parameter $Q(x)$ is equal to 0.53 for a 25-mg/cm^2 thick gold foil and $CdR(0)$ is the cadmium ratio of the detector (gold foil) corrected for the flux depression and the self-shielding effects [3]. Here 'x' represents the foil thickness in mg/cm^2.

The measured cadmium ratio has been utilized in the determination of the epithermal-neutron fluence rate. The ratio of the thermal flux to the resonance or epithermal flux in terms of the effect on the particular detector is given by:

$$\frac{\sigma_a \phi_{th}}{RI \, \phi_{epi}} = \frac{CdR}{1.02} - 1 \qquad (3)$$

where RI is the resonance integral, defined as:

$$RI = \int_{0.4 \, eV}^{\infty} \sigma(E) \frac{dE}{E} \qquad (4)$$

and ϕ_{epi} is the epithermal fluence rate per energy decade [5].

The epithermal resonance of the gold foil at 4.9 eV has been determined and the resonance integral (RI) has a value of 1558 barns. By using a thermal activation cross section (σ_a) of 98.8 barns, the epithermal-neutron fluence per energy decade was measured.

The fast-neutron fluence was measured in accordance with the In(n,n') reaction. Indium foils of thickness 0.127 mm with a mass of 98 mg/cm^2 were shielded by an enriched-^{10}B sphere to eliminate the 1.45 eV indium resonance. Counting was done using a high purity Ge detector for which the efficiency was determined using a standard ^{152}Eu source. Table One includes the various components of the measured neutron fluence rates at the epithermal port of the BMRR for a power level of 1 MW.

IONIZATION CHAMBERS

A prerequisite for biomedical studies is that energy dissipation in irradiated material be determined with a sufficient degree of accuracy and precision. This is a difficult task

Table One

Measured Parameters at the BMRR (at 1 MW power)
for the Al_2O_3/Al Filtered Epithermal Beam.

i)	Epithermal-neutron fluence rate from 0.4 eV to 10 keV	6×10^8 n/cm^2-s
ii)	Thermal-neutron fluence rate up to 0.4 eV	2.4×10^7 n/cm^2-s
iii)	Fast-neutron fluence rate for energy >1.2 MeV	3.2×10^7 n/cm^2-s
iv)	Cadmium ratio	1.025
v)	Fast-neutron dose rate for soft tissue, free in air, using TE ionization chambers	1.75 rad/min
vi)	Gamma-dose rate for soft tissue, free in air, using graphite ionization chambers and TLDs	0.4 rad/min
vii)	Peak thermal-neutron fluence rate generated inside a cylindrical head phantom filled with tissue-equivalent liquid	8.5×10^8 n/cm^2-s
viii)	Relative biological effectiveness (RBE) obtained with V-79 Chinese hamster cells	2.2

considering the complexity of dose determination due to inhomogeneities in tissue composition and density. In addition, specification of the absorbed dose does not account for the microscopic distribution of energy deposition, i.e. the radiation quality. However, this microscopic distribution must be considered because the degree of damage depends not only on the total amount of energy deposited but also on the spatial distribution of the energy deposition, i.e., on LET. Because of the differences in relative biological effectiveness of the various radiation components, it is necessary to determine the neutron absorbed dose from fast neutrons, D_N, the gamma absorbed dose, D_γ, and the absorbed dose from the $^{14}N(n,p)^{14}C$ reaction, D_p, in the mixed field separately.

The use of a calibrated A-150 plastic tissue-equivalent (TE) ionization chamber filled with TE gas is generally considered to be the most satisfactory method for determining the total absorbed dose in mixed neutron/gamma fields for biomedical applications. Ionization chambers respond to both neutrons and gammas. Thus, one of the principal problems in neutron dosimetry with chambers is the separation of the absorbed dose contributed by each of these radiations. This separation was accomplished by using a tissue-equivalent chamber with tissue-equivalent gas (methane-based) to measure the total absorbed dose and by using a neutron-insensitive graphite wall chamber filled with CO_2 gas to essentially measure the gamma component. Finally, by subtracting the gamma component from the total absorbed dose, the neutron absorbed dose was determined [6].

The fast neutron and gamma dose to tissue, free in air, for the epithermal beam at the port surface of the BMRR was determined using Far West Technology ionization chambers.* Charge was measured using a Keithley 614 electrometer.** Two different size chambers were used. These were: (a) a tissue-equivalent chamber and a graphite chamber both with a volume of 0.1 cc (shown in Figure 1) and (b) a tissue-equivalent chamber with a volume of 1 cc and a graphite chamber with a volume of 2.5 cc. The tissue-equivalent methane-based gas and CO_2 gas were obtained from Matheson.*** During the course of the measurements, the 25 cm x 25 cm port was shielded by a ^6Li metal plate to reduce the thermal-neutron component. All chambers were calibrated by Far

---

* Far West Technology Inc., Coleta, CA.
** Keithley Instruments Inc., Cleveland, OH.
*** Matheson Gas Co., Twinsburg, OH.

West Technology (FWT) and also by BNL staff in the BNL calibration facility. There was an excellent agreement between the calibrations. Calibration was performed at a fixed pressure and temperature, thus necessitating a correction for these two parameters for different pressure and temperature conditions. These will be discussed with other correction factors later.

The fundamental equation for the determination [7] of the absorbed dose rate, $\dot{D}_m$, in the wall material adjacent to the cavity of the chamber is:

$$\dot{D}_m = \frac{\dot{Q}}{M} \cdot \frac{\overline{w}}{e} \cdot s_{m,g} \tag{5}$$

where $\dot{Q}$ is the ionization current,
       M is the mass of the cavity gas,
       $\overline{w}$ is the average energy expended to produce an ion pair,
       e is the fundamental electronic charge, and
      $s_{m,g}$ is the ratio of average mass stopping power of the wall of
          the chamber (A–150 plastic) to that of the gas.

The absorbed dose rate $\dot{D}_m$ in the unknown field relative to the absorbed dose rate, $\dot{D}_m^c$, in the calibration field is:

$$\frac{\dot{D}_m}{\dot{D}_m^c} = \frac{\dot{Q}}{\dot{Q}^c} \cdot \frac{M^c}{M} \cdot \frac{\overline{w}}{\overline{w}^c} \cdot \frac{s_{m,g}}{s_{m,g}^c} \tag{6}$$

The mass ratio can be replaced by pressure and temperature because the chamber volume and gas composition are constant. At a point in the mixed (n,γ) field, where $\dot{D}_m^n$ and $\dot{D}_m^\gamma$ are the absorbed dose rates in tissue from neutrons and gammas respectively, we get:

$$\frac{\dot{Q}}{\dot{Q}^c} \frac{M^c}{M} = \frac{\overline{w}^c \, s_{m,g}^c \, \dot{D}_m^n}{\overline{w}^n \, s_{m,g}^n \, \dot{D}_m^c} + \frac{\overline{w}^c \, s_{m,g}^c \, \dot{D}_m^\gamma}{\overline{w}^\gamma \, s_{m,g}^\gamma \, \dot{D}_m^c} \tag{7}$$

where $\overline{w}^n$ and $\overline{w}^\gamma$ are the average energy expended to produce an ion pair by the neutron and the gamma components of the mixed field [8]. The quantities $s_{m,g}^n$ and $s_{m,g}^c$ are the ratio of the average stopping powers of the wall material and the gas for each of the radiation components (neutron and gamma) [9].

The calibration fields of the ^{60}Co and ^{137}Cs sources at BNL are calibrated in units of the exposure rate, $\dot{X}^c$, free in air. We can obtain the absorbed dose rate $\dot{D}_m^c$ to the wall of the chamber from the exposure rate $\dot{X}^c$, using the following relation:

$$\dot{D}_m^c = f \, \overline{\left(\mu/\rho\right)}_{energy}^{m,air} \cdot \lambda \cdot \dot{X}^c \tag{8}$$

where f is a conversion factor from exposure (roentgen) to absorbed dose (rads) and is equal to 0.873 rad/R,

$\overline{\left(\mu/\rho\right)}_{energy}^{m,air}$ is the average ratio of the mass energy absorption coefficient in the chamber wall and in the air of the calibration facility [10],

λ is a factor that takes into account the attenuation of the calibration field by the wall of the chamber, and

$$\dot{X}^c \qquad \text{is the exposure rate, free in air, of the calibration field.}$$

Now, by combining equations (7) and (8), we obtain:

$$\dot{D}_m^n = 0.873 \, \overline{\left(\mu/\rho\right)}_{energy}^{m,air} \cdot \lambda \cdot \dot{X}^c \cdot \left[\frac{\dot{Q}\,M^c}{\dot{Q}^c\,M} - \frac{\overline{w}^c}{\overline{w}^\gamma} \frac{s_{m,g}^c}{s_{m,g}^\gamma} \frac{\dot{D}_m^\gamma}{\dot{D}_m^c} \right] \cdot \frac{\overline{w}^n \; s_{m,g}^n}{\overline{w}^c \; s_{m,g}^c} \qquad (9)$$

where $\dot{Q}$ and $\dot{Q}^c$ are the measured values of the currents in the two fields.

The entrance absorbed dose rate to the tissue, free in air, due to the fast-neutron component of the mixed field $\dot{D}_{s.t.}^n$ can be calculated as shown below:

$$\dot{D}_{s.t.}^n = \lambda \cdot \frac{k_{s.t.}^n}{k_m^n} \cdot \dot{D}_m^n \qquad (10)$$

Here, the quantity $k_{s.t.}^n/k_m^n$ is the ratio of the KERMA in soft tissue and the wall material weighted by energy over the neutron spectrum.

The quantity $(\overline{w}^c \cdot s_{m,g}^c \cdot \dot{D}_m^\gamma)/(\overline{w}^\gamma \cdot s_{m,g}^\gamma \cdot \dot{D}_m^c)$ is usually written as h_T and defined as the sensitivity of the chamber to the mixed-field gamma rays relative to the sensitivity of the radiation used for the calibration. A graphite chamber filled with CO_2 gas has a lower sensitivity for neutrons and thus was used to determine the gamma component of the mixed field.

Finally, by substituting the values of the various parameters in Equation (7) for the TE and graphite chambers, we obtain:

$$\dot{Q}^{n,\gamma} = 1.035 \, \dot{D}_m^\gamma + 0.92 \, \dot{D}_m^n \qquad \text{for the TE chamber, and} \qquad (11)$$

$$\dot{Q}^{n,\gamma} = 1.036 \, \dot{D}_m^\gamma + 0.088 \, \dot{D}_m^n \qquad \text{for the Graphite chamber.} \qquad (12)$$

Here, the KERMA values over the neutron spectrum, were taken from ICRU Report 26. $\dot{Q}^{n,\gamma}$ is the total registered current converted to dose rate using the calibration constant of the chambers. These two coupled equations have been solved to give the fast-neutron and gamma-dose components of the mixed-field radiation.

Although ionization chambers are the best devices for measuring the fast-neutron and gamma components of the mixed field, results obtained from these measurements are susceptible to large errors. As has been pointed out earlier, dose determination requires accurate values of w and the KERMA factors. Calculation of fast-neutron dose also requires a precise knowledge of the neutron spectrum to avoid large errors. Other errors can be minimized by applying an appropriate correction factor.

Ionization chamber results are determined by measuring the ionizations produced in the cavity by charged particles created in the wall, central electrode, and gas. The relative contributions from the wall material and gas are dependent on the neutron energy spectrum. The wall of a chamber must be thick enough to establish secondary charged particle equilibrium. The minimum thickness is determined by the maximum range of the secondaries. However, walls also cause attenuation of the primary radiation. Under conditions of charged particle equilibrium, absorbed dose is equal to KERMA. A 5-mm thick TE plastic cup reduces the response by 10%. For the determination of KERMA, free-in-air, the readings made from wall thicknesses in excess of the minimum value required for the establishment of charged particle equilibrium are generally extrapolated to zero wall thickness and so in this case a correction factor has been applied [11].

The calibration factors supplied with the Far West Technology ionization chambers are normalized to 760 mm Hg at 22 °C. However, the calibration undertaken at the Brookhaven calibration facility was performed under pressure and temperature conditions different from those in the chambers that were used to measure dose from the mixed-field radiations. Therefore, a correction was applied as follows:

$$\text{Factor } F = \frac{273 + t}{273 + T} \cdot \frac{P}{p} \tag{13}$$

where T and P are the temperature and pressure at the time of calibration, and the quantities t and p are true respective values of the temperature and pressure at the time of the mixed-field measurement.

Other factors which were incorporated in the final evaluation of the dose to soft tissue, free in air, were:

    i)   saturation effects,
    ii)   polarity effects,
    iii)   gas flow dependence,
    iv)   offset and leakage current,
    v)   influence of the ^{6}Li metal cover, and
    vi)   stem scattering.

The fast-neutron dose and gamma dose for soft tissue, free in air, for the epithermal beam at the BMRR was measured using FWT ionization chambers. The results, after incorporating the appropriate correction factors, are shown in Table One.

THERMOLUMINESCENT DOSIMETERS

Thermoluminescent devices have been used extensively for the dosimetry of X, γ, and β-rays. In neutron dosimetry, the main use of thermoluminescent dosimeters (TLD) has been for the measurement of the associated gamma rays. Thermoluminescence (TL) is a phenomenon exhibited by a variety of organic and inorganic materials. The latent luminescence induced by the ionizing radiation (basically by trapped electrons in host material) is released at high temperature. The proportionality between the released thermoluminescent light and the absorbed dose is the basis for this dosimetric technique.

The incident gamma-ray component of the total dose rate in soft tissue, free in air, at the epithermal port of the BMRR was determined by using LiF-TLD rods (TLD-700), obtained from the Harshaw Chemical Corp.† The dimensions of these rods were 1 mm x 1 mm x 6 mm, as shown in Figure 1. During reactor irradiations, the TLD dosimeters were placed in a plastic bottle containing ^{6}Li-enriched (~95.6% ^{6}Li) LiF powder and were positioned on the central axis of a cylindrical teflon base, 14-mm high x 7.2 mm in diameter. In all cases, the plastic bottle was placed on its horizontal axis of symmetry, thus assuring that the minimum thickness of shielding for the TLD was 0.5 g/cm^2 of ^{6}LiF, which is equivalent to a 2.1-mm thickness of ^{6}Li metal. In order to verify that the TLDs were well protected from the thermal neutrons, gold wires were inserted between the teflon base and the plastic inner bottle. The induced activity of the gold wires corresponded to 10^6 thermal neutrons/cm^2-s at 1 MW. It has been determined for this batch of TLDs that a fluence of 10^{10} thermal neutrons/cm^2 at 1 MW would induce a signal equivalent to ≈ 1.2 rads in soft tissue. Therefore, it was concluded that the TLD rods were adequately protected from the thermal-neutron beam, and that they produced negligible signal. In a similar fashion, it was assumed that the contribution to the TL signal from high-energy

---

† Harshaw Chemical Corp., Solon, OH.

neutrons was insignificant because of their inability to induce a TL response. (This assumption is currently under question, because of inadequate knowledge of the neutron spectrum.)

Standard heat treatment was applied to the TLD-700. The one-hour pre-irradiation annealing at 400 °C in a stainless-steel base was followed by a slow cooling to room temperature (15-min furnace-cool, followed by a 30-min air-cool). The post-irradiation annealing (100 °C in air for 15 min) in a teflon base was followed with a 45-min air cooling. This post-irradiation treatment helps in eliminating peaks 2 and 3 from the spectrum of the glow curve of the LiF-700 dosimeters. The emitted TL signal (integrated between room temperature and 350 °C) was detected using Harshaw's model 2000 A and B and also Harshaw's model 4000 thermoluminescence reader in 30 seconds. The glow curve of each dosimeter was recorded on paper as well as on computer mass media storage. The high temperature limit of 350 °C was selected to reduce the black body radiation contribution. There are several problems that complicate the interpretation of the glow curve results. For example, after irradiation with neutrons or gammas, the glow curves sometimes have different shapes. This is one of the reasons that the entire glow curve was integrated to ensure proper interpretation of the results.

The gamma-dose rate component $\dot{D}_\gamma^{s.t.}$ in an elemental mass of soft tissue, free in air, was determined using the following relation:

$$\dot{D}_\gamma^{s.t.} = \overline{(\mu/\rho)}_{energy}^{s.t./tef.} \cdot \lambda \cdot C \cdot \dot{D}_{Co}^{LiF} \tag{14}$$

where $\overline{(\mu/\rho)}_{energy}^{s.t./tef.}$ is the ratio of mass energy absorption coefficients of soft tissue and Teflon, λ is the linear attenuation coefficient for gamma rays in Teflon, C is the cavity-theory correction for gamma rays, and $\dot{D}_{Co}^{LiF}$ is the dose rate in the TLDs located in the Teflon base and irradiated with standard calibrated ^{60}Co and ^{137}Cs sources. Teflon promotes the production of charged particle equilibrium and provides a base for placement of the TLD rods on the reactor port. Teflon was used rather than hydrogenous materials because it does not produce any contaminating radiation. Calculations were made assuming that the gamma component from the reactor has an energy of 2.2 MeV. At present, we have inadequate information on the gamma spectrum.

At least eight dosimeters were used to determine the background (electronic noise of the instrument, black body radiation, etc.) components during each run. Cobalt-60 and cesium-137 sources from the BNL calibration facility were used to calibrate TLDs for each run. The same Teflon block and cylinders were used throughout the calibration. Care was exercised in converting the exposure rate, free in air, to the dose rate in the dosimeter material that was placed in the Teflon block. The gamma-dose rate in soft tissue, free in air, was determined by comparing the total light output to that of the calibrated TLDs and then converting the dose rate to soft tissue.

The absorbed dose from the ^{14}N(n,p)^{14}C reaction was calculated as follows:

$$^{14}N_D = 1.602 \times 10^{-8}\, EFN\sigma\Phi \tag{15}$$

where $^{14}N_D$ is the absorbed dose in rads due to nitrogen,

$^{14}N_D$	is the absorbed dose in rads due to nitrogen,
1.602×10^{-8}	is the conversion factor, MeV to 100 ergs,
E	is the energy released, in MeV per event,
F	is the fraction by weight of the nitrogen in tissue,
N	is the number of atoms per gram of tissue of the nitrogen,
σ	is microscopic absorption cross section in $cm^2 \cdot$ event/atom, and
Φ	is the thermal-neutron fluence in neutrons/cm^2.

314

For the ^{14}N(n,p)^{14}C reaction, the energy released per event E is 0.585 MeV and the microscopic absorption cross section σ is 1.85 x 10^{-24} cm^2. Taking the nitrogen content in brain tissue by weight fraction to be 1.84%, the nitrogen-thermal neutron dose has been calculated using the following equation:

$$^{14}N_D = 0.745 \times 10^{-9} \, F\Phi \quad \text{in rads.} \tag{16}$$

In a similar fashion, the ^{10}B(n,α)^{7}Li dose has been calculated at various depths in the head phantom by measuring the thermal-neutron fluence rate using gold and copper foils. The ^{10}B dose to the tissue is, assuming uniform distribution of boron over the tissue volume, given by:

$$^{10}B_D = 1.602 \times 10^{-8} \, EFN\sigma\Phi. \tag{17}$$

For the ^{10}B reaction, the energy released per event is 2.34 MeV and the microscopic absorption cross section, σ, used in calculating the dose has the value 3838 x 10^{-24} cm^2. Finally, the equation used to calculate the dose is:

$$^{10}B_D = 8.66 \times 10^{-6} \, F\Phi \quad \text{in rads.} \tag{18}$$

V-79 Chinese hamster cells were irradiated in suspension as detailed elsewhere [12]. Neutron irradiations were carried out at the epithermal port of the BMRR at a distance of 1.2 cm from the port face. X-irradiation was carried out at the 250-kVp GE Maxitron 250, 60 cm from the source, using 0.5-mm copper and 1.0-mm aluminum filters. Following the irradiation, cells were plated for survival assay. The ratio of the D_0 values obtained from the linear portion of each survival curve demonstrated an RBE of ~2.2.

ACKNOWLEDGMENTS

This research was supported in part by the U.S. Department of Energy under Contract No. DE-AC02-76CH00016. Accordingly, the U.S. Government retains a nonexclusive, royalty-free license to publish or reproduce the published form of this contribution, or allow others to do so, for U.S. Government purposes.

REFERENCES

1.   R. G. Fairchild, J. Kalef-Ezra, S. K. Saraf, S. Fiarman, E. Ramsey, L. Wielopolski, B. H. Laster, and F. J. Wheeler, "Installation and Testing of an Optimized Epithermal Neutron Beam at the Brookhaven Medical Research Reactor (BMRR)." (These Proceedings.)

2.   R. G. Fairchild, J. A. Kalef-Ezra, S. Fiarman, L. Wielopolski, J. Hanz, S. Mussolino, and F. Wheeler, "Optimization of an Epithermal Neutron Beam for NCT at the Brookhaven Medical Research Reactor (BMRR)," <u>Strahlenther. Onkol.</u>, 165(2/3):84 (1989).

3.   R. G. Fairchild, "Development and Dosimetry of an 'Epithermal' Neutron Beam for Possible Use in Neutron Capture Therapy," <u>Phys. Med. Biol.</u>, 10(4):491 (1965).

4.   S. Pearlstein and E. V. Weinstock, "Scattering and Self-Shielding Corrections in Cadmium-Filtered Gold, Indium, and 1/v Foil Activation Measurements," <u>Nucl. Sci. Eng.</u>, 29(1):28 (1967).

5.   D. J. Hughes, in <u>Pile Neutron Research</u>, Addison-Wesley, Cambridge, MA (1953).

6.   H. H. Rossi and G. Failla, "Tissue-Equivalent Ionization Chambers," <u>Nucleonics</u>, 14(2):32 (1956).

7.   "Neutron Dosimetry for Biology and Medicine," International Commission on Radiation Units and Measurements, Bethesda, MD, <u>ICRU Report 26</u> (1977).

8.   "Average Energy Required to Produce an Ion Pair," International Commission on Radiation Units and Measurements, Bethesda, MD, <u>ICRU Report 31</u> (1979).

9.   "Stopping Powers for Electrons and Positrons," International Commission on Radiation Units and Measurements, Bethesda, MD, <u>ICRU Report 37</u> (1984).

10.   J. H. Hubell, "Photon Mass Attenuation and Energy-Absorption Coefficients from 1 keV to 20 MeV," <u>Int. J. Appl. Radiat. Isot.</u>, 33:1269 (1982).

11.   J. J. Broerse and J. Zoetelief, <u>Advances in Dosimetry for Fast Neutrons and Heavy Charged Particles for Therapy Applications</u>, IAEA (1984).

12.   D. Gabel, R. G. Fairchild, H. G. Borner, and B. Larsson, <u>Radiat. Res.</u>, 98:307 (1984).

A BEAM-MODIFICATION ASSEMBLY FOR EXPERIMENTAL NEUTRON CAPTURE THERAPY OF BRAIN TUMORS

D. N. Slatkin,[1] J. A. Kalef-Ezra,[2] S. K. Saraf,[1] and D. D. Joel[1]

[1] Medical Department, Brookhaven National Laboratory
Upton, NY
[2] Laboratory of Medical Physics, School of Medicine
University of Ioannina, Ioannina, Greece

Recent attempts to treat intracerebral rat gliomas by boron neutron capture therapy (BNCT) have been somewhat disappointing [1,2], perhaps in part because of excessive whole-body and nasopharyngeal irradiation. Intracerebral rat gliomas were treated by BNCT with more success using a new beam-modification assembly (Figure 1). Rats were infused with the sulfhydryl borane dimer $Na_4{}^{10}B_{24}H_{22}S_2$ intraperitoneally before irradiation at the rate of ~2 mg ^{10}B per kg body weight per hour for 72 hours [3]. Boron-10 concentrations measured by the neutron-induced prompt-gamma technique [4] in 0.3 ml aliquots of blood sampled from each rat several minutes after the end of infusion averaged 35 µg per gram (range 24-42 µg/g). Boron-10 concentrations in similar rat brain tumors after 72 hours of identical infusion were twenty-five percent lower than blood ^{10}B concentrations, on the average.

An 18 mm-thick, 25.3 cm x 25.3 cm-square ^6LiF powder/epoxy plastic (1:1 v/v) collimator was molded with a central 11.5 mm diameter therapy aperture at the apex of a truncated conical (30° half-angle) neutron-restrictive air column. The collimator was apposed to the 25.7 cm x 25.7-cm square, thickly coated* bismuth plate at the face of the thermal neutron port of the Brookhaven Medical Research Reactor. For thermal-neutron irradiation of the head in the brain tumor area, rats were anesthetized and placed supine in a 5.0 cm-diameter, 9.9 cm-long cylindrical air space at atmospheric pressure and ambient temperature. This space is partially shielded from adventitious radiations by a 5.7 cm inner-diameter, 7.5 cm outer-diameter, 9 mm-thick cylinder of ^6LiF powder that fills an ~3.5 mm-thick-walled, cylindrical, water-tight plastic case, the inner wall of which is TFE-Teflon and the outer wall of which is transparent Lexan. This cylindrical assembly is surrounded by a water-filled, 25.2 cm-wide x 25.5 cm-high x 9.9 cm-deep box, the walls of which are made of transparent, 1.7 cm-thick Lexan. Rat heads were cradled in the adjustable TFE-Teflon gantry (Figure 1) to center the tumor zone at the 11.5 mm-diameter therapy aperture.

Thermal neutron fluences measured with 0.25 mm-diameter,** 3 mm-long segments of gold wire positioned in the cerebrum antero-posteriorly using a surgical technique, at the same depths as are the centers of similarly implanted rat gliomas (about 5 mm beneath the skin), were ~2 x 10^{12} cm^{-2}. Abdominal, thoracic, and cervical doses from

---

*   Krylon Acrylic Spray Coating, No. 1303, Borden, Inc., Columbus, Ohio 43215
**   Johnson Matthey, Danvers, Massachusetts 01923

Neutron Beam Design, Development, and Performance for Neutron Capture Therapy
Edited by O. K. Harling *et al.*
Plenum Press, New York, 1990

Fig. 1. Anesthetized rat held supine in the body radiation shield of the beam-modification assembly. The gantry-supported head is photographed through a Lucite plate that is a surrogate for the neutron collimator to depict the orientation of the rat head at the therapy aperture. The position of the circular aperture and several horizontal fiducial lines are marked on the plate. Similar lines on the surface of the plastic collimator facilitate similar placement of the head at the reactor.

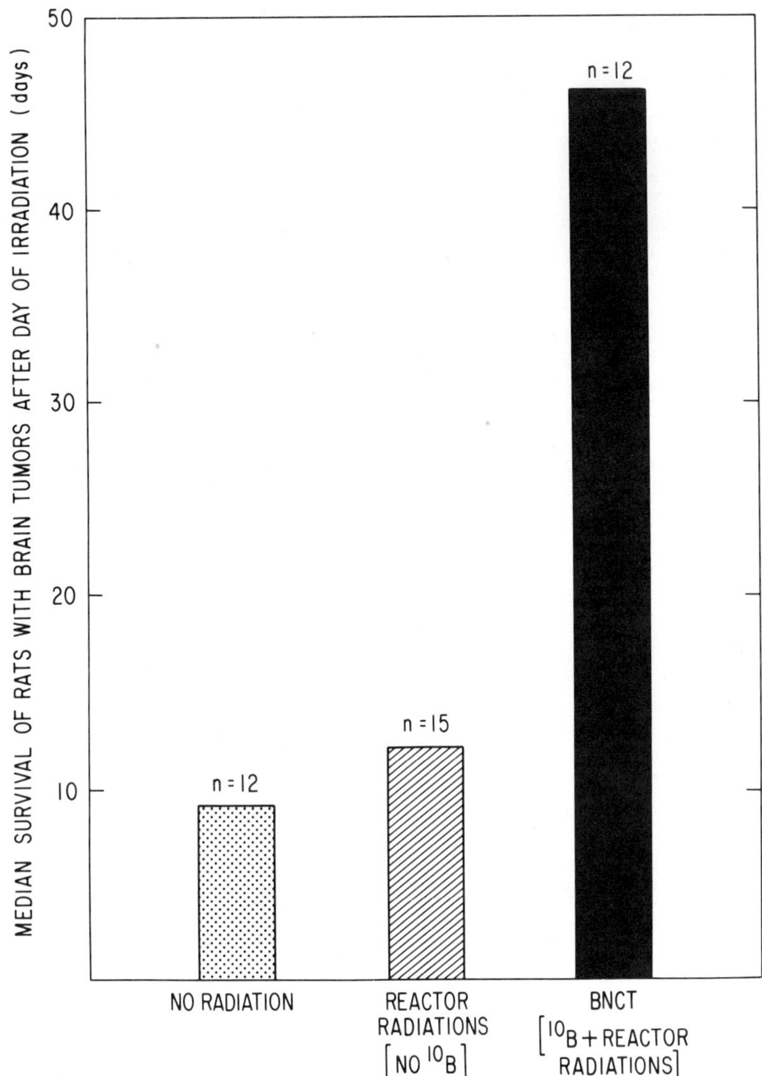

Fig. 2.   Malignant, non-metastasizing gliomas were initiated in the left frontal lobes of the cerebra of twenty-seven rats 14 days before the day of irradiation (middle and right bars). Similar gliomas were initiated on the same day in a group of twelve untreated rats (left bar). When rats either died or became moribund with a demonstrable life expectancy of less than two days, necropsy revealed progressive growth of tumor around the site of initiation of the tumor to be the sole cause of lethality. Spread of tumor via cerebrospinal fluid to leptomeninges or to the ventricles was not observed.

fast neutrons and gamma photons were in the ranges ~0.5-1.0 Gy and ~0.4-0.5 Gy, respectively. The results of this study, as indicated by median post-irradiation survivals, are shown in Figure 2. The range of the post-irradiation survival for the reactor-irradiation-only group (15 rats) was 7 to 54 days (all but one, 7 to 25 days), and the corresponding range for the BNCT group (12 rats) was 14 to 374 days*** (all but two, 14 to 78 days). The range of survival of the untreated group (12 rats) after the day of irradiation of the other groups (i.e., after the fourteenth day following tumor initiation) was 6 to 66 days (all but one, 6 to 21 days). According to the two-tailed, two-sample (non-parametric) Wilcoxon test [5], BNCT prolonged post-irradiation survival longer than did the treatment by reactor radiations alone at a probability level of P << 0.01. This study shows that important parameters of BNCT for brain tumors can be evaluated in small animals in vivo. However, the feasibility of doing so is particularly dependent on a suitable configuration for each component of the body radiation shielding and of the beam-modification assembly and on appropriate nuclide compositions of these components. [Editors' Note: This paper contains data obtained since the original presentation at the Workshop.]

ACKNOWLEDGMENTS

This research was supported by the U.S. Department of Energy under Contract No. DE-AC02-76CH00016. Accordingly, the U.S. Government retains a nonexclusive, royalty-free license to publish or reproduce the published form of this contribution, or allow others to do so, for U.S. Government purposes. The research described in this report involved animals maintained in animal care facilities fully accredited by the American Association for Accreditation of Laboratory Animal Care.

REFERENCES

1.   D. N. Slatkin, D. D. Joel, R. G. Fairchild, P. L. Micca, M. M. Nawrocky, B. H. Laster, J. A. Coderre, G. C. Finkel, C. E. Poletti, and W. H. Sweet, "Distributions of Sulfhydryl Borane Monomer and Dimer in Rodents and Monomer in Humans: Boron Neutron Capture Therapy of Melanoma and Glioma in Boronated Rodents," in Proc. 1988 Workshop on Clinical Aspects of Neutron Capture Therapy, R. G. Fairchild, V. P. Bond, and A. D. Woodhead, eds., Basic Life Sciences Series, Vol. 50, Plenum Press, New York, p. 179 (1989).

2.   N. R. Clendenon, R. F. Barth, J. H. Goodman, A. E. Staubus, W. A. Gordon, M. L. Moeschberger, F. Alam, A. H. Soloway, R. G. Fairchild, D. N. Slatkin, and J. A. Kalef-Ezra. "Enhanced Survival in a Rat Glioma Model Following BNCT," Strahlenther. Onkol., 165(2/3):222 (1989) .

3.   D. Joel, D. Slatkin, R. Fairchild, P. Micca, and M. Nawrocky, "Pharmacokinetics and Tissue Distribution of the Sulfhydryl Boranes (Monomer and Dimer) in Glioma-Bearing Rats," Strahlenther. Onkol., 165(2/3):167 (1989) .

4.   R. G. Fairchild, D. Gabel, B. H. Laster, D. Greenberg, W. Kiszenick, and P. L. Micca, "Microanalytical Techniques for Boron Analysis Using the $^{10}B(n,\alpha)^7Li$ Reaction," Med. Phys., 13:50 (1986) .

5.   P. G. Marshall, M. E. Miller, S. Grand, P. L. Micca, and D. N. Slatkin, "Toxicities of $Na_2B_{12}H_{11}SH$ and $Na_4B_{24}H_{22}S_2$ in Mice," in Proc. 1988 Workshop on Clinical Aspects of Neutron Capture Therapy, R. G. Fairchild, V. P. Bond, and A. D. Woodhead, eds., Basic Life Sciences Series, Vol. 50, Plenum Press, New York, p. 333 (1989).

*** Two of the twelve animals in this group are alive and symptom-free 374 days after BNCT, as of December 11, 1989.

BIOMEDICAL IRRADIATION SYSTEM FOR BORON NEUTRON CAPTURE THERAPY AT THE KYOTO UNIVERSITY REACTOR

T. Kobayashi, K. Kanda, Y. Ujeno, and M. R. Ishida

Kyoto University Research Reactor Institute
Osaka, Japan

ABSTRACT

Physics studies related to radiation source, spectroscopy, beam quality, dosimetry, and biomedical applications using the Kyoto University Reactor Heavy Water Facility are described. Also, described are a Nickel Mirror Neutron Guide Tube and a Super Mirror Neutron Guide Tube that are used both for the measurement of boron concentration in phantom and living tissue and for precise measurements of neutron flux in phantom in the presence of both light and heavy water. Discussed are: (1) spectrum measurements using the time of flight technique, (2) the elimination of gamma rays and fast neutrons from a thermal neutron irradiation field, (3) neutron collimation without producing secondary gamma rays, (4) precise neutron flux measurements, dose estimation, and the measurement of boron concentration in tumor and its periphery using guide tubes, (5) the dose estimation of boron-10 for the first melanoma patient, and (6) special-purpose biological irradiation equipment. Other related subjects are also described.

INTRODUCTION

Physics studies concerning both neutron fields and dosimetry for boron neutron capture therapy (BNCT) have been performed at the Kyoto University Research Reactor Institute (KURRI). The Institute's facilities have been made available to many biological and medical researchers both inside and outside KURRI.

The Kyoto University Reactor (KUR) is a tank-type reactor that operates at 5MW. It has a heavy water thermal neutron facility, which was originally designed for biomedical irradiations and the production of a standard field of thermal neutrons [1]. By using a bismuth scatterer and a LiF collimator, a low gamma-ray neutron field has been realized [2]. For the convenience of biological experimenters, a conventional irradiation tube was installed for use during continuous KUR operation [3]. A sophisticated irradiation box, which controls temperature, atmosphere, and the gamma-ray dose contamination rate, has been constructed for biological experiments at the cell level. Also, the heavy water facility can be used for clinical purposes, but because use of the facility must be scheduled six months in advance, it is generally limited to the treatment of critically-ill patients.

In addition to the heavy water facility described above, a Nickel Mirror Neutron Guide Tube or NMNGT [4] is used for basic reactor physics studies such as the precise determination of neutron flux distributions both in phantoms [5,6,7] and human bodies, as well as the measurement of boron concentrations in them using prompt gamma rays [8]. The availability of the NMNGT greatly facilitated the estimation of boron concentrations and dose for the first melanoma patient [9].

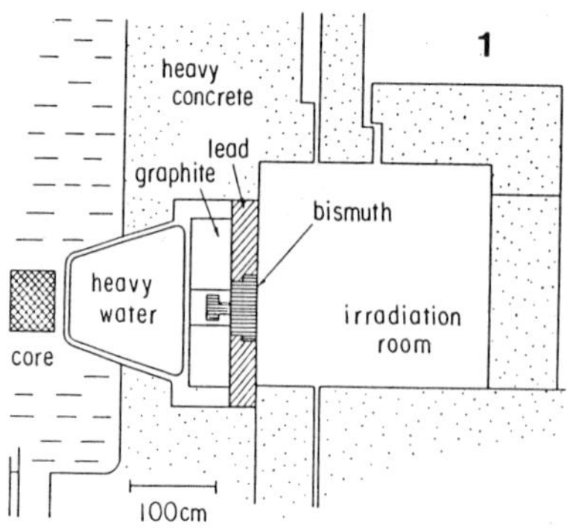

Figure 1. Heavy Water Facility of the KUR.

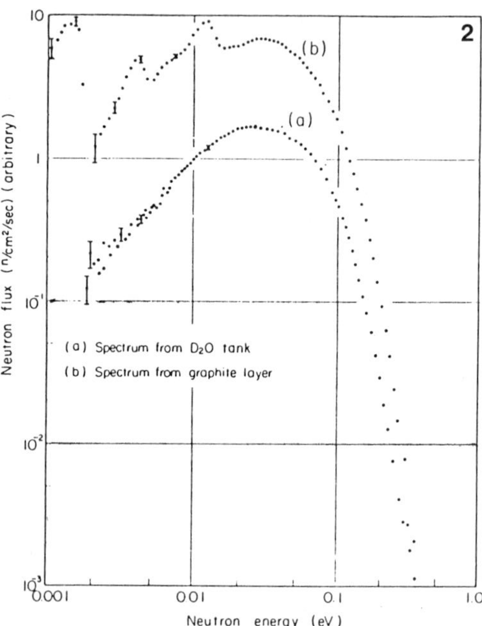

Figure 2. Neutron Spectra from Heavy Water and Graphite Layers.

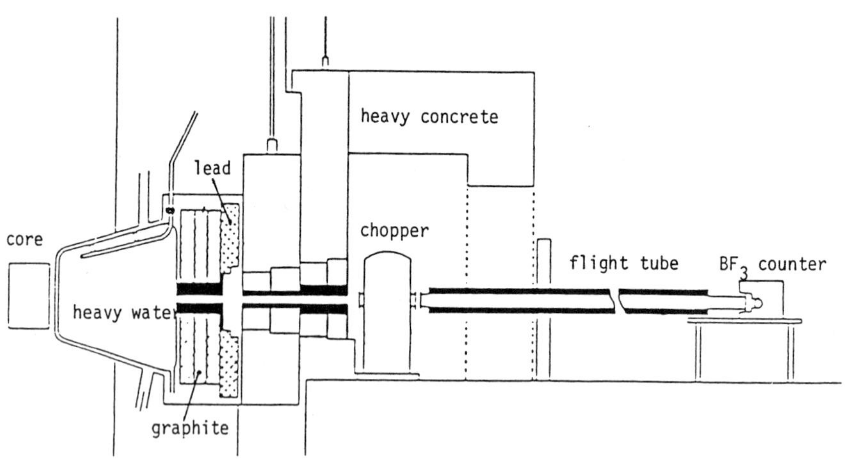

Figure 3. Experimental Arrangement of the Time-of-Flight Apparatus.

Another guide tube attached to the KUR is the Super Mirror Neutron Guide Tube or SMNGT. This is a new facility with performance characteristics superior to the NMNGT [10]. The SMNGT's characteristics have recently been measured and it has now been opened to general researchers including those in biomedicine.

KUR HEAVY WATER FACILITY

The KUR's Heavy Water Facility produces a pure thermal neutron field. This facility and its measured spectrum are shown in Figures 1 and 2 respectively. The facility was approved as a means of generating a standard thermal neutron field by the International Atomic Energy Agency (IAEA) in 1977. It is the only facility in the world to have been so designated. The neutron spectrum from the heavy water tank was measured using the time-of-flight technique. The cognizant apparatus is shown in Figure 3.

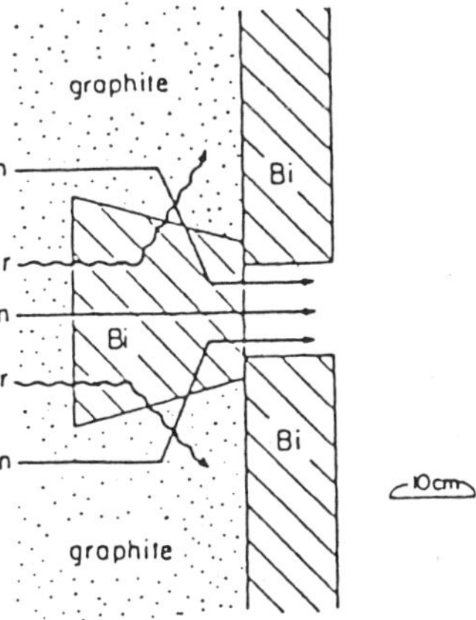

Figure 4.   Basic Principle of the Bismuth Scatterer Method.

ELIMINATION OF GAMMA RAYS

Gamma-ray contamination was reduced by the bismuth scatterer technique. The basic principle of the bismuth scatterer method is illustrated schematically in Figure 4. The method is based on the fact that the diffusion length of thermal neutrons and the mean free path of gamma rays are different in moderator and shielding materials. This scatterer method entails the introduction of a piece of bismuth of sufficient thickness to reduce the gamma-ray dose. The neutron fluence distribution within the bismuth is largely dictated by that in the graphite. Hence, the neutron fluence is reduced only slightly through the introduction of a small piece of bismuth within the graphite. Bismuth is expensive, but the total volume used in this device is not large. Thus, a neutron field with a low gamma-ray background is achieved without incurring a large reduction in the neutron fluence rate. For the initial standard design, the KUR heavy water thermal-neutron facility contains two tons of heavy water, a 50 cm graphite layer, and a 15 cm bismuth layer.

Experiments were performed on this heavy water thermal-neutron facility for various scattering geometries. The distribution of the neutron fluence rate and the gamma-ray dose rate in relation to the bismuth scatterer were determined experimentally.

LITHIUM FLUORIDE NEUTRON SHIELD

The usual shielding materials for thermal neutrons are cadmium metal and boron composites, such as borated polyethylene. However, gamma rays are produced from these materials as the result of neutron capture reactions. To achieve a thermal neutron field with a low gamma-ray component, Li-6 is the most useful material because it has an adequate absorption cross section and the gamma-ray production in Li-6 by thermal neutron capture is almost zero. Among the possible lithium compounds, lithium-fluoride (LiF) was selected because of its handling stability and high lithium density. The LiF shielding materials described below were manufactured [11].

Tile: The LiF tiles used in the experiments were of two standard sizes, 5 x 5 x 1 cm and 10 x 10 x 1 cm. These are now commercially manufactured.[†] The advantage of a LiF tile over one made of boron or cadmium for reducing the gamma-to-neutron ratio is shown in Figure 5. Note that if LiF powder were to be sheathed in an aluminum or stainless steel clad, the net effect would not be advantageous because a considerable number of gamma rays would be produced from neutron capture in the clad materials.

Flexible Sheet: For the fine adjustment of the neutron collimator, we developed a flexible sheet containing LiF. After several trials, we succeeded in producing the desired material in sheet form by using ethylene/vinylacetate copolymer. The manufacturing

---

† Nippon Kaghaku Togyo Co. Ltd., c/o Wako Shoken Bldg., 6-22 Kitahama 2-chome, Chuo-ku, Osaka 541, Japan, Phone: 81-6-202-4686, Fax: 81-6-222-6214.

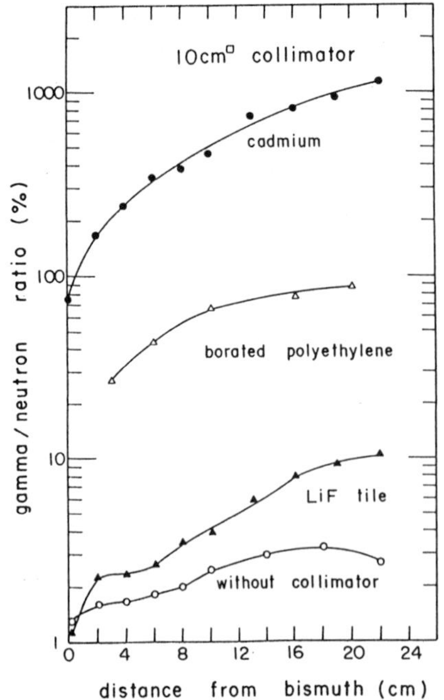

Figure 5. Gamma-to-Neutron Ratio for
Different Collimating Materials.

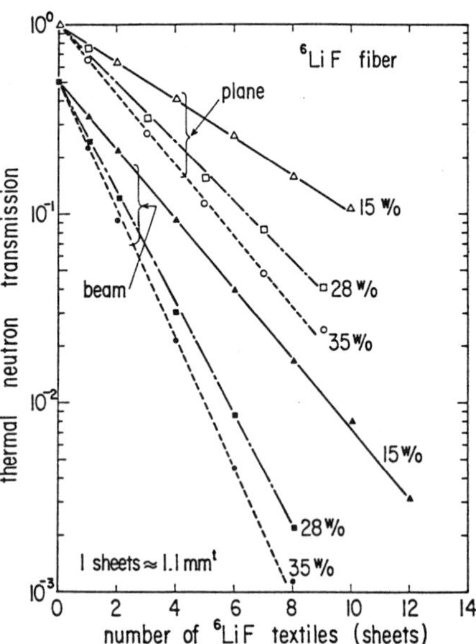

Figure 6. Transmission of Thermal
Neutrons Measured for LiF.

process is described in [11]. The standard size sheets are 30 x 30 x 0.5 cm and 60 x 90 x 0.5 cm. These sheets can be cut by a knife. When the sheets have been irradiated for a long time, they become discolored and lose both flexibility and adhesion.

Textile: In treating patients we need a highly flexible and sterile neutron shielding material in the form of a gauze. After several trials, we succeeded in making a fiber containing LiF by the melt spinning method using polyethylene. The characteristics of this material are listed in Table One. The fiber can be woven into various textiles, including woven or knitted fabrics, which can in turn be used as gauze, clothing, or curtains. The transmission of thermal neutrons measured for LiF containing textile is shown in Figure 6.

Other: Other useful materials, such as Teflon and thermo-plastics, were also manufactured.

Besides LiF, materials containing B_4C and BN were also developed. Boron has the disadvantage of a substantial gamma-ray yield. But it has the advantages of a large neutron absorption cross section and low cost. Both lithium and boron shielding materials are used depending on the objectives at hand. The shielding capability for thermal neutrons of the above developed materials was measured by the transmission method using the KUR Heavy Water Facility. The results are given in Table Two.

NEUTRON GUIDE TUBES

Neutron guide tubes attached to the KUR are listed in Table Three and their neutron spectra are shown in Figure 7 [4,10]. The NMNGT facility is frequently used not only for boron-10 analysis in tumor and tissue, but also for basic physics studies related to BNCT.

Table One

Characteristics of Neutron Shielding Fibers

	LiF	B_4C	BN
Contents (w/o)	40	40	30
Diameter (μm)	29	30	33
Density (g/cm^3)	1.23	1.27	1.16
Denier (d†)	5.4	6.4	8.0
Tensile Strength (g/d)	2.1	2.5	1.5

† One denier (d) is the fineness of a yarn having one gram for each 9000 meters.

Table Two

Thermal Neutron Shielding Ability for Various Materials [11]

	Thickness (mm)	Density (g/cc)	Thermal Neutron Transmission
LiF Tile			
Natural	10	2.51	2.51×10^{-2}
Enriched	2	2.41	1.52×10^{-3}
Sheet			
40 w/o ^{6}LiF	5	1.36	7.30×10^{-3}
40 w/o B_4C	5	1.36	2.69×10^{-4}
Textile			
40 w/o ^{6}LiF	13.0^a	*	1.45×10^{-2}
40 w/o B_4C	12.5^b	*	1.25×10^{-2}
30 w/o BN	11.7^c	*	3.40×10^{-2}

[a] 0.49 g/cm^2,　[b] 0.43 g/cm^2,　[c] 0.47 g/cm^2,　* depends on weaving

Table Three

Neutron Guide Tubes at KUR [10]

Name	Beam Hole at KUR	Characteristic Wavelength (Å)	Neutron Mirror	Length of NGT (m)	Beam Size (cm)	Flux (n/cm^2-s)	Year
NMNGT	E-3	2.85	Nickel	10.8	1x7.4	2.5 E6	1973
SMNGT	B-4	1.17	Super Mirror	11.7	1x7.4	5.5 E7	1984

FINE NEUTRON FLUX MEASUREMENT

By using the NMNGT facility, fine neutron distributions in a polyethylene phantom 15 cm in diameter and in a water phantom 5 cm in diameter were measured. These measured data were then compared with theoretical calculations [5,7]. The experimental arrangement for the polyethylene phantom is shown in Figure 8. A comparison between the experimental and calculated results is shown in Figure 9. The latter were performed using DOT 3.5. Figure 10 compares the experimental result for the water phantom of 5 cm in diameter with TWOTRAN calculations, in which non-scattered neutrons were separately calculated as shown in Figure 11. This technique is very effective as a means of calculating neutron flux distributions in a heavy water phantom. An experiment was performed to evaluate the effectiveness of using heavy water for deep tumor treatment. One of the experimental and calculational results is shown in Figure 12 [12].

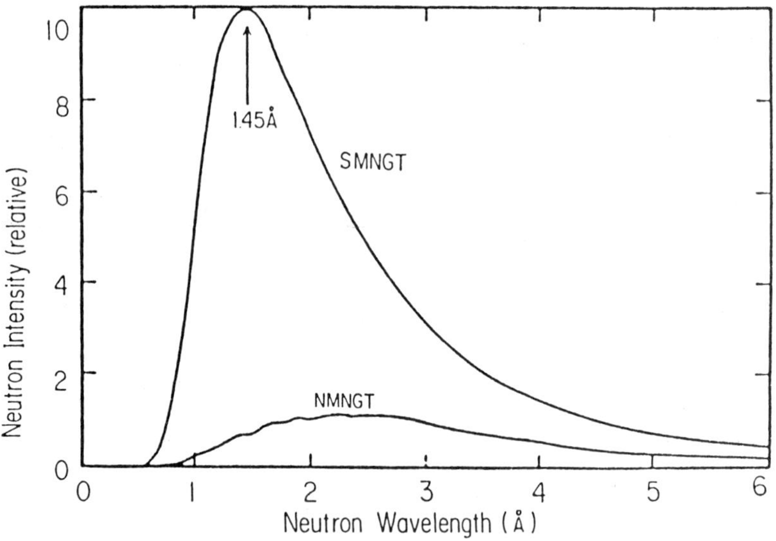

Figure 7.   Calculated Neutron Spectra of the SMNGT and the NMNGT.

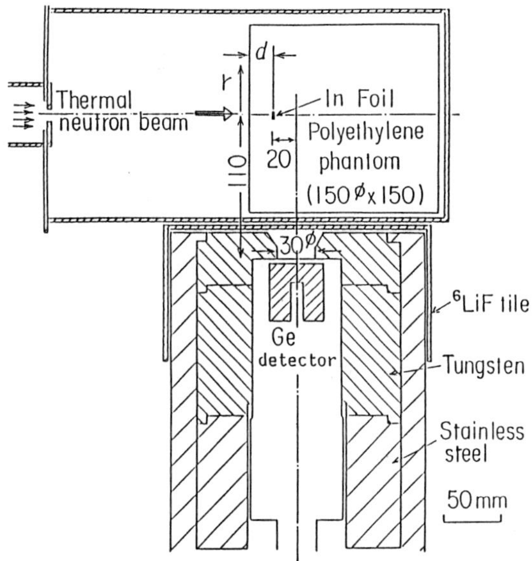

Figure 8. Experimental Arrangement for Polyethylene Phantom.

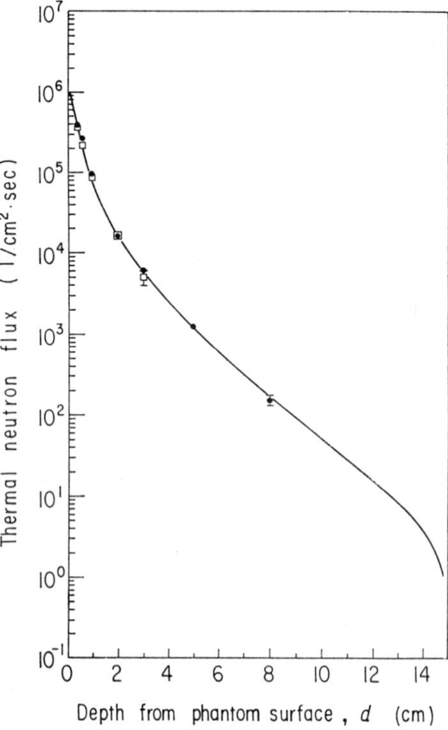

Figure 9. Thermal Neutron Flux Distribution along Center Axis of Polyethylene Phantom.

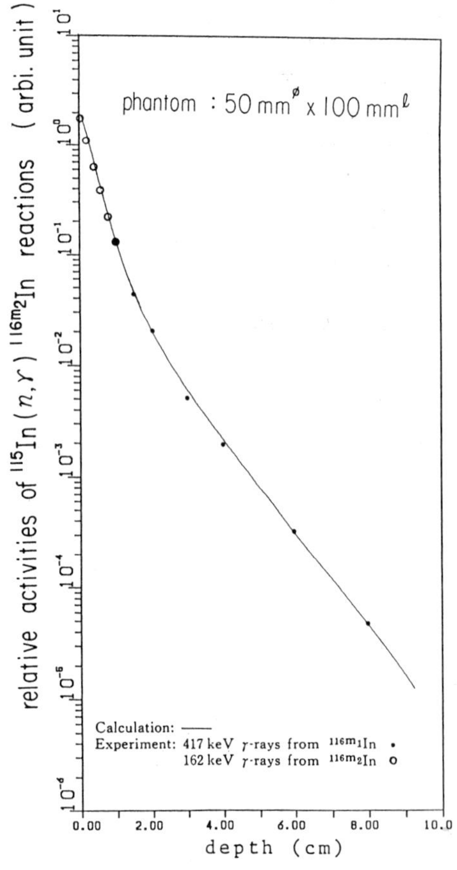

Figure 10. Measured and Calculated Axial Neutron Distribution for the Water Phantom, 5 cm Diameter (normalized at depth of 10 mm).

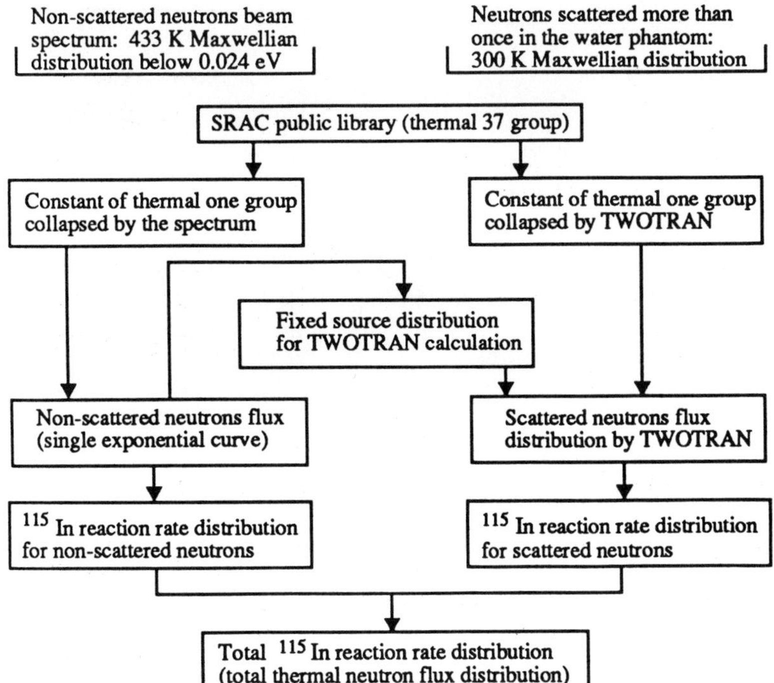

Figure 11. Flow Diagram of TWOTRAN Calculation.

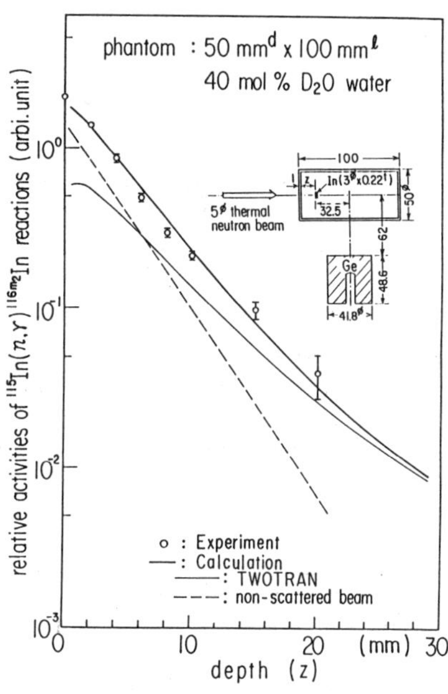

Figure 12. Measured and Calculated Axial Neutron Distribution near the Injection Site for the Heavy Water Phantom, 5 cm in Diameter (normalized at depth of 4 mm).

Figure 13.   Boron-10 Movement in a Hamster.

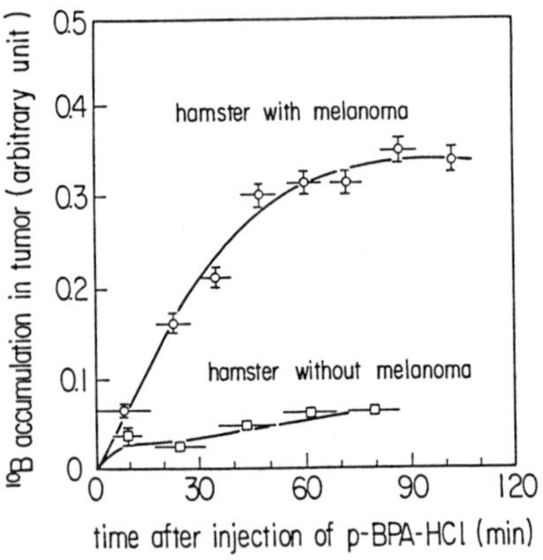

IN-VIVO MEASUREMENT OF BORON CONCENTRATIONS

By using the NMNGT facility, a prompt gamma-ray spectroscopy capability was developed. This was made possible by the high quality thermal neutron beam and Li-6 shielding materials [8]. Both in-vitro and in-vivo measurements were made.

First, the in-vitro measurements are described. The principle is as follows. Concentrations of boron and hydrogen can be assumed to be homogeneous in tissue. Hence, prompt gamma rays emitted from hydrogen can be used for normalization. In the thermal energy region, cross sections of both the B-10(n,α)Li-7 and the H(n,γ)D reactions exhibit the characteristic '1/v' behavior. The branching ratio of Li* is constant as follows:

$$^{10}\text{B} + \text{n} \nearrow {^7\text{Li}} + \alpha \qquad (6.3\%)$$
$$\longrightarrow {^7\text{Li*}} + \alpha \qquad (93.7\%)$$
$$\longrightarrow {^7\text{Li}} + \gamma \qquad (478 \text{ keV}) \qquad \text{H.T.} = 7.3 \times 10^{-14} \text{ s}$$

At any point in the sample, the ratio of gamma rays emitted from the B-10(n,α)Li-7 reaction to that from the H(n,γ)D reaction is independent of the thermal neutron irradiation conditions (i.e., neutron fluence rate, neutron energy spectrum, or neutron distribution) as long as the distributions of B-10 and H in the samples are homogeneous. This means that the B-10 concentration in an unknown sample can be determined by comparison with known sample data. This method has the following advantages:

1) Thermal neutron irradiation conditions (i.e., neutron fluence rate, neutron distribution, and neutron energy spectrum in the sample) are not affected.

2) Counting efficiency of the germanium detector is not affected.

3) Quantity and dimension of sample are not restricted.

The quantity of the sample is, in theory, not essential in this method. However, the error in the gamma ray counting statistics is inversely proportional to sample volume. Hence, the quantity of the sample is important for practical measurements.

In-vivo applications of the above method for measuring boron concentration are now described [13]. Figure 13 shows an example of the measurement using hamsters both with and without tumor. Using only one hamster, we can obtain a time-dependent B-10 concentration in either tumor or normal tissue. We assumed that both the B-10 and H are

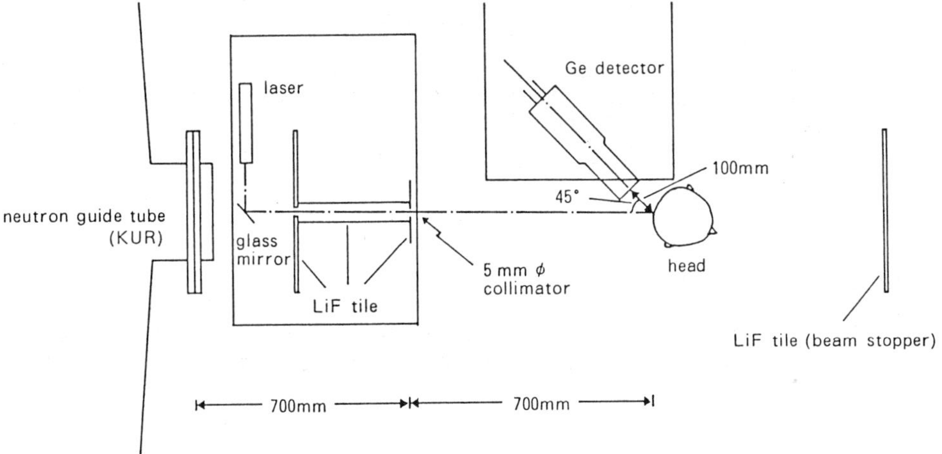

Figure 14.   Non-Destructive System for Measuring ^{10}B Concentrations in Tissue Using the KUR Neutron Guide Tube.

homogeneously distributed in the small volume of tissue, e.g., 5 mm in diameter. This method has the advantage that time-dependent B-10 concentrations can be measured. If employing other methods, such as chemical analysis, we would have had to sacrifice several animals to obtain such a curve. For this measurement, the neutron fluence rate for tumor is 2×10^6 n/cm^2-s and the gamma-ray dose rate is about 10 mrem/hr. These results are also germane to the development of new compounds such as monoclonal antibodies containing boron-10 complex.

DOSE ESTIMATION IN THE FIRST CLINICAL TRIAL TO TREAT MELANOMA

Prior to the first clinical irradiation of a patient, the B-10 concentrations in both the melanoma tumor and its peripheral parts in a patient were measured with the prompt gamma-ray spectroscopy technique using neutrons from the NMNGT facility [9]. A hypodermic injection of boron compound was made in ten fractions, 4 cm from the boundary of the tumor. The B-10 concentrations were measured at 15 points in the X-Y direction. Each was 1.5 - 2 cm apart. The measuring time was 7 minutes for each point. The B-10 concentration at the skin surface up to 4 mm into the tumor was 3 - 5 ppm and that in tumor was about 20 ppm. This was determined by comparison with the result of the phantom experiments.

For the clinical irradiation, a neutron collimator consisting of ^{6}LiF tile and sheet, sized to have the same dimensions as those of the tumor (elliptical form 6.5 x 6.7 cm), was used. The collimator size was determined according to the results of in situ B-10 concentration measurements. The total irradiation time was determined from the neutron fluence data observed during the first 30 minutes of irradiation. In this case, the maximum fluence was adjusted to be 1×10^{13} nvt at the skin surface.

IN SITU MEASUREMENT OF ^{10}B CONCENTRATIONS

The measuring system using the NMNGT facility is shown in Figure 14. The thermal neutron beam from the guide tube, for which the neutron flux is about 2×10^6 n/cm^2-s, was collimated into a beam 5 mm in diameter by the ^{6}LiF tile neutron shielding material. The patient's skin surface was irradiated using the thermal neutron beam at the proper angle.

Figure 15 shows the B-10/H gamma-ray counting ratio for the first clinical trial. The horizontal axis shows the distance from the tumor center, [1+0] denotes one injection of B-10 compound 17 hours before measurement, while [1+1] denotes two injections 17 and 4 hours before measurement. The solid curve is the estimated B-10 concentration in the skin which is assumed to be 4 mm thick as judged by X-ray CT imaging. The overall distribution of B-10 concentration along the skin surface is considered to be a function of the diffusion of the boron compounds. The B-10 within each measured spot (5 x 5 x 4 mm) is assumed to be uniformly distributed. The dotted lines denote the limits of accuracy for the estimation. From these data, we concluded that the size of the neutron collimator should be the same as the tumor size because the B-10 concentration at the periphery was higher than that in the tumor.

Figure 16 shows the results of a series of phantom experiments that were performed in order to determine the B-10 concentration both in and surrounding the tumor.

Figure 15. Distribution of the ^{10}B/H Gamma-Ray Counting Ratio on and around the Tumor.

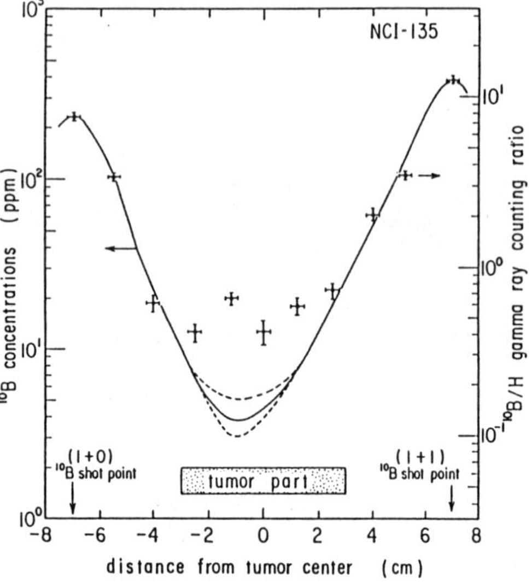

Figure 16. Phantom Experiments to Determine Distribution of B-10 Concentration in Skin.

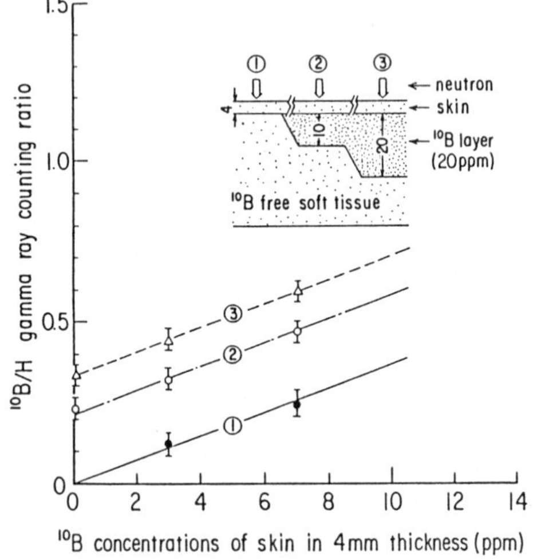

332

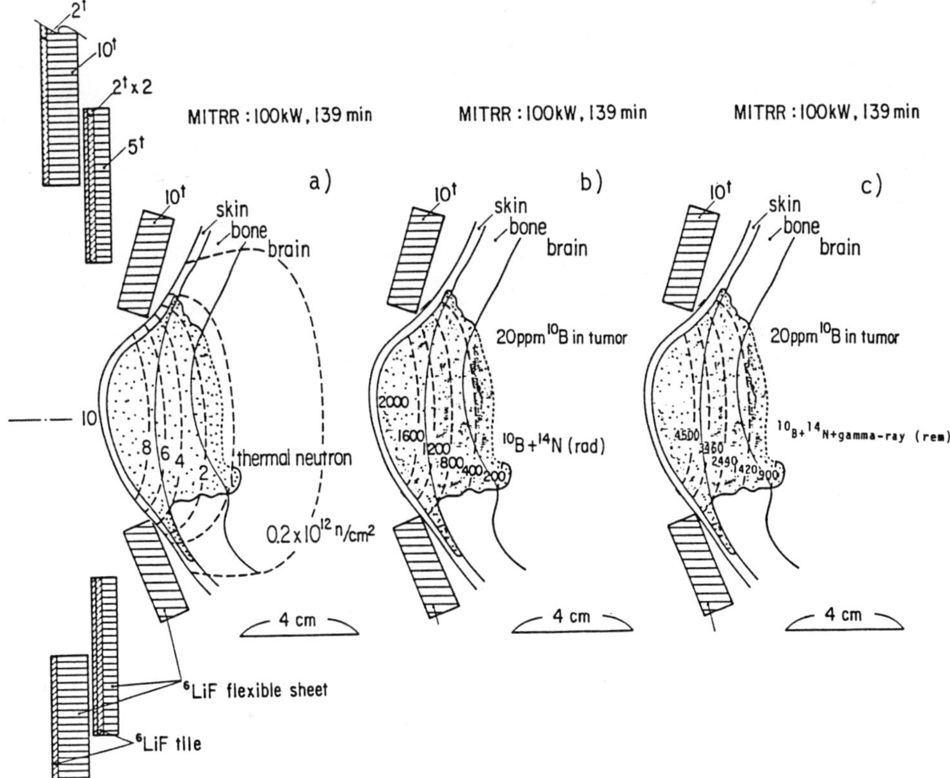

Figure 17.   Collimator Arrangement and the Dose Distribution in the Tumor.
a) Thermal neutron fluence.
b) Absorbed dose in rad caused by B-10 and N-14.
c) Absorbed dose in rems from thermal neutron and gamma rays
(the RBEs of B-10 and N-14 assumed 2.5).

In this case, the B-10 concentration in the tumor was fixed at 20 ppm distributed homogeneously. The relation between the measured B-10/H gamma-ray counting ratio and the B-10 concentration in 4 mm thick skin was measured. From Figures 15 and 16, the B-10 concentration in the skin up to 4 mm into the tumor was estimated to be 3 - 5 ppm.

CLINICAL IRRADIATION AND DOSE ESTIMATION

Gold foils 0.02 mm in thickness and 3 mm in diameter and a gold wire 0.25 mm in diameter were used for the estimation of the neutron flux distribution on the skin surface of the tumor and its peripheral parts. The thermal neutron distribution in the tumor and tissue were estimated from several phantom experiments. The gamma ray dose rate was measured by handmade TLDs of Mg_2SiO_4(Tb) sheathed with polyethylene: 2 mm in outer diameter and 1 mm in inner diameter and 10 mm in length. Figure 17 shows the collimator arrangement and the contours of both the dose estimation for the clinical irradiation at the Musashi Institute of Technology Reactor (MuITR) and the thermal neutron fluence in the tumor [9].

BIOLOGICAL IRRADIATION EQUIPMENT

For basic studies related to BNCT, equipment for cell irradiations is important [14,15]. The equipment prepared by us is described.

The first item of equipment was a conventional exposure tube that faced the bismuth layer of the KUR Heavy Water Facility. Figure 18a shows two irradiation exposure tubes that were used simultaneously. One is of polyethylene and the other of Teflon. Figure 18b shows the corresponding dose distributions. Between the two tubes was a 5-cm thick bismuth layer. It was so located to preclude interference of secondary gamma rays.

The second item of equipment was a neutron direction converter which was made for the convenience of biologists who wanted to utilize thermal neutrons vertically. Figure 19 shows its design and characteristics.

The third item was the cell irradiation box, which controls atmospheric gas and its temperature as shown in Figures 20 and 21.

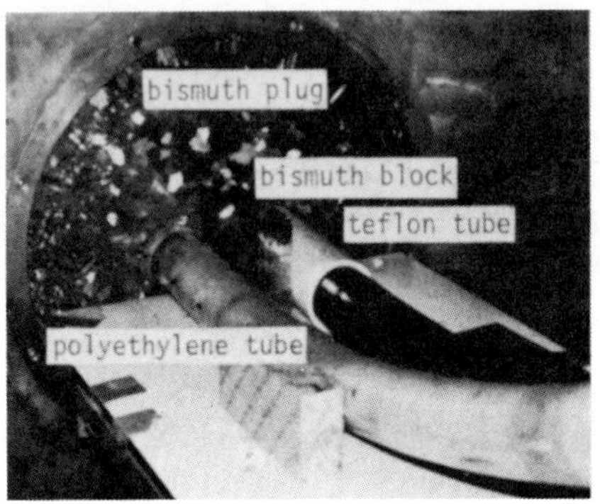

Figure 18a. Photo of Exposure Tubes. (Upper tube is Teflon, the lower one is polyethylene.)

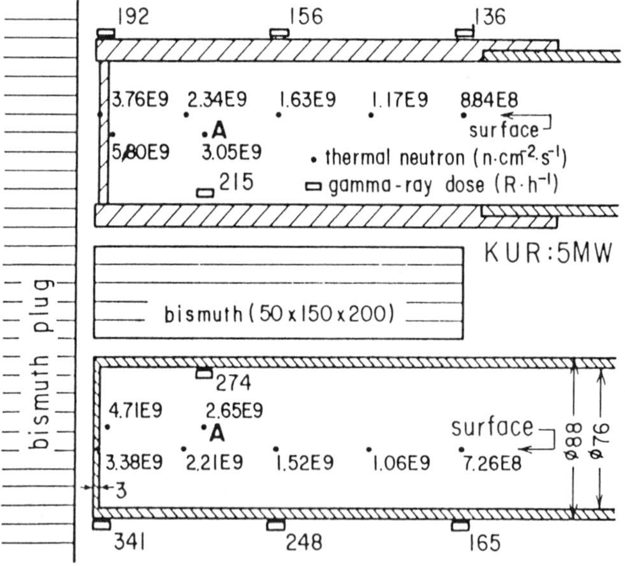

Figure 18b. Dose Distributions in the Exposure Tubes.

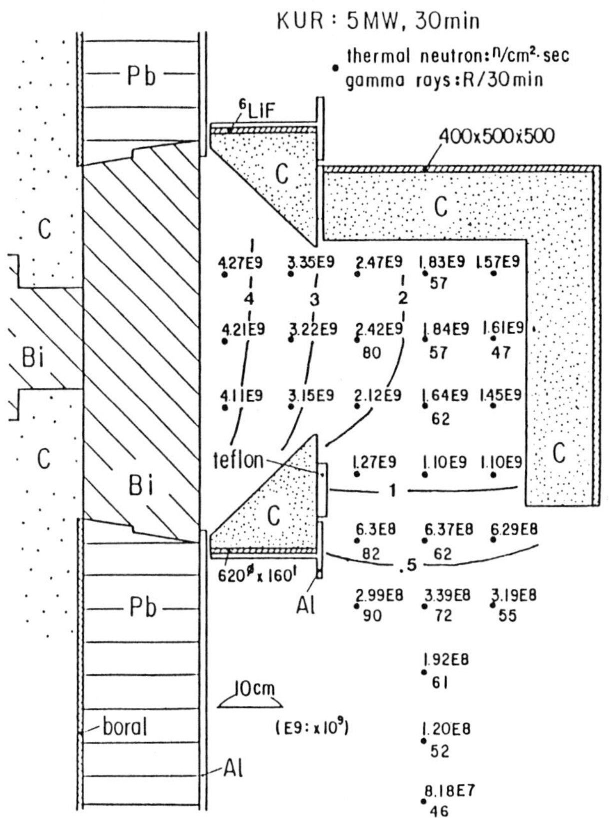

Figure 19.   Dose Distributions for Thermal Neutrons and Gamma Rays in the Direction Converter.

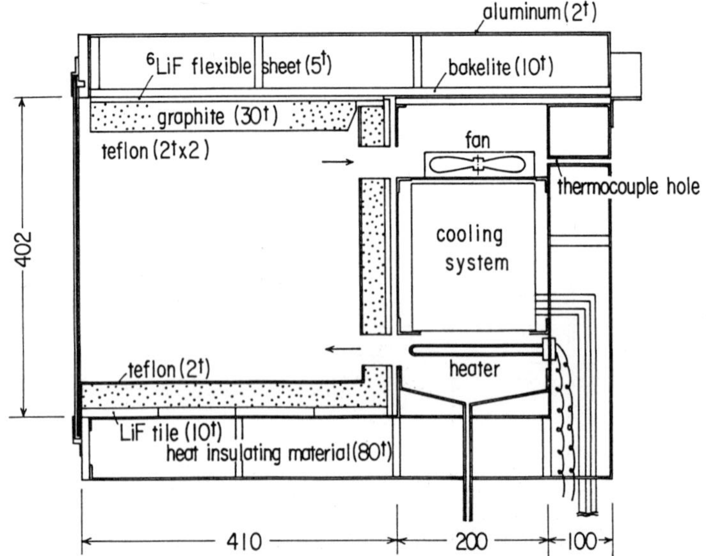

Figure 20. Biological Irradiation Box with Thermo-Control System for Thermal Neutrons with a Low Gamma-Ray Component.

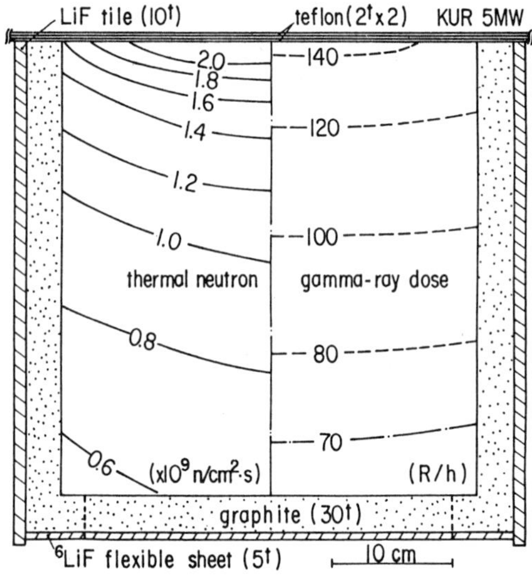

Figure 21. Distribution of Thermal Neutrons and Gamma Rays in the Irradiation Box.

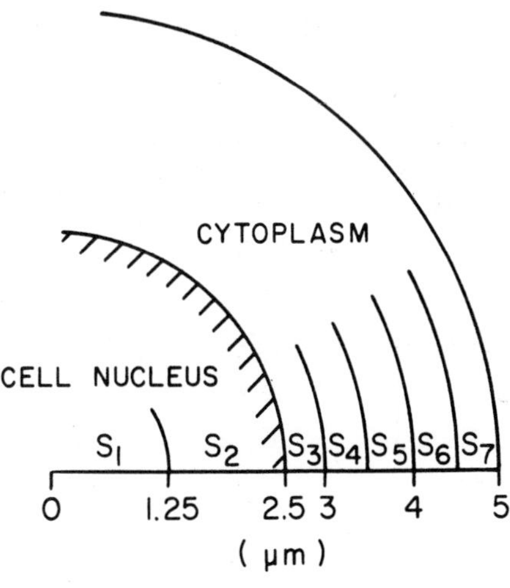

Figure 22.   Cell Structure Assumed for
Calculating BSDRs.

THEORETICAL APPROACH TO EFFECTIVE NEUTRON DOSE ON CELL

Irradiation with thermal neutrons may cause damage to spherical cell nuclei in tissue as the result of producing heavy charged particles. Among the possible effects that can occur, the following were evaluated by means of analytical calculations [16]:

1) The absorbed energy in the cell nucleus.

2) The number of particles hitting and or passing through the cell nucleus.

3) The integrated particle track length in the cell nucleus.

These quantities depend on the change of particle range in the cell, the range-LET relation, and the cell nucleus radius. The distribution of B-10 in the cell is one of the most important factors required to estimate the absorbed dose in the cell nucleus. This dose is needed to explain the biological experiment results well. For reference, as a relevant physics study of BNCT, the selective coefficient of B-10 in cell was quantitatively estimated by defining the Boron Selective Dose Ratio, BSDR [17], as the following:

$$BSDR = \frac{\text{Absorbed energy in cell nucleus from heavy charged particles in tissue containing } ^{10}B}{\text{Absorbed energy in cell nucleus from heavy charged particles in normal tissue}}$$

A cell is assumed to be of spherical shape and divided into several regions (refer to Figure 22 and Table Four). The detailed calculation of the above is described in [16, 17].

ACKNOWLEDGMENTS

The authors would like to acknowledge Prof. Y. Mishima of Kobe University School of Medicine and Prof. H. Hatanaka of Teikyo University Hospital for invaluable discussions. They are indebted to Prof. S. Okamoto and his group for their offer of neutron guide tubes for the present study. This work was partially supported by a Grant-in-Aid of the Ministry of Education, Science, and Culture of Japan.

Table Four

BSDRs for a Typical Cell.

Region	Radius of Region Containing ^{10}B (μm)	^{10}B Concentration in Region (ppm)	BSDR	Average ^{10}B Concentration at BSDR=2 (ppm)
S_1	0 - 1.25	$64.00 \cdot c$[†]	$1+1.42 \cdot c$	0.70
S_2	1.25 - 2.5	$9.14 \cdot c$	$1+1.05 \cdot c$	0.95
S_3	2.5 - 3.0	$10.99 \cdot c$	$1+0.443 \cdot c$	2.26
S_4	3.0 - 3.5	$7.87 \cdot c$	$1+0.239 \cdot c$	4.18
S_5	3.5 - 4.0	$5.92 \cdot c$	$1+0.139 \cdot c$	7.19
S_6	4.0 - 4.5	$4.61 \cdot c$	$1+0.0901 \cdot c$	11.1
S_7	4.5 - 5.0	$3.69 \cdot c$	$1+0.0608 \cdot c$	16.4
S_1+S_2	0 - 2.5	$8.00 \cdot c$	$1+1.10 \cdot c$	0.91
S_3++S_7	2.5 - 5.0	$1.14 \cdot c$	$1+0.148 \cdot c$	6.75
S_1++S_7	0 - 5.0	$1.00 \cdot c$	$1+0.268 \cdot c$	3.73

† c: Average ^{10}B Concentration in Cell (ppm).

REFERENCES

1. K. Kanda, K. Kobayashi, S. Okamoto, and T. Shibata, "Thermal Neutron Standard Field with a Maxwellian Distribution Using the KUR Heavy Water Facility," Nucl. Instr. Meth., 148:535 (1978).

2. K. Kanda, T. Kobayashi, K. Ono, T. Sato, T. Shibata, Y. Ujeno, Y. Mishima, H. Hatanaka, and Y. Nishiwaki, "Elimination of Gamma Rays from a Thermal Neutron Field for Medical and Biological Irradiation Purposes," IAEA-SM-193/68, IAEA, Vienna, Austria, March 10-14, 1975, Biomedical Dosimetry, p. 205.

3. T. Kobayashi, T. Kozuka, H. Chatani, K. Kanda, and T. Shibata, "Experimental Study on Increase of Thermal Neutron Flux in the KUR Heavy Water Facility for Effective Use of Exposure Tubes and a U-235 Fission Converter," Annu. Rep. Res. Reactor Inst. Kyoto Univ., 18:133 (1985).

4. S. Okamoto, T. Akiyoshi et al., "KUR Neutron Guide Tube," Kyoto University Research Reactor Institute (1974).

5. K. Aoki, T. Kobayashi, K. Kanda, and I. Kimura, "Flux Distribution in Phantom for Biomedical Use of Beam-Type Thermal Neutrons," J. Nucl. Sci. Technol., 22:949 (1985).

6. K. Aoki, T. Kobayashi, and K. Kanda, "Phantom Experiment and Analysis for in vivo Measurement of Boron-10 Concentrations in Melanoma for Boron Neutron Capture Therapy," J. Nucl. Sci. Technol., 21:647 (1984).

7. M. Ono, T. Kobayashi, and K. Kanda, "Improvement of Calculation Technique on Flux Distribution in a Water Phantom Caused by Narrow Thermal Neutron Beam for Neutron Capture Therapy," Annu. Rep. Res. Reactor Inst. Kyoto Univ., 21:102 (1988).

8.   T. Kobayashi and K. Kanda, "Microanalysis System of ppm-order ^{10}B Concentrations in Tissue for Neutron Capture Therapy by Prompt Gamma-Ray Spectrometry," <u>Nucl. Instr. Meth.</u>, 204:525 (1983).

9.   T. Kobayashi, S. Hatta, C. Honda, K. Yamamura, T. Akiyoshi, K. Kanda, and Y. Mishima, "Estimation of Absorbed Dose in Human Malignant Melanoma Treated by Neutron Capture Therapy: With Special References to Vertical Direction," <u>Pigment Cell Research</u>, 2:361 (1989).

10.   T. Ebisawa, T. Akiyoshi, S. Tasaki, T. Kawai, N. Achiwa, M. Uturo, and S. Okamoto, "Nickel Mirror and Super Mirror Neutron Guide Tubes at the Kyoto University Research Reactor," in <u>Thin-Film Neutron Optical Devices</u>, SPIE, Vol. 983, p. 54 (1988).

11.   K. Kanda, T. Kobayashi, M. Takeuchi, and S. Ouchi, "Development of Neutron Shielding Material Using LiF," <u>Proc. Sixth Int. Conf. on Radiation Shielding</u>, Vol. II, JAERI, Tokyo, Japan, p. 1258 (1983).

12.   T. Kobayashi, M. Ono, and K. Kanda, "Measurement and Analysis on Neutron Flux Distributions in a Heavy Water Phantom Using the KUR Neutron Guide Tube for BNCT," <u>Strahlenther. Onkol.</u>, 165(2/3):101 (1989).

13.   T. Hamada, K. Aoki, T. Kobayashi, and K. Kanda, "The in vivo Measurement of the Time-Dependent ^{10}B Movement in Tumor of Hamsters," <u>Annu. Rep. Res. Reactor Inst. Kyoto Univ.</u>, 16:112 (1983).

14.   Y. Ujeno, 0. Niwa, K. Takimoto, K. Kanda, T. Kobayashi, and K. Ono, "Distributions of Thermal Neutrons and Gamma Rays in CO_2 Incubators Set in Thermal Neutron Field for Medical and Biological Irradiation Purposes," in <u>Proc. Sixth Int. Congress of Radiat. Res.</u>, Tokyo, Japan, A-30-1 (1979).

15.   M. R. Ishida, K. Kanda, T. Kobayashi, and Y. Ujeno, "Organization and Facilities for Boron Neutron Capture Therapy in KURRI," in <u>Proc. Second Japan-Australia Int. Workshop on Thermal Neutron Capture Therapy for Malignant Melanoma</u>, Kobe, Japan, III-2 (1987).

16.   T. Kobayashi and K. Kanda, "Analytical Calculation of Boron-10 Dosage in Cell Nucleus for Neutron Capture Therapy," <u>Radiat. Res.</u>, 91:77 (1982).

17.   T. Kobayashi and K. Kanda, "Boron-10 Dosage in Cell Nucleus for Neutron Capture Therapy - Boron Selective Dose Ratio," in <u>Boron Neutron Capture Therapy for Tumors</u>, H. Hatanaka, ed., Chapter XXII, Nishimura Co., Ltd., Niigata, Japan, p. 293 (1986).

Thursday Morning - Session I

O. K. Harling
Massachusetts Institute of Technology
Cambridge, Massachusetts, USA

Thursday Morning - Session II

R. G. Fairchild
Brookhaven National Laboratory
Upton, New York, USA

Thursday Afternoon - Session III

J. L. Russell, Jr.
Theragenics Corporation
Norcross, Georgia, USA

Thursday Afternoon - Session IV

R. M. Brugger
University of Missouri
Columbia, Missouri, USA

Friday Morning - Session V

R. G. Zamenhof
Tufts - New England Medical Center
Boston, Massachusetts, USA

Friday Morning - Session VI

F. J. Wheeler
Idaho National Engineering Laboratory
Idaho Falls, Idaho, USA

Friday Afternoon - Session VII

G. Constantine
Harwell Laboratory, UKAEA
Oxfordshire, England, UK

PARTICIPANTS

AIZAWA, Otohiko
Atomic Energy Research Laboratory
Musashi Institute of Technology
Ozenji 971, Kawasaki 215
Japan

AMOLS, H.
Department of Radiation Oncology
Columbia University
New York, NY 10032
USA

BECKER, G. K.
Idaho National Engineering Laboratory
EG & G Idaho, Inc.
Idaho Falls, ID 83415-3519
USA

BERNARD, John A.
Nuclear Reactor Laboratory
Massachusetts Institute of Technology
Cambridge, MA 02139
USA

BLUE, James W.
Cleveland Clinic
Cleveland, OH 44106
USA

BLUE, Thomas E.
2901 Robinson Laboratory
Ohio State University
Columbus, OH 43210-1107
USA

BRENNER, John F.
Department of Radiation Oncology
Tufts - New England Medical Center
Boston, MA 02111
USA

BROWNELL, Gordon L.
Physics Research Laboratory
Massachusetts General Hospital
Boston, MA 02114
USA

BRUGGER, Robert M.
Nuclear Engineering Department
University of Missouri
Columbia, MO 65211
USA

BURIAN, Jiri
Nuclear Research Institute
CSSR 250, 68 Rez near Prague
Czechoslovakia

CLEMENT, Stephen D.
Nuclear Reactor Laboratory
Massahusetts Institute of Technology
Cambridge, MA 02139
USA

CHOI, J. Richard
Nuclear Reactor Laboratory
Massachusetts Institute of Technology
Cambridge, MA 02139
USA

COLE, Donald
Human Health and Assessments Division
U. S. Department of Energy
Washington, D.C. 20545
USA

CONSTANTINE, Geoffrey
Materials Physics and Metallurgy Division
Harwell Laboratory
Harwell, Oxon
United Kingdom

CRAWFORD, J. F.
Paul Scherrer Institute
CH-5234 Villingen
Switzerland

CROCKER, J. G.
Idaho National Engineering Laboratory
EG & G Idaho, Inc.
Idaho Falls, ID 83415-3519
USA

CSOM, Gyula
Institute of Nuclear Technics
Technical University of Budapest
Budapest
Hungary

DURAN, James
Eagle-Picher Research Laboratories
Miami, OK 74355
USA

FAIRCHILD, Ralph G.
Medical Department
Brookhaven National Laboratory
Upton, NY 11973
USA

FIARMAN, Sidney
Medical Department
Brookhaven National Laboratory
Upton, NY 11973
USA

GRUSELL, Erik
Department of Radiation Sciences
Uppsala University
Uppsala S-751 21
Sweden

HARKER, Yale D.
Idaho National Engineering Laboratory
P.O. Box 1625
Idaho Falls, Idaho 83415
USA

HARLING, Otto K.
Nuclear Reactor Laboratory
Massachusetts Institute of Technology
Cambridge, MA 02139
USA

HARRINGTON, Baiba
Australian Nuclear Science &
Technology Organization
Lucas Heights Research Laboratories
New Illawarra Road
Lucas Heights, NSW
Australia

HATANAKA, Hiroshi
Department of Neurosurgery
Teikyo University Medical School
2-11-1 Kaga, Itabashi-ku
Tokyo 173
Japan

HOPKINS, George R.
Nuclear Reactor Laboratory
Massachusetts Institute of Technology
Cambridge, MA 02139
USA

KOBAYASHI, Tooru
Research Reactor Institute
Kyoto University
Kumatori-cho, Sennan-gun
Osaka 590-04
Japan

KWOK, Kwan S.
Nuclear Reactor Laboratory
Massachusetts Institute of Technology
Cambridge, MA 02139
USA

MADOC-JONES, Hywel
Department of Radiation Oncology
Tufts - New England Medical Center
Boston, MA 02111
USA

McCALMONT, Samuel A.
Callery Chemical
Pittsburgh, PA 15230
USA

MILL, A. J.
CEGB
Berkeley Nuclear Laboratories
Berkeley Glos GL139PB
United Kingdom

MOSS, Raymond L.
Commission of the European Communities
Joint Research Center
Institute of Advanced Materials
Petten Establishment
1755 ZG Petten
The Netherlands

MUSOLINO, Stephen V.
Safety-Environmental Protection Division
Brookhaven National Laboratory
Upton, NY 11973
USA

NEUMAN, William A.
Idaho National Energy Laboratory
EG & G Idaho, Inc.
Idaho Falls, ID 83415-3519
USA

NIGG, David W.
Idaho National Engineering Laboratory
EG & G Idaho, Inc.
Idaho Falls, ID 83415-3519
USA

RAMSEY, Eric B.
Department of Radiation Oncology
State University of New York
Stony Brook, NY 11794
USA

REINSTEIN, Lawrence E.
Department of Radiation Oncology
University Hospital at Stony Brook
Stony Brook, NY 11794
USA

ROBERTS, Kevin
Nuclear Reactor Laboratory
Massachusetts Institute of Technology
Cambridge, MA 02139
USA

RORER, David
Brookhaven National Laboratory
Upton, NY 11973
USA

RUSSELL, John L., Jr.
Theragenics Corporation
Norcross, GA 30093
USA

SCHOFIELD, P.
UKAEA - Harwell Laboratory
Theoretical Physics Division
Oxfordshire OXll ORA
United Kingdom

SOLOWAY, Albert H.
College of Pharmacy
Ohio State University
Columbus, OH 43210-1291
USA

SARAF, Sharad K.
Medical Department
Brookhaven National Laboratory
Upton, NY 11973
USA

SHEFER, R. E.
Science Research Laboratory
Somerville, MA 02143
USA

SLATKIN, Daniel N.
Medical Department
Brookhaven National Laboratory
Upton, NY 11973
USA

SOLARES, Guido
Nuclear Reactor Laboratory
Massachusetts Institute of Technology
Cambridge, MA 02139
USA

STECHER-RASMUSSEN, F.
Netherlands Energy Research Foundation
NL-1755 ZG Petten
The Netherlands

UJENO, Yowri
Research Reactor Institute
Kyoto University
Kumatori-cho, Sennan-gun
590-04 Osaka
Japan

WANG, C.-K. Chris
Department of Nuclear Engineering
Kansas State University
Manhattan, KS 66506-2503
USA

WASHIO, Takashi
Nuclear Reactor Laboratory
Massachusetts Institute of Technology
Cambridge, MA 02139
USA

WAZER, David E.
Department of Radiation Oncology
Tufts New England Medical Center
Boston, MA 02111
USA

WHEELER, F. J.
Idaho National Engineering Laboratory
EG&G Idaho, Inc.
Idaho Falls, ID 83415-3515
USA

WHITTEMORE, William L.
General Atomics
San Diego, CA 92138
USA

YANCH, Jacquelyn C.
Radiological Sciences Program
Massachusetts Institute of Technology
Cambridge, MA 02139
USA

ZAMENHOF, Robert G.
Department of Radiation Oncology
Medical Physics Division
Tufts - New England Medical Center
Boston, MA 02111
USA